$$e^{ikr\cos\theta} = \sum_{l=0}^{\infty} i^l(2l+1)j_l(kr)P_l(\cos\theta)$$

$$j_l(\rho) \underset{\rho\to 0}{\to} \frac{\rho^l}{(2l+1)!!} \qquad j_l(\rho) \underset{\rho\to\infty}{\to} \frac{\sin(\rho - l\pi/2)}{\rho}$$

$$n_l(\rho) \underset{\rho\to 0}{\to} -\frac{(2l-1)!!}{\rho^{l+1}} \qquad n_l(\rho) \underset{\rho\to\infty}{\to} -\frac{\cos(\rho - l\pi/2)}{\rho}$$

$$e^{ik\rho\cos\theta} = J_0(k\rho) + 2\sum_{m=1}^{\infty} i^m J_m(k\rho)\cos(m\theta)$$

$$J_m(\rho) \underset{\rho\to 0}{\to} \frac{\left(\frac{1}{2}\rho\right)^m}{m!} \qquad J_m(\rho) \underset{\rho\to\infty}{\to} \sqrt{\frac{2}{\pi\rho}}\cos\left(\rho - \frac{m\pi}{2} - \frac{\pi}{4}\right)$$

$$N_m(\rho) \underset{\rho\to 0}{\to} -\frac{(m-1)!}{\pi\left(\frac{1}{2}\rho\right)^m} \qquad N_m(\rho) \underset{\rho\to\infty}{\to} \sqrt{\frac{2}{\pi\rho}}\sin\left(\rho - \frac{m\pi}{2} - \frac{\pi}{4}\right)$$

$$j_l(\rho) = \sqrt{\frac{\pi}{2\rho}}\,J_{l+\frac{1}{2}}(\rho)$$

$$[\nabla_r^2 + k^2]\frac{e^{ik|r-\vec{r}'|}}{|r-\vec{r}'|} = -4\pi\,\delta(r-\vec{r}')$$

$$\left(\frac{d^2}{dx^2} + k^2\right)e^{ik|x-x'|} = 2ik\,\delta(x-x')$$

$$f(x) = \sum_{m=0}^{\infty}\left[a_m\cos\left(\frac{2m\pi x}{L}\right) + b_m\sin\left(\frac{2m\pi x}{L}\right)\right] \qquad \text{if} \quad f(x) = f(x+L)$$

$$\text{where } a_m = \frac{2}{L(1+\delta_{m,0})}\int_0^L \cos\left(\frac{2m\pi x}{L}\right)f(x)\,dx$$

$$b_m = \frac{2}{L}\int_0^L \sin\left(\frac{2m\pi x}{L}\right)f(x)\,dx$$

$$\int_{-\infty}^{\infty} e^{ik(x-x')}\,dk = 2\pi\,\delta(x-x')$$

$$f(x) = \frac{1}{\sqrt{2\pi}}\int_{-\infty}^{\infty} e^{i\omega x}g(\omega)\,d\omega, \qquad g(\omega) = \frac{1}{\sqrt{2\pi}}\int_{-\infty}^{\infty} e^{-i\omega x}f(x)\,dx$$

$$f(x+\varepsilon) = \sum_{n=0}^{\infty}\frac{\varepsilon^n}{n!}\frac{d^n f(x)}{dx^n} = \exp\left(\varepsilon\frac{d}{dx}\right)f(x)$$

$$f(r+\varepsilon) = \sum_{n=0}^{\infty}\frac{1}{n!}(\vec{\varepsilon}\cdot\vec{\nabla}_{\vec{r}})^n f(\vec{r}) = \exp(\vec{\varepsilon}\cdot\vec{\nabla}_r)f(\vec{r})$$

$$\int f(x_1, x_2, \ldots)\,dx_1\,dx_2\ldots = \int f(x_1(y_1, y_2, \ldots), x_2(y_1, y_2, \ldots), \ldots)\left|\frac{\partial(x_1 x_2 \ldots)}{\partial(y_1 y_2 \ldots)}\right| dy_1\,dy_2\ldots$$

$$\int_0^{\infty} x^n e^{-\lambda x}\,dx = \frac{\Gamma(n+1)}{\lambda^{n+1}} \qquad \Gamma(n+1) = n\Gamma(n)$$

$$\int_0^{\infty} x^n e^{-\lambda x^2}\,dx = \frac{1}{2}\frac{\Gamma\left(\frac{n+1}{2}\right)}{\lambda^{\left(\frac{n+1}{2}\right)}} \qquad \Gamma(1) = 1,\ \Gamma\left(\tfrac{1}{2}\right) = \sqrt{\pi}$$

A REVIEW OF
UNDERGRADUATE PHYSICS

A REVIEW OF
UNDERGRADUATE PHYSICS

BENJAMIN F. BAYMAN
University of Minnesota

MORTON HAMERMESH
University of Minnesota

JOHN WILEY & SONS
New York / Chichester / Brisbane / Toronto / Singapore

Library of Congress Cataloging in Publication Data:

Bayman, Benjamin F., 1930–
  A review of undergraduate physics.

  Includes indexes.
  1. Physics.   I. Hamermesh, M. (Morton), 1915–
II. Title.
QC21.2.B38 1986   530   85-26577
ISBN 0-471-81684-1 (pbk)

Printed in the United States of America

10 9 8 7 6 5 4 3 2 1

PREFACE

In this book we have tried to present a concise summary of most of the material covered in an undergraduate program in physics. Each topic is developed from fundamental principles and then applied to the solution of illustrative problems. These problems are of the type used by American graduate schools in their comprehensive physics examinations and in the Graduate Record Examination. This book should therefore be especially useful to someone who is preparing for such a comprehensive examination. We hope it will also be useful to students who are currently in an undergraduate physics program, and to engineers and scientists who are interested in more advanced treatments of subjects they encountered in their introductory physics courses.

We have tried to make our presentation as self-contained as possible. Of course, each of our chapters is too brief to be considered as a replacement for a monograph or textbook on its subject. However, if the goal is a review of a wide variety of physical ideas and applications in a reasonable amount of time, then brevity is necessary. Furthermore, by treating different subject areas of physics within the same volume, we have emphasized the important basic ideas that are common to these different areas. This makes the review process more efficient and deepens our appreciation of the unity of physics.

We believe that a book of this sort is most useful if its size and cost are both kept reasonably small. We have therefore included very little factual material of the kind that would be covered in introductory courses in atomic, nuclear, or solid-state physics. This factual material is an important component of a physics education, but it is not easily summarized. Moreover, in the interest of brevity we have assumed that the reader has a good understanding of vector algebra and calculus, and of the elementary properties of differential equations.

Most of this text uses the cgs Gaussian system of units. This is the system used in most graduate-level work in physics and in most of the research literature. Of course, the physical description of any system should be independent of units. Thus, a student who prefers to work in a different system of units should be able to transcribe all our expressions into his or her units with no change in essential physical content.

The idea for this book developed out of an informal seminar offered during the past ten years to help first-year physics graduate students at the University of Minnesota prepare for our Graduate Written Examination. Most of our illustrative examples are taken from previous University of Minnesota examinations. We have also included problems from the comprehensive examinations given at several other universities.

B. F. B.
M. H.

CONTENTS

A REVIEW OF
UNDERGRADUATE PHYSICS

CHAPTER 1

CLASSICAL MECHANICS

Books that attempt to survey all of physics traditionally begin with classical mechanics. There are several good reasons for following this tradition. Most of our physical intuition is based on mechanical models, most of the important concepts of physics have their simplest realization in mechanical systems, and the newer ideas of relativity and quantum mechanics are perhaps best appreciated in terms of their contrast with the views of classical mechanics.

1.1 NEWTON'S SECOND LAW OF MOTION

If the force $\mathbf{f}$ acts on a point particle of mass m, then

$$\mathbf{f} = \frac{d\mathbf{p}}{dt} = m\frac{d\mathbf{v}}{dt} = m\frac{d^2\mathbf{r}}{dt^2} \tag{1.1a}$$

Here $\mathbf{p}$ and $\mathbf{v}$ are the momentum and velocity of the particle relative to an inertial frame of reference, and $\mathbf{r}$ is a vector from a fixed point 0 in that frame of reference to the location of the particle. From (1.1a) we can derive

$$\boldsymbol{\tau} = \mathbf{r} \times \mathbf{f} = m\mathbf{r} \times \frac{d\mathbf{v}}{dt} = \frac{d}{dt}(m\mathbf{r} \times \mathbf{v})$$

$$= \frac{d}{dt}(\mathbf{r} \times \mathbf{p}) = \frac{d\mathbf{l}}{dt} \tag{1.1b}$$

$\boldsymbol{\tau}$ is the torque on the particle and $\mathbf{l} = \mathbf{r} \times \mathbf{p}$ is its angular momentum, both defined relative to the point 0.

Now consider a system of particles. The force $\mathbf{f}_i$ on particle i can be written as

$$\mathbf{f}_i = \mathbf{f}_i^{\text{ext}} + \sum_{j \neq i} \mathbf{f}(j \text{ on } i) \tag{1.2}$$

$\mathbf{f}_i^{\text{ext}}$ is the external force on particle i, and $\mathbf{f}(j \text{ on } i)$ is the force on particle i due to particle j. Newton's third law of motion asserts that

$$\mathbf{f}(j \text{ on } i) = -\mathbf{f}(i \text{ on } j) \tag{1.3}$$

If we now sum (1.1a) and (1.1b) over all the particles and use (1.2) and (1.3), we find that

$$\mathbf{F} \equiv \sum_i \mathbf{f}_i^{\text{ext}} = \frac{d}{dt} \sum_i \mathbf{p}_i = \frac{d}{dt} \mathbf{P}_{\text{tot}} \tag{1.4a}$$

$$\boldsymbol{\tau} \equiv \sum_i \boldsymbol{\tau}_i = \frac{d}{dt} \sum_i \mathbf{l}_i = \frac{d}{dt} \mathbf{L}_{\text{tot}} \tag{1.4b}$$

The derivation of (1.4b) also requires that we assume that $\mathbf{f}(j \text{ on } i)$ is directed along the line joining particles i and j. Both the total torque $\boldsymbol{\tau}$ and the total angular momentum $\mathbf{L}_{\text{total}}$ in (1.4b) must be defined with respect to the same point. This point may be any point fixed in an inertial frame of reference (i.e., at rest in such a frame or moving with uniform velocity relative to it), or it may be the point that moves with the mass center of the system,[1] located at

$$\mathbf{R}_{\text{CM}} = \frac{\sum_i m_i \mathbf{r}_i}{\sum_i m_i} = \frac{\sum_i m_i \mathbf{r}_i}{M_{\text{tot}}} \tag{1.5}$$

Usually we need to relate $\mathbf{P}_{\text{tot}}$ and $\mathbf{L}_{\text{tot}}$ to the motion of the system. For $\mathbf{P}_{\text{tot}}$ we have

$$\mathbf{P}_{\text{tot}} = \sum_i \mathbf{p}_i = \sum_i m_i \frac{d\mathbf{r}_i}{dt} = \frac{d}{dt} \sum_i m_i \mathbf{r}_i = \frac{d}{dt}(M_{\text{tot}} \mathbf{R}_{\text{CM}}) = M_{\text{tot}} \frac{d}{dt} \mathbf{R}_{\text{CM}} \tag{1.6a}$$

The relationship between angular momentum and angular velocity is more complicated. It is discussed in Section 1.8 below. Our present considerations will be limited to uniform rigid bodies rotating about an axis about which the body has rotational symmetry, or an axis which is perpendicular to a plane of reflection symmetry. In these cases we can write

$$L = I\omega \tag{1.6b}$$

Here ω is the angular speed of the body (in radians per unit time), and I is the moment of inertia of the body about the rotation axis, defined by

$$I = \int dm \, s^2 \tag{1.7}$$

The integration goes over every mass element dm of the body, and s is the perpendicular distance of the mass element dm from the rotation axis. Equations (1.4) and (1.6) can be combined to yield

$$\mathbf{F} = M_{\text{tot}} \frac{d\mathbf{V}_{\text{CM}}}{dt} = M_{\text{tot}} \frac{d^2 \mathbf{R}_{\text{CM}}}{dt^2} \tag{1.8a}$$

$$\tau = I \frac{d\omega}{dt} = I \frac{d^2\theta}{dt^2} \tag{1.8b}$$

where $\omega = d\theta/dt$. We purposely avoid writing (1.8b) as a vector equation since it applies only to the case of rotation about a principal axis (see Section 1.8).

---

[1]Another valid (but less useful) choice is any point accelerating toward or away from the mass center.

1.2 SOME COMMONLY ENCOUNTERED FORCES

1.2.1 Friction

Suppose that an object is in contact with a surface. The force that the surface exerts on the object can be resolved into a perpendicular component $\mathbf{N}$ and a tangential component $\mathbf{f}$. If the object slides along the surface, it is often a good approximation to assume that the magnitudes of $\mathbf{f}$ and $\mathbf{N}$ are related by

$$f = \mu_k N \tag{1.9a}$$

μ_k is called the coefficient of kinetic or sliding friction. The direction of $\mathbf{f}$ is usually assumed to be opposite to the velocity of the object relative to the surface. If the object is at rest on the surface the value of f depends on the other forces acting, but cannot exceed a critical value given by

$$f \leq \mu_s N \tag{1.9b}$$

μ_s is called the coefficient of static friction. The values of μ_s and μ_k are usually assumed to depend only on the nature of the surfaces in contact, and to be independent of the area of contact and the magnitude of $\mathbf{N}$. If a problem refers to a "smooth" surface, this implies that $\mu_s = \mu_k = 0 = f$, so that the force that such a surface exerts on an object is exactly perpendicular to the surface.

1.2.2 Gravitation

The gravitational force on a point mass m_1 due to another point mass m_2 is

$$\mathbf{F}_{(\text{on 1 due to 2})} = Gm_1m_2\frac{(\mathbf{r}_2 - \mathbf{r}_1)}{|\mathbf{r}_2 - \mathbf{r}_1|^3} = -Gm_1m_2\frac{\mathbf{r}}{r^3} = -Gm_1m_2\frac{\hat{r}}{r^2} \tag{1.10}$$

where G is the fundamental gravitational constant and $\mathbf{r}$ is the vector from mass m_2 to mass m_1. We can also use (1.10) to find the force on a point mass m_1 due to a *spherically symmetric* mass distribution. In this case $\mathbf{r}$ is the vector to m_1 from the center of the continuous mass distribution, and m_2 is the total mass within a distance r from the center (see Figure 1.1). In particular, if m_1 is wholly outside the continuous distribution, m_2 is the total spherical mass.

Now suppose the continuous mass distribution is the earth (assumed spherical), and we want the gravitational force on a point particle m_1 slightly above its surface. Then $\mathbf{r}$ points from the center of the earth, so an observer near the particle would say that the force on m_1 is vertically downward. If the height of

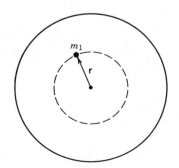

FIGURE 1.1 The gravitational force on m_1 is that of a point particle of mass m_2 located at the center of the sphere. The total mass within the dotted sphere of radius r is m_2.

the particle above the earth is small compared to the radius of the earth, (1.10) becomes

$$\mathbf{F} = -m_1\left(\frac{Gm_2}{R_e^2}\right)\hat{r} = -m_1 g\hat{r} \equiv m_1\mathbf{g} \tag{1.11}$$

Thus, we can describe the gravitational field near the surface of the earth as uniform, of magnitude g ($= 32.2$ ft/s^2 $= 9.8$ m/s^2) and directed vertically downward. The total gravitational force and torque on a finite object are given by

$$\mathbf{F}_{\text{grav}} = \int dm\,\mathbf{g} = \mathbf{g}\int dm = M_{\text{tot}}\mathbf{g} \tag{1.12a}$$

$$\boldsymbol{\tau}_{\text{grav}} = \int \mathbf{r} \times dm\,\mathbf{g} = \left(\int dm\,\mathbf{r}\right) \times \mathbf{g} = M_{\text{tot}}\mathbf{R}_{\text{CM}} \times \mathbf{g}$$

$$= \mathbf{R}_{\text{CM}} \times (M_{\text{tot}}\mathbf{g}) = \mathbf{R}_{\text{CM}} \times \mathbf{F}_{\text{grav}} \tag{1.12b}$$

Equation (1.12b) shows that we get the correct value of the gravitational torque on an object if we assume that the entire gravitational force (weight) acts at a single point of the object, its mass center. This implies that the gravitational torque on an object, defined with respect to its mass center, is zero. This is true in general only for a uniform gravitational field.

1.2.3 Hooke's Law Springs

Suppose that a Hooke's law (or ideal) spring has an unstretched length l_0 and spring constant k. If the spring is stretched or compressed to length l it exerts a restoring force of magnitude

$$F = k|l - l_0| \tag{1.13}$$

Now consider the situation shown in Figure 1.2. Since the mass in Figure 1.2*b* is in equilibrium, the upward force due to the spring must equal the downward

FIGURE 1.2 Three springs suspended from a ceiling: (a) the unstretched spring; (b) mass *m* in equilibrium under the combined forces of gravity and the spring; and (c) the mass displaced a distance *x* from equilibrium.

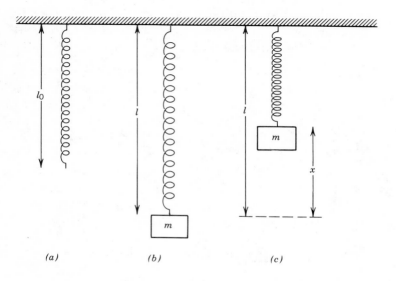

(a)   (b)   (c)

force due to gravity. Thus,

$$k(l - l_0) - mg = 0$$

In Figure 1.2c the mass m has been given an additional upward displacement x, so that the length of the spring is now $l - x$. The upward force due to the spring is now $k(l - x - l_0)$ while the downward gravitational force is still mg. Thus, the net upward force on the mass is

$$F = k(l - x - l_0) - mg = -kx \tag{1.14}$$

The minus sign in $-kx$ implies that an upward displacement of the mass results in a downward net force on the mass, and vice versa. We see that k governs the restoring force for oscillations about equilibrium. The equation of motion of the mass is

$$F = m\ddot{x} = -kx \tag{1.15a}$$

whose general solution

$$x(t) = A \sin\left(\sqrt{\frac{k}{m}}\, t + \phi\right), \qquad (A, \phi, \text{constants}) \tag{1.15b}$$

describes oscillations about equilibrium ($x = 0$), with constant amplitude A and initial phase ϕ. A and ϕ depend on the initial conditions under which the mass is set into oscillation. The circular frequency

$$\omega = \sqrt{\frac{k}{m}} \tag{1.15c}$$

depends only on the materials of which the system is made. In particular, it is independent of the amplitude of the oscillations.

PROBLEM 1.2.1   A heavy object, when placed on a rubber pad that is to be used as a shock absorber, compresses the pad by 1 cm. If the object is given a vertical tap, it will oscillate. Ignoring the damping, estimate the oscillation frequency.

   Let k be the spring constant of the rubber, and let x_0 ($= 1$ cm) be the equilibrium displacement. At equilibrium the upward force on the object is kx_0, and the downward force is mg. Thus, $kx_0 = mg$, $k = mg/x_0$. The circular frequency of small oscillations about equilibrium is $\omega = \sqrt{k/m} = \sqrt{g/x_0} = \sqrt{980/1}$ rad/s. Thus, the frequency is $(1/2\pi)\sqrt{980}$ cycles/s $= 4.98$ Hz.

PROBLEM 1.2.2   An automobile, with nobody inside, has a mass of 1000 kg, and has ground clearance 18 cm. After four persons with total mass 300 kg get into the car, the ground clearance is only 12 cm. They drive off. At what speed will the car, with its four passengers, bounce in resonance while moving along a road that is straight, level, and smooth, except for a transverse tar patch every 15 m? For simplicity assume that the shock absorbers are ineffective, and also that the fore and aft suspensions have the same bouncing frequency.

   Adding passenger weight of $300g$ Newtons causes a 6-cm deflection. Thus, $k = 300$ kg $\times$ 9.8 m/s²$/.06$ m $= 5000 \times 9.8$ kg/s². Since the total mass of the loaded car is 1300 kg, the circular frequency is

$$\omega = \sqrt{k/M} = \sqrt{5000 \times 9.8 \text{ kg/s}^2/1300 \text{ kg}} = \sqrt{5000 \times 9.8/1300} \text{ s}^{-1}.$$

Thus, the period of the oscillations is

$$\tau = \frac{1}{\nu} = \frac{2\pi}{\omega} = 2\pi\sqrt{\frac{1300}{5000 \times 9.8}} \text{ s.}$$

If the car has a speed of

$$v = \frac{15 \text{ m}}{\tau} = \frac{15}{2\pi}\sqrt{\frac{5000 \times 9.8}{1300}} \text{ m/s} = 14.7 \text{ m/s} = 52.8 \text{ km/h}$$

the impulses due to the tar patch will be at the resonant frequency.

PROBLEM 1.2.3 A stick of length l is held so that one end rests on a smooth plane, making an angle θ with the plane. The stick is then released. How far will the left end of the stick have moved by the time the stick hits the plane?

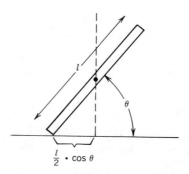

$\frac{l}{2} \cdot \cos\theta$

The external forces acting on the stick (gravity and the surface contact force) are both vertical. Thus, $\mathbf{F}_{\text{ext}}$ has no horizontal component, and the acceleration of the mass center of the stick is vertical. Since the horizontal component of the velocity of the mass center is initially zero, it remains zero as the stick falls. This implies that the mass center of the stick falls vertically, so that by the time the stick is horizontal the left-hand end will have moved by $(l/2)[1 - \cos\theta]$.

PROBLEM 1.2.4 A thin stick of length L and mass m is supported at its ends by vertical strings so as to be in a horizontal position. One of the strings is cut at time t.

(a) Find the downward acceleration of the center of the stick at time $t + \delta$ (where $\delta \to 0$).

At time $t + \delta$, the external forces acting on the stick are shown in the free-body diagram (Figure 1.3b). The total external torque about the left end is $mgL/2$. Thus, the angular acceleration, α, of the stick about its left end is

$$\alpha = \frac{\tau}{I} = \frac{mgL/2}{(1/3)mL^2} = \frac{3}{2}\frac{g}{L}$$

The linear acceleration, a, of the center of the stick is then

$$a = \frac{\alpha L}{2} = \frac{3}{2}\frac{g}{L}\frac{L}{2} = \frac{3}{4}g$$

vertically downward.

(b) Find the sideward acceleration of the center of the stick. At time $t + \delta$, all the external forces acting on the stick are vertical. Thus, the total external force has no horizontal component, and the sideward acceleration of the stick is zero.

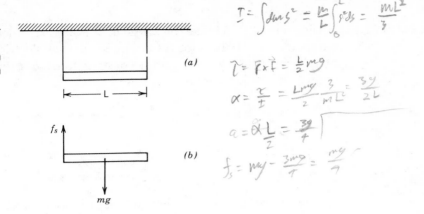

FIGURE 1.3 (a) The stick immediately after the right-hand string has been cut. (b) The forces acting on the string at that instant.

(a)

f_s

(b)

mg

$dm = \frac{M \, ds}{L}$

$I = \int dm \, s^2 = \frac{m}{L} \int_0^L s^2 ds = \frac{mL^2}{3}$

$\tau = \bar{F} \times \bar{r} = \frac{L}{2} mg$

$\alpha = \frac{\tau}{I} = \frac{Lmg}{2} \cdot \frac{3}{mL^2} = \frac{3g}{2L}$

$a = \alpha \frac{L}{2} = \frac{3g}{4}$

$f_s = mg - \frac{3mg}{4} = \frac{mg}{4}$

(c) Find the tension f_s in the remaining string. Newton's second law applied to the stick gives

$$mg - f_s = ma = m \cdot \tfrac{3}{4}g$$

$$f_s = \tfrac{1}{4}mg$$

PROBLEM 1.2.5   An hourglass with vertical sides is placed on a critically damped balance, the sand trickling through the hole. What does the balance read? Discuss the direction of deflection of the balance during all stages of the flow.

Let the mass of the hourglass plus sand be M, and let F_b be the upward force that the balance pan exerts on the hourglass. According to Newton's third law, F_b is also the force that the hourglass exerts on the balance pan and thus F_b determines the reading on the balance scale. If y is the height of the center of mass of the hourglass plus sand, then

$$F_b - Mg = M\ddot{y}$$

Thus, if $\ddot{y} = 0$, $F_b = Mg$, but if $\ddot{y}$ is positive (negative), F_b will be greater (less) than Mg.

Suppose that all the sand is at rest in the upper portion of the hourglass for $t < t_0$. At $t = t_0$ the sand starts to fall, and reaches a steady stream at $t = t_1$. The steady stream continues until $t = t_2$, when the flow starts to wane and comes to a stop at $t = t_3$. Thus, for $t < t_0$ and $t > t_3$ the sand is at rest, $\dot{y} = 0 = \ddot{y}$, and $F_b = Mg$. Between $t = t_1$ and $t = t_2$, $\dot{y} < 0$, but $\ddot{y} = 0$ (since the sand is falling at a constant rate) so that F_b still equals Mg. Between $t = t_0$ and $t = t_1$ we are going from a situation in which $\dot{y} = 0$ to one in which $\dot{y} < 0$. Thus, $\ddot{y} < 0$ between t_0 and t_1, so that $F_b < Mg$. Conversely, between $t = t_2$ and $t = t_3$ we are going from a situation in which $\dot{y} < 0$ to one in which $\dot{y} = 0$. Thus, $\ddot{y} > 0$ between t_2 and t_3, so that $F_b > Mg$. To summarize: the balance reads

$$
\begin{array}{ll}
Mg & \text{for } t \leq t_0 \\
< Mg & \text{for } t_0 < t < t_1 \\
Mg & \text{for } t_1 \leq t \leq t_2 \\
> Mg & \text{for } t_2 < t < t_3 \\
Mg & \text{for } t \geq t_3
\end{array}
$$

The transitions between the different flow conditions described above will be smooth, since the momentum flux will not change discontinuously.

PROBLEM 1.2.6   A right circular cylinder has a density that is a function of distance from the symmetry axis. It rests on a frictionless surface. A string is wrapped around the periphery of the cylinder and a constant force F is applied to the string for a time T, in the horizontal direction.

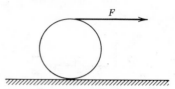

(a) Describe qualitatively the translational and rotational motion of the object.

Since the surface is frictionless, the force that it exerts on the cylinder has no horizontal component. Thus, the horizontal component of the total external force is F when $0 \leq t \leq T$, and zero when $T \leq t$. The axis of the cylinder, therefore, has acceleration F/M when $0 \leq t \leq T$, and zero when $T \leq t$, so that its speed, v, is given by

$$v = \frac{F}{M}t, \qquad \text{for } 0 \leq t \leq T$$

$$= \frac{F}{M}T, \qquad \text{for } T \leq t$$

In the interval $0 \leq t \leq T$, the external torque about the axis is FR, so the angular acceleration is $\alpha = FR/I$ for $0 \leq t \leq T$, and zero for $T \leq t$. Thus, the angular speed of rotation of the cylinder about its axis is

$$\omega = \frac{FR}{I}t, \qquad \text{for } 0 \leq t \leq T$$

$$= \frac{FR}{I}T, \qquad \text{for } T \leq t$$

(b) Find a specific geometry for the object so that the kinetic energy is equally divided between translational and rotational motion.

The translational part of the kinetic energy is

$$\frac{1}{2}Mv^2 = \frac{1}{2}M\left(\frac{F}{M}t\right)^2$$

The rotational part is

$$\frac{1}{2}I\omega^2 = \frac{1}{2}I\left(\frac{FR}{I}t\right)^2$$

If these are to be equal, we must have

$$\frac{1}{M} = \frac{R^2}{I}, \qquad I = MR^2$$

The only way this can occur is if all the mass of the cylinder is at distance R from the axis. Thus, the cylinder must consist of a thin layer of material around an empty core.

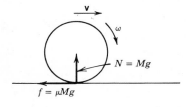

Handwritten margin notes (top left):
$\tau = r \mu mg$ $\alpha = \dfrac{r\mu mg}{I}$

$w = \int \dfrac{r\mu mg}{I} \, dt = \dfrac{r\mu mg}{rI} t$

Stop condition: $v = \omega r$

Handwritten margin notes (left column):
$E_0 = \tfrac{1}{2} m v_0^2$

$V = V_0 - \mu g t$

$\omega = \dfrac{r\mu mg}{I} t$

$V_0 - \mu g t = \dfrac{r^2 \mu mg}{I} t$

$t = \dfrac{V_0}{\left(\dfrac{\mu mg r^2 + \mu g}{I}\right)}$

$t = \dfrac{V_0}{\mu g\left(\dfrac{mr^2 + I}{I}\right)}$

$= \dfrac{V_0 I}{\mu g (mr^2 + I)}$

The I to equal $\tfrac{2}{5} m r^2$

$t = \dfrac{V_0 \, 2mr^2 5}{5\mu g (7mr^2)} = \dfrac{V_0 \cdot 2}{7\mu g}$

$V = V_0 - \mu g \dfrac{V_0 \cdot 2}{7\mu g}$

$= V_0 \left(1 - \tfrac{2}{7}\right)$

$= \dfrac{5 V_0}{7}$

$I = \int dm \, s^2$

PROBLEM 1.2.7   A bowling ball of mass M and radius R is thrown onto a surface with speed v_0. The coefficient of kinetic friction between the ball and the surface is μ. Initially, the ball is sliding without rolling. What will be its speed when it rolls without sliding?

The friction force f slows the speed of the mass center of the ball and increases the angular speed of the ball around its mass center:

$$M\frac{dv}{dt} = -Mg\mu \qquad \text{(horizontal component of external force)}$$

$$\frac{d}{dt}I\omega = Mg\mu R \qquad \text{(torque of external force about mass center)}$$

Thus,

$$v = v_0 - g\mu t$$

$$\omega = \omega_0 + \frac{Mg\mu Rt}{I} = \frac{Mg\mu Rt}{I}$$

Pure rolling will occur when $v = \omega R$, because then the point of contact of the ball with the surface will have zero speed relative to the surface. This occurs at a time t satisfying

$$v_0 - g\mu t = \frac{Mg\mu R^2}{I}t$$

$$t = \frac{v_0}{g\mu\left[1 + \dfrac{MR^2}{I}\right]}$$

The speed of the mass center at this time is

$$v = v_0 - g\mu\frac{v_0}{g\mu\left[1 + \dfrac{MR^2}{I}\right]} = v_0\frac{MR^2}{I + MR^2}$$

To complete the solution we need the moment of inertia of a uniform sphere about a diameter. Suppose that the center of the sphere is at the origin of a rectangular coordinate system. The moment of inertia of the sphere about the z axis is

$$I = \int dm \left(x^2 + y^2\right)$$

The symmetry of the sphere implies that

$$\int dm\, x^2 = \int dm\, y^2 = \int dm\, z^2 = \frac{1}{3}\int dm \left(x^2 + y^2 + z^2\right)$$

Handwritten note (center, near time equation):
M around center, not around axis.

$\tfrac{4\pi r^5}{5}\left(\dfrac{3}{4\pi r^3} M\right) = \dfrac{3Mr^2}{5}$

FIGURE 1.4 The center of the ball moves with velocity **v**, while the ball rotates about its center with angular speed ω. **N** and **f** are components of the force that the surface exerts on the ball.

so that we can write

$$I = \frac{2}{3} \int dm \left(x^2 + y^2 + z^2 \right) = \frac{2}{3} \frac{M}{(4/3)\pi R^3} \int_{\substack{r=0 \to R \\ \theta=0 \to \pi \\ \phi=0 \to 2\pi}} r^2 \sin\theta \, dr \, d\theta \, d\phi \cdot r^2$$

$$= \frac{2}{3} \frac{M}{(4/3)\pi R^3} 4\pi \int_0^R r^4 \, dr = \frac{2}{5} MR^2$$

Finally, the mass-center speed when the ball begins to roll without sliding is given by

$$v = v_0 \frac{MR^2}{\frac{2}{5}MR^2 + MR^2} = \frac{5}{7}v_0$$

PROBLEM 1.2.8   A ball of mass m and radius r rests on a sled of mass M, which in turn rests on a frictionless plane. If the coefficient of friction between the ball and sled is μ, what is the maximum horizontal force F that can be exerted on the sled without causing the ball to slide?

Let A be the acceleration (positive to the right) of the sled relative to the plane, and a the acceleration (positive to the right) of the ball also relative to the plane. Free-body diagrams of the ball and sled indicate all the forces acting on them. Newton's second law applied to each object yields

$$f = ma$$

$$F - f = MA$$

We can get one more relation by noting that fr is the total external torque on the

FIGURE 1.5 (a) The sled being pulled to the right with the ball on top. (b) Free-body diagram of the ball. (c) Free-body diagram of the sled.

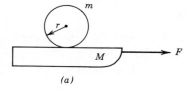

(a)

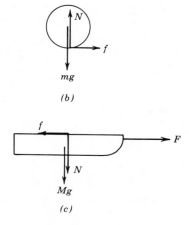

(b)

(c)

ball about its mass center, so that

$$fr = I\alpha = \tfrac{2}{5}mr^2\alpha$$

where α is the angular acceleration of the ball about its mass center. If the ball rolls on the sled, the points of the ball and sled that are in contact must be at rest relative to each other, so that

$$a + \alpha r = A$$

$$fr = \frac{2}{5}mr^2\left(\frac{A - a}{r}\right), \qquad f = \frac{2}{5}m(A - a)$$

and when this is combined with $f = ma$, we see that $A = \tfrac{7}{2}a$, and

$$F = ma + MA = \left(\tfrac{7}{2}M + m\right)a$$

But we also know that $f \leq \mu N = \mu mg$. Thus, $ma \leq \mu mg$, $a \leq \mu g$, and

$$F \leq \left(\tfrac{7}{2}M + m\right)\mu g$$

Thus, $F = \left(\tfrac{7}{2}M + m\right)\mu g$ is the maximum horizontal pull on the sled consistent with pure rolling (no sliding) of the ball.

1.3 IMPULSIVE FORCES

We can rewrite (1.8) in integrated form as follows:

$$\mathscr{I}_{\text{ext}} \equiv \int_{t_1}^{t_2} \mathbf{F}_{\text{ext}}\, dt = \mathbf{P}_{\text{tot}}(t_2) - \mathbf{P}_{\text{tot}}(t_1) \tag{1.16a}$$

$$\mathbf{J}_{\text{ext}} \equiv \int_{t_1}^{t_2} \boldsymbol{\tau}_{\text{ext}}\, dt = \mathbf{L}_{\text{tot}}(t_2) - \mathbf{L}_{\text{tot}}(t_1) \tag{1.16b}$$

Thus, the external *linear impulse*, $\mathscr{I}_{\text{ext}}$, equals the change in total linear momentum, and the external *angular impulse*, $\mathbf{J}_{\text{ext}}$, equals the change in total angular momentum.

Now suppose that the time interval $\Delta t \equiv t_2 - t_1$ approaches zero, but that the time average values of $\mathbf{F}_{\text{ext}}$ and $\boldsymbol{\tau}_{\text{ext}}$ become infinitely large, in such a way that the impulses (1.16a) and (1.16b) remain finite. Then the changes in linear and angular momentum will be finite. However, the linear and angular *displacements* that occur during Δt involve an additional power[2] of Δt, and so they vanish as $\Delta t \to 0$. Forces and torques that act so powerfully but so briefly that they produce finite changes in linear and angular momenta, while the system undergoes negligible displacement, are said to be *impulsive*. They are usually simple to handle, since their only effect is to produce discontinuous changes in linear or angular momenta.

---

[2]For example, if the force is constant during Δt, we have

$$\Delta P = F\Delta t$$

$$\Delta x = \frac{1}{2}a(\Delta t)^2 = \frac{1}{2}\left(\frac{F\Delta t}{m}\right)\Delta t$$

Thus, if $F \cdot \Delta t$ is finite as $\Delta t \to 0$, $(F \cdot \Delta t) \cdot \Delta t$ will vanish as $\Delta t \to 0$.

PROBLEM 1.3.1   A long narrow uniform stick of length l and mass M rests on a smooth horizontal surface. At time $t = 0$ it is struck a sharp horizontal blow near one end, in a direction perpendicular to the length of the stick, giving it a linear impulse $\mathcal{J}$. Describe its subsequent motion.

Before $t = 0$ the stick was at rest, so $\mathbf{P}_{tot}$ and $\mathbf{L}_{tot}$ were zero. Immediately after $t = 0$, $\mathbf{P}_{tot}$ equals $\mathcal{J}$. Since the horizontal surface is smooth, no further horizontal forces act on the stick, so that $\mathbf{P}_{tot}$ remains equal to $\mathcal{J}$. Hence, the mass center of the stick moves with constant velocity

$$\mathbf{V}_{CM} = \frac{\mathbf{P}_{tot}}{M} = \frac{\mathcal{J}}{M}$$

in the direction of the impulse. The angular impulse relative to the mass center of the stick has magnitude $\mathcal{J}l/2$ and is directed perpendicular to the surface. It equals the angular momentum of the stick relative to its mass center after the impulse. Thus, the angular speed of the stick about its mass center is given by

$$\omega = \frac{L}{I} = \frac{\mathcal{J}l/2}{I} = \frac{\mathcal{J}l/2}{\frac{1}{12}Ml^2} = \frac{6\mathcal{J}}{Ml}$$

The motion of the stick after the impulse consists of a uniform translation of the center of the stick with linear speed $\mathcal{J}/M$, while the stick rotates about its center with angular speed $6\mathcal{J}/Ml$.

PROBLEM 1.3.2   A proton of mass M moves with constant speed V along a nearly straight-line orbit that passes a distance b from an electron of mass m. Estimate the amount of kinetic energy acquired by the electron during the encounter, and the angle of deflection Θ of the proton's velocity. You may assume that Θ will be small, and that the movement of the electron is negligible during the period when most of the energy is transferred.

Let us use a dimensional argument to anticipate the dependence of Θ on m, M, b, V, and the charge e. Since the motion of the electron is neglected, m is irrelevant. The deflection angle Θ is dimensionless. Since the only dimensionless quantity that can be formed from M, b, V, and e is $(e^2/b)/(MV^2)$, we conclude that Θ must be a function of this combination of variables.

Since we are told that Θ will be small, we will calculate the momentum transfer to the electron by assuming that the orbit of the proton is perfectly straight (Figure 1.6). When the proton is a distance x from the origin, the Coulomb force it exerts on the electron is

$$\mathbf{F}\left(\begin{array}{c}\text{on electron,}\\\text{due to proton}\end{array}\right) = e^2 \frac{x\hat{x} - b\hat{y}}{|x\hat{x} - b\hat{y}|^3}$$

FIGURE 1.6 The stationary electron is on the *y* axis, while the proton moves along the *x* axis with speed *V*.

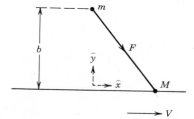

[cf. (1.10)]. The impulse felt by the electron is

$$\mathscr{I} = \int_{-\infty}^{\infty} \mathbf{F}\, dt = e^2 \int_{-\infty}^{\infty} \frac{x\hat{x} - b\hat{y}}{\left[x^2 + b^2\right]^{3/2}}\, dt = e^2 \int_{-\infty}^{\infty} \frac{x\hat{x} - b\hat{y}}{\left[x^2 + b^2\right]^{3/2}} \frac{dx}{V}$$

$$= -\frac{be^2}{V}\hat{y} \int_{-\infty}^{\infty} \frac{dx}{\left[x^2 + b^2\right]^{3/2}} = -\frac{2e^2}{Vb}\hat{y}$$

The x integral is conveniently done using the substitution $x = b \tan\theta$. Since the initial electron momentum was zero, the final electron momentum is $\mathbf{p} = \mathscr{I} = -(2e^2/Vb)\hat{y}$, and the final electron kinetic energy is

$$K.E. = \frac{1}{2}mv^2 = \frac{p^2}{2m} = \frac{2e^4}{mV^2b^2}$$

This result is useful in the theory of energy loss of charged particles moving through matter.

The impulse on the proton is opposite to the impulse on the electron. Thus, the final velocity of the proton is approximately $V\hat{x} + (2e^2/MVb)\hat{y}$. This implies that the direction of motion of the proton is deflected through an angle

$$\Theta = \arctan\left(\frac{2e^2}{MV^2b}\right)$$

which must be small if our initial assumption of a straight-line orbit is to be accurate. In Section 1.12, we will see that the exact result in this case is

$$\Theta = 2\arctan\left(\frac{e^2}{MV^2b}\right)$$

These two expressions for Θ are approximately equal when Θ is small.

1.4 CONSERVATION OF LINEAR MOMENTUM, ANGULAR MOMENTUM, AND MECHANICAL ENERGY

If the external force and/or torque are equal to zero, (1.4a) and (1.4b) tell us that the linear and/or angular momentum are independent of time.

From Newton's second law for a point particle we can derive

$$\mathbf{f} \cdot \mathbf{v} = m\frac{d\mathbf{v}}{dt} \cdot \mathbf{v} = \frac{m}{2}\frac{d}{dt}(\mathbf{v} \cdot \mathbf{v}) = \frac{d}{dt}\left(\frac{1}{2}mv^2\right) \tag{1.17a}$$

$\mathbf{f} \cdot \mathbf{v}$ is the rate at which $\mathbf{f}$ does work on the particle, and (1.17a) shows that it equals the rate of change of the particle's kinetic energy. If we integrate (1.17a) from $t = t_a$ to $t = t_b$, we get

$$\int_{t_a}^{t_b} \mathbf{f} \cdot \mathbf{v}\, dt = \int_{\mathbf{r}_a}^{\mathbf{r}_b} \mathbf{f} \cdot d\mathbf{r} = \int_{t_a}^{t_b} \frac{d}{dt}\left(\frac{1}{2}mv^2\right) dt$$

$$= \frac{1}{2}mv^2(t_b) - \frac{1}{2}mv^2(t_a) \tag{1.17b}$$

and the work done on the particle when it moves from $\mathbf{r}_a$ to $\mathbf{r}_b$ equals its change in kinetic energy.

If the system is *conservative*, the total work done on a particle when it moves around any closed curve is zero:

$$\oint_{\substack{\text{any closed} \\ \text{curve}}} \mathbf{f} \cdot d\mathbf{r} = 0 \qquad (1.18a)$$

Now consider two points $(\mathbf{r}_a, \mathbf{r}_b)$, and any two paths connecting them. We can combine these two paths to make a closed curve, to which (1.18a) applies. This means that

$$\underset{\substack{\mathbf{r}_a \\ (\text{path 1})}}{\int^{\mathbf{r}_b}} \mathbf{f} \cdot d\mathbf{r} + \underset{\substack{\mathbf{r}_b \\ (\text{path 2})}}{\int^{\mathbf{r}_a}} \mathbf{f} \cdot d\mathbf{r} = 0$$

$$\underset{\substack{\mathbf{r}_a \\ (\text{path 1})}}{\int^{\mathbf{r}_b}} \mathbf{f} \cdot d\mathbf{r} + \left(- \underset{\substack{\mathbf{r}_a \\ (\text{path 2})}}{\int^{\mathbf{r}_b}} \mathbf{f} \cdot d\mathbf{r} \right) = 0$$

$$\underset{\substack{\mathbf{r}_a \\ (\text{path 1})}}{\int^{\mathbf{r}_b}} \mathbf{f} \cdot d\mathbf{r} = \underset{\substack{\mathbf{r}_a \\ (\text{path 2})}}{\int^{\mathbf{r}_b}} \mathbf{f} \cdot d\mathbf{r}$$

so that the work done on the particle when it moves from $\mathbf{r}_a$ to $\mathbf{r}_b$ is independent of the path from $\mathbf{r}_a$ to $\mathbf{r}_b$. In this case we choose a reference point $\mathbf{r}_0$, and define the *potential* energy at point $\mathbf{r}_a$ by

$$U(\mathbf{r}_a) = - \int_{\mathbf{r}_0}^{\mathbf{r}_a} \mathbf{f} \cdot d\mathbf{r} \qquad \left(\begin{array}{l} \text{independent of the} \\ \text{path between } \mathbf{r}_0 \text{ and } \mathbf{r}_a \end{array} \right) \qquad (1.18b)$$

Then we can write

$$\int_{\mathbf{r}_a}^{\mathbf{r}_b} \mathbf{f} \cdot d\mathbf{r} = \int_{\mathbf{r}_0}^{\mathbf{r}_b} \mathbf{f} \cdot d\mathbf{r} - \int_{\mathbf{r}_0}^{\mathbf{r}_a} \mathbf{f} \cdot d\mathbf{r} = U(\mathbf{r}_a) - U(\mathbf{r}_b)$$

and (1.17b) becomes

$$\tfrac{1}{2}mv^2(t_a) + U(\mathbf{r}_a) = \tfrac{1}{2}mv^2(t_b) + U(\mathbf{r}_b) \qquad (1.19)$$

Thus, the *mechanical energy*, defined as the sum of the kinetic and potential energies, is independent of time.

If we know the potential energy function $U(\mathbf{r}_a)$, we can recover the force field $\mathbf{f}(\mathbf{r}_a)$ by calculating the gradient of (1.18b) with respect to $\mathbf{r}_a$,

$$\mathbf{f}(\mathbf{r}_a) = -\nabla_{\mathbf{r}_a} U(\mathbf{r}_a) \qquad (1.20)$$

If we change the reference point from $\mathbf{r}_0$ to $\mathbf{r}_0'$, the effect on $U(\mathbf{r}_a)$ in (1.18b) is the addition of $U(\mathbf{r}_0') - U(\mathbf{r}_0)$, which is independent of $\mathbf{r}_a$. This has no effect on the gradient of $U(\mathbf{r}_a)$ with respect to $\mathbf{r}_a$, which verifies that the force field $\mathbf{f}(\mathbf{r}_a)$ does not depend on the reference point $\mathbf{r}_0$ chosen for defining the potential energy.

A particle moving in a central force field experiences a force whose position dependence is given by

$$\mathbf{f}(\mathbf{r}) = g(r)\hat{r} \qquad (1.21)$$

Here $\mathbf{r}$ is the vector from the "force center" to the location of the particle. (The gravitational force on particle 1 exerted by particle 2, given by (1.10), can be regarded as a central force with the location of particle 2 as the force center.) We can easily verify that every central force is conservative, since

$$\int_{\mathbf{r}_0}^{\mathbf{r}_a} g(r)\hat{r} \cdot d\mathbf{r} = \int_{r_0}^{r_a} g(r)\, dr$$

which depends only on $\mathbf{r}_0$ and $\mathbf{r}_a$, and not on the integration path from $\mathbf{r}_0$ to $\mathbf{r}_a$. For the particular case of the gravitational force (1.10), $g(r)$ is given by

$$g(r) = -G\frac{m_1 m_2}{r^2}$$

and (1.18b) becomes

$$U(\mathbf{r}_a) = -\int_{r_0}^{r_a}\left(-G\frac{m_1 m_2}{r^2}\right) dr = -Gm_1 m_2 \left[\frac{1}{r_a} - \frac{1}{r_0}\right]$$

In this case it is usual to choose the reference point $\mathbf{r}_0$ to be at infinity, so that $1/r_0 = 0$ and

$$U(\mathbf{r}_a) = -G\frac{m_1 m_2}{r_a} \tag{1.22}$$

It can be easily verified that if we apply (1.20) to (1.22), we recover the original gravitational force field (1.10).

If $\mathbf{f}$ is the force on a particle due to a uniform gravitational field (1.11), the potential energy is

$$U(\mathbf{r}_a) = -\int_{\mathbf{r}_0}^{\mathbf{r}_a} m\mathbf{g} \cdot d\mathbf{r} = mg\int_{\mathbf{r}_0}^{\mathbf{r}_a} \hat{z} \cdot d\mathbf{r} = mg(z_a - z_0) \tag{1.23}$$

Thus, the change in potential energy equals mg times the change in height. We can also use (1.23) for a finite object, if we interpret m as its total mass and $z_a - z_0$ as the change in height of the mass center.

Any one-dimensional force field $f(x)$ is conservative if the force depends on x alone. In the particular case of a harmonic restoring force (1.14), we have

$$U(x_a) = -\int_{x=0}^{x=x_a} (-kx)\, dx = \frac{1}{2}kx_a^2 \tag{1.24}$$

Here we have followed the usual practice of using the equilibrium position, $x = 0$, as the zero of potential energy.

Conservation of mechanical energy is especially useful when we need to solve for $v(t)$ or for the elapsed time, $\int dr/v(r)$. To calculate the kinetic energy of a finite object, it is convenient to use the formula

$$T = \frac{1}{2}\int dm(\dot{\mathbf{R}} + \dot{\mathbf{s}})^2 = \frac{1}{2}M(\dot{\mathbf{R}})^2 + \frac{1}{2}\int dm(\dot{\mathbf{s}})^2 \tag{1.25}$$

where $\mathbf{R}$ is the vector to the mass center of the object, and $\mathbf{s}$ locates each particle of the object relative to its mass center. The second term in (1.25) would be the kinetic energy of the object as seen by an observer who moves with the mass center. If the motion of the object about its mass center consists of rigid-body rotation with angular speed ω about a principal axis with moment of inertia I,

FIGURE 1.7 Locating a point in an object relative to its mass center.

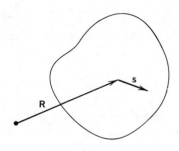

then (1.25) becomes

$$T = \tfrac{1}{2}M(\dot{\mathbf{R}})^2 + \tfrac{1}{2}I\omega^2 \qquad (1.26)$$

PROBLEM 1.4.1   A massless ice skater is holding two weights each 50 cm from his body, and rotating four times a second around a vertical axis. He then extends each of the weights to 1 m from his body. How many times a second does he rotate?

Since no external torque acts, angular momentum is conserved:

$$L_1 = I_1\omega_1 = L_2 = I_2\omega_2$$

$$\omega_2 = \omega_1 I_1/I_2 = \omega_1\left(\frac{0.5}{1.0}\right)^2 = \frac{\omega_1}{4}$$

so his final angular speed is one revolution per second.

PROBLEM 1.4.2   A simple pendulum of length L suddenly doubles its length as it passes through its vertical position. Find the change in amplitude and frequency of motion.

The external forces acting on the pendulum are gravity and the support force at the pivot. The support force has zero torque about the pivot. Furthermore, when the pendulum is vertical the line of action of the gravitational force also passes through the pivot, so that gravity also has zero torque about the pivot. Thus, during the extension of the pendulum there is no external torque about the pivot, so the angular momentum about the pivot is conserved.

Before $t = 0$, $\theta(t) = A\sin\omega t$, with $\omega = \sqrt{g/L}$. The angular momentum at $t = 0_-$ is $ML^2\dot{\theta}(t = 0_-) = AML^2\omega = AM\sqrt{gL^3}$. Immediately after $t = 0$, we have $\theta(t) = A'\sin\omega't$, with $\omega' = \sqrt{g/2L}$. The angular momentum at $t = 0_+$ is $M(2L)^2\dot{\theta}(t = 0_+) = 4A'ML^2\omega' = A'M\sqrt{8gL^3}$. Thus, angular momentum conservation implies that $AM\sqrt{gL^3} = A'M\sqrt{8gL^3}$, so we conclude that

$$A' = \frac{A}{2\sqrt{2}}, \qquad \omega' = \frac{\omega}{\sqrt{2}}$$

Note that the total mechanical energy of the pendulum is not conserved during the extension of the pendulum, since the forces that act on the pendulum during the extension do some work.

PROBLEM 1.4.3 Consider the stick in Problem 1.2.3. What is the speed of the right end at the instant it strikes the table?

The mechanical energy of the stick is conserved as it falls, since the force that the surface exerts on the left end is perpendicular to the velocity of that end, so that $\mathbf{f} \cdot \mathbf{r} = 0$ and no work is done by $\mathbf{f}$. Let us choose the condition of zero potential energy to occur when the stick lies horizontally on the surface. Initially, the kinetic energy is zero and the potential energy is $mg(l/2)\sin\theta$. Thus, the initial (and constant) mechanical energy is $mg(l/2)\sin\theta$. When the stick hits the surface its potential energy is zero, so that the mechanical energy equals the kinetic energy

$$mg\frac{l}{2}\sin\theta = \frac{m}{2}v_{\mathrm{CM}}^2 + \frac{1}{2}I\omega^2$$

At the instant the stick hits the surface, its motion consists of pure rotation about its stationary left end. Thus, $v_{\mathrm{CM}} = l\omega/2$, and we have

$$mg\frac{l}{2}\sin\theta = \frac{m}{2}\left(\frac{l\omega}{2}\right)^2 + \frac{1}{2} \cdot \frac{1}{12}ml^2\omega^2 = \frac{ml^2}{6}\omega^2$$

$$\omega = \sqrt{\frac{3g}{l}\sin\theta}$$

and the speed of the right end of the stick is

$$v_{\mathrm{re}} = l\omega = \sqrt{3gl\sin\theta}$$

PROBLEM 1.4.4 A yo-yo of mass m and moment of inertia I (about its spin axis) falls from rest. As it falls, the string unwinds from the inner shaft of radius b. Assuming no energy dissipation, find the speed v of the mass center of the *yo-yo* as a function of the vertical distance of fall.

If the yo-yo falls a distance s, its potential energy decreases by mgs, and so its kinetic energy increases by mgs (the string does no work on the yo-yo). But the kinetic energy is given by

$$T = \frac{1}{2}mv^2 + \frac{1}{2}I\omega^2 = \frac{1}{2}mv^2 + \frac{1}{2}I\left(\frac{v}{b}\right)^2 = \frac{1}{2}\left(m + \frac{I}{b^2}\right)v^2$$

Thus, when the yo-yo has fallen a distance s,

$$\frac{1}{2}\left(m + \frac{I}{b^2}\right)v^2 = mgs, \qquad v = \sqrt{\frac{2gs}{1 + (I/mb^2)}}$$

FIGURE 1.8 The string is attached to the inner shaft of the falling rotating yo-yo.

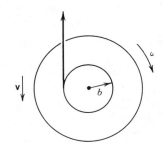

PROBLEM 1.4.5   A uniform stick of length l is pivoted at one end, and released from rest in a horizontal position. How long will it take for it to swing down to the vertical position?

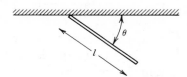

We use energy conservation to calculate the angular speed as a function of θ. Take the condition of zero potential energy to occur when the stick is horizontal. Since the stick is released from rest, its kinetic energy is also zero when the stick is horizontal, so the total mechanical energy is zero. When the stick has fallen to angle θ, its potential energy is $-mg(l/2)\sin\theta$, which implies that its kinetic energy is $+mg(l/2)\sin\theta$. We equate this to $(1/2)I\dot\theta^2$, where I is the moment of inertia of the stick about its pivoted end:

$$mg\frac{l}{2}\sin\theta = \frac{1}{2}I\dot\theta^2 = \frac{1}{2}\left(\frac{1}{3}ml^2\right)\dot\theta^2$$

$$\frac{d\theta}{dt} = \sqrt{\frac{3g}{l}\sin\theta}, \qquad dt = \sqrt{\frac{l}{3g}}\frac{d\theta}{\sqrt{\sin\theta}}$$

Thus, the time T required to go from $\theta = 0$ to $\theta = 90°$ is

$$\int_{t=0}^{T} dt = T = \sqrt{\frac{l}{3g}}\int_{\theta=0}^{\pi/2}\frac{d\theta}{\sqrt{\sin\theta}}$$

$$\approx 2.622\sqrt{\frac{l}{3g}}$$

This integral over θ can be expressed in terms of a complete elliptic integral of the first kind.

PROBLEM 1.4.6   Because of increasing air pollution, the mean air temperature on earth will probably rise, making for longer growing seasons and resulting in taller shrubs and trees, with a consequent increase in the moment of inertia of the earth.

(a) Assume that this growth is equivalent to raising a thin ring of mass m to a height h above the ground. Calculate the fractional change $\Delta T/T$ in the length of the day. Assume that the earth is a sphere and that the vegetation is concentrated around the equator.

Let the original moment of inertia about the earth's axis be I_0. When we raise a mass m from R to $R + h$, we increase the moment of inertia by

$$m(R + h)^2 - mR^2 \simeq 2mRh$$

Thus if ω_i and ω_f are the initial and final angular speeds, angular momentum conservation requires that

$$L = I_0\omega_i = (I_0 + 2mRh)\omega_f$$

If T_i and T_f are the initial and final rotation periods,

$$T_f = 2\pi/\omega_f, \qquad T_i = 2\pi/\omega_i$$

$$\frac{T_f - T_i}{T_i} = \frac{2\pi/\omega_f - 2\pi/\omega_i}{2\pi/\omega_i} = \frac{\omega_i}{\omega_f} - 1 = \frac{I_0 + 2mRh}{I_0} - 1$$

$$\Delta T/T = 2mRh/I_0$$

FIGURE 1.9 A loop of chain resting in a horizontal plane on a smooth cone of half-angle α.

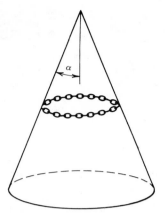

(b) Does the total kinetic energy increase, decrease, or remain the same?

Initially, the kinetic energy is $(1/2)I_0\omega_i^2 = L^2/(2I_0)$. Finally, it is $(1/2)(I_0 + 2mRh)\omega_f^2 = L^2/(2(I_0 + 2mRh))$. Since angular momentum L is conserved, we have

$$\frac{(K.E.)_i}{(K.E.)_f} = \frac{L^2/2I_0}{L^2/(2(I_0 + 2mRh))} = \frac{I_0 + 2mRh}{I_0} > 1$$

and the kinetic energy *decreases* as a result of the increase in moment of inertia.

PROBLEM 1.4.7   A single closed loop of chain of mass m and length l rests on the surface of a smooth, frictionless cone. The chain lies in a horizontal plane. The half-angle of the cone is α. Determine the tension in the chain.

Let the tension be T. If the chain length increased by dl, the elastic potential energy would increase by $T\,dl$. The radius would increase by $dl/2\pi$, the height would decrease by $dl/(2\pi \tan \alpha)$, so the gravitational potential energy would decrease by $mg\,dl/(2\pi \tan \alpha)$. Thus, the total potential energy change associated with the length change dl is

$$dU = \left(T - \frac{mg}{2\pi \tan \alpha} \right) dl$$

At equilibrium this should be zero, since otherwise stretching or contracting the chain would be associated with an increase in kinetic energy. Thus,

$$T = \frac{mg}{2\pi \tan \alpha}$$

1.5 COLLISIONS BETWEEN PARTICLES

In collision problems it is assumed that the colliding particles interact for such a short time that the impulse due to external forces is negligible. Thus, the total linear momentum of the system is the same immediately before and after the collision. The simplest way to impose the constraint of constant total linear momentum is to work in a coordinate system in which the total linear momentum is zero, the mass-center system. If the total kinetic energy of the particles is unchanged by the collision, the collision is said to be elastic. For an inelastic

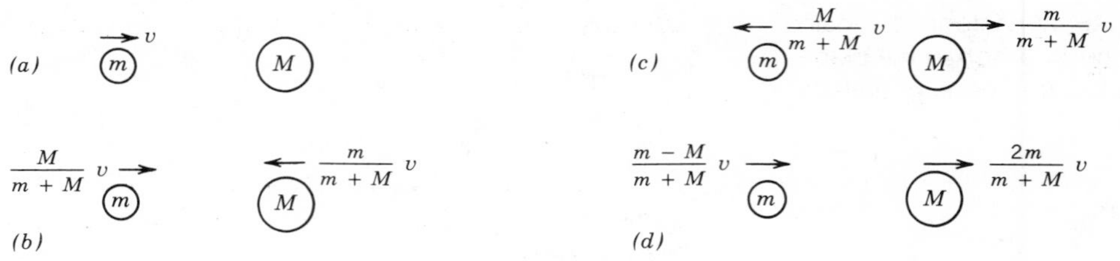

FIGURE 1.10 Two masses colliding in one dimension.

(a) $\overrightarrow{m}$ v M

(c) $\overleftarrow{\dfrac{M}{m+M}v}$ m M $\overrightarrow{\dfrac{m}{m+M}v}$

$\dfrac{M}{m+M}v$ $\overrightarrow{}$ m M $\overleftarrow{\dfrac{m}{m+M}v}$

(b)

$\dfrac{m-M}{m+M}v$ $\overrightarrow{}$ m M $\overrightarrow{\dfrac{2m}{m+M}v}$

(d)

collision the Q-value is defined to be the increase in total kinetic energy (so $Q = 0$ for an elastic collision).

In an elastic collision in one dimension, an incident particle with speed v and mass m strikes a stationary particle of mass M. What fraction of the initial kinetic energy of the incident particle is transferred to the target particle? Figure 1.10a shows the situation before the collision, as viewed from the laboratory. The mass center of the two particles moves to the right, relative to the laboratory, with speed $m/(m + M) \cdot v$. Figure 1.10b shows this same situation, but viewed by an observer moving to the right with the speed of the mass center. Note that the relative velocity of the two particles still equals v, but the total momentum is now zero. The total kinetic energy in this mass-center system is

$$\frac{1}{2}m\left(\frac{M}{m+M}v\right)^2 + \frac{1}{2}M\left(\frac{m}{m+M}v\right)^2 = \frac{1}{2}\frac{mM}{m+M}v^2$$

After the collision, the total momentum in the mass-center system is still zero, so the speeds of m and M are still in the ratio M/m, and their motion is oppositely directed. If their relative speed is v', their total kinetic energy is $1/2 \times mM(v')^2/(m + M)$. Since the collision is elastic, it must be that $(v')^2 = v^2$. The solution $v' = v$ would correspond to the initial situation, or the particles would move past each other without interacting. The only other way of conserving momentum and energy is to set $v' = -v$. This is depicted in Figure 1.10c. Finally, we make the transition back to the laboratory system by adding the mass-center velocity $(m/(m + M) \cdot v$ to the right) to each velocity of Figure 1.10c. The result is shown in Figure 1.10d. Thus, the final kinetic energy of the target particle is $(1/2)M(2mv/(m + M))^2$, and the fraction of the energy of the incident particle transferred to the target particle is

$$\frac{\dfrac{1}{2}M\left(\dfrac{2m}{m+M}v\right)^2}{\dfrac{1}{2}mv^2} = \frac{4mM}{(m+M)^2}$$

Inspection of Figure 1.10d shows that in all cases the target particle moves to the right, but whether the incident particle moves to the right or to the left after the collision depends on whether its mass is greater or less than the mass of the target particle. In particular, if the incident and target particles have the same mass, the speed of the incident particle after the collision is zero, and it transfers all of its kinetic energy to the target particle.

PROBLEM 1.5.1 A ball of mass m is dropped down an elevator shaft. Just before it hits the top of the elevator, it has downward speed v and the elevator has upward speed V, both measured relative to the shaft. What is the upward speed of the ball (relative to the shaft) immediately after it makes an elastic collision with the top of the elevator?

Let us view the ball from the elevator. Immediately before the collision it has downward speed $v + V$. Immediately after the collision it has upward speed $v + V$. Thus, the upward speed immediately after the collision, as seen from the shaft, is $(v + V) + V = v + 2V$.

PROBLEM 1.5.2 The ball in the previous problem is dropped at the instant when the top of the rising elevator is below it by a distance h. How high will the ball rebound?

Suppose that it takes time t for the ball to hit the elevator. During this time the ball falls a distance $(1/2)gt^2$ and the elevator rises a distance Vt, and so $(1/2)gt^2 + Vt = h$. The $t > 0$ solution of this equation is

$$t = \frac{\sqrt{V^2 + 2gh} - V}{g}$$

Then the downward speed, v, of the ball just before it hits the top of the elevator is

$$v = gt = \sqrt{V^2 + 2gh} - V$$

According to the result of Problem 1.5.1, the upward speed of the ball after the collision is

$$v + 2V = \sqrt{V^2 + 2gh} + V$$

so the increase in mechanical energy of the ball, as a result of the collision, is

$$\frac{m}{2}\left[\sqrt{V^2 + 2gh} + V\right]^2 - \frac{m}{2}\left[\sqrt{V^2 + 2gh} - V\right]^2 = 2mV\sqrt{V^2 + 2gh}$$

Thus, the ball will rebound to a height of

$$z = \frac{2mV\sqrt{V^2 + 2gh}}{mg} = \frac{2V\sqrt{V^2 + 2gh}}{g}$$

above the point at which it was released.

PROBLEM 1.5.3 A uniform rod of length l and mass M lies on a smooth horizontal surface. A bullet of mass m, traveling with speed v in a direction perpendicular to the rod, strikes the end of the rod and becomes embedded there. Describe the subsequent motion of the system.

Let us view the collision in the mass-center system, which moves with speed $m/(m + M) \cdot v$. In Figure 1.11a (which is analogous to Figure 1.10b), the rod and bullet approach each other. Of course, in this mass-center system, the mass center appears to be stationary. After the bullet comes to rest in the rod, the combined system rotates about the still stationary mass center (Figure 1.11b). Since the external angular impulse is negligible during the collision, the total angular momentum is conserved. Before the collision, its magnitude is

$$L = m\left(\frac{M}{m + M}\right)v\left(\frac{M}{m + M}\frac{l}{2}\right) + M\left(\frac{m}{m + M}\right)v\left(\frac{m}{m + M}\frac{l}{2}\right)$$

$$= \frac{mM}{m + M}v\frac{l}{2}$$

FIGURE 1.11 (a) Before the collision, as seen in the mass-center system. The symbol ⊗ locates the stationary center of mass. (b) immediately after the collision, the system rotates about the still stationary mass center.

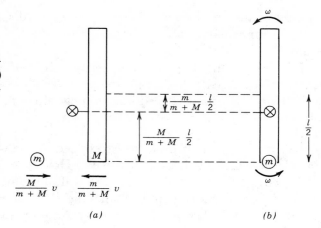

After the collision, $L = I\omega$, where ω is the angular speed with which the system rotates about its mass center, and I is the total moment of inertia. This is composed of two parts:

$$I(\text{bullet}) = m\left(\frac{M}{m + M}\frac{l}{2}\right)^2$$

$$I(\text{stick}) = \frac{1}{12}Ml^2 + M\left(\frac{m}{m + M}\frac{l}{2}\right)^2$$

This last expression is an example of the use of the parallel axis theorem.[3] Altogether, $I = 1/12 \times M(M + 4m)l^2/(M + m)$, and

$$\omega = \frac{\dfrac{mM}{m + M}v\dfrac{l}{2}}{\dfrac{1}{12}M\left(\dfrac{M + 4m}{m + M}\right)l^2} = \frac{6mv}{(M + 4m)l}$$

In the laboratory, it appears as if the mass center of the system moves with constant velocity (speed $m/(m + M) \cdot v$), while the system rotates with constant angular speed ω about the mass center.

1.6 PROBLEMS IN WHICH THE MOTION IS SPECIFIED AND THE FORCES MUST BE DETERMINED

According to (1.1a) the force exerted on an object equals the rate at which its momentum changes. If we know the motion of the object, we can calculate the rate of change of momentum and, hence, the force acting on it.

---

[3] The parallel axis theorem states that I_A, the moment of inertia of an object about an axis A, is given by

$$I_A = I_{\text{CM}} + Md^2$$

where I_{CM} is the moment of inertia of the object about an axis parallel to A through the mass center, d is the perpendicular distance between the two axes, and M is the total mass of the object. This theorem is easily derived from the definition (1.7) of the moment of inertia.

FIGURE 1.12 The water approaches the wall perpendicular to it, and ends up moving parallel to it.

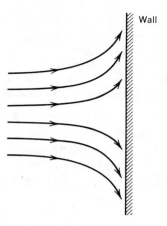

Wall

PROBLEM 1.6.1   A stream of water of constant cross-sectional area A and speed v strikes a wall and spreads out sideways. What pressure does it exert on the wall?

In time dt the mass of water striking the wall is $v\,dt \cdot A \cdot \rho$, where ρ is the density. Before it strikes, the component of momentum of this water perpendicular to the wall is $v\,dt\,A\rho \cdot v$. After it strikes, the water moves parallel to the wall, so the perpendicular component of its momentum is zero. Thus, in time dt, the wall changes the momentum of the water by $\rho A v^2\,dt$. The rate at which the wall delivers momentum to the water is $\rho A v^2$, and this is the force that the wall exerts on the water. According to Newton's third law, it is also the force that the water exerts on the wall. Thus, the pressure is ρv^2.

PROBLEM 1.6.2   A chain of mass m and length l is suspended above a platform scale, with the lowest link just touching the platform. The chain is released. What is the scale reading when the last link hits the platform? Assume that each link falls independently of the others.

When the last links hit the platform, their speed is $\sqrt{2gl}$. Consider a time interval dt. The mass of chain hitting during dt is $(m/l) \cdot \sqrt{2gl}\,dt$ and the momentum delivered to the scale is $\sqrt{2gl} \cdot (m/l) \cdot \sqrt{2gl}\,dt = 2mg\,dt$. Thus, the

FIGURE 1.13 The chain at the instant it is released.

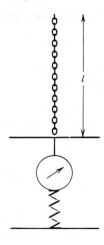

l

impact force as the last link hits is $2mg$. At this time the weight of chain on the scale is mg, so that the total scale reading is $3mg$.

1.7 USE OF NON-INERTIAL REFERENCE FRAMES

In Section 1.1 we stated that Newton's second law of motion is valid if we locate the particle relative to a point fixed in an inertial frame. Sometimes, however, the forces and constraints acting on the system under study are most conveniently expressed in a noninertial reference frame. In these cases it may be advantageous to use this noninertial frame, even though (1.1) must then be replaced by something somewhat more complicated.

1.7.1 Uniformly Accelerated Reference Frames

In Figure 1.14, 0 is a fixed point in an inertial reference frame. Point P undergoes constant acceleration relative to 0. Thus,

$$\frac{d^2\mathbf{s}}{dt^2} = \mathbf{a} \qquad \text{(constant)}$$

$$\frac{d^2\mathbf{r}}{dt^2} = \frac{d^2}{dt^2}(\mathbf{r}' + \mathbf{s}) = \frac{d^2\mathbf{r}'}{dt^2} + \mathbf{a}$$

Substituting this into (1.1), we get

$$\mathbf{F} - m\mathbf{a} = m\frac{d^2\mathbf{r}'}{dt^2} \tag{1.27}$$

We can interpret (1.27) as telling us that we can still use Newton's second law of motion when the masses are located relative to a uniformly accelerated coordinate system, provided that we add a fictitious force $-m\mathbf{a}$ to the real force $\mathbf{F}$ acting on each particle. This would be the effect of a uniform gravitational field whose strength and direction are given by $-\mathbf{a}$ [cf. (1.11)]. Thus, we can use Newton's second law with a uniformly accelerated reference frame provided that we add the fictitious uniform gravitational field $-\mathbf{a}$ to any real fields acting on the system.

PROBLEM 1.7.1   A helium-filled balloon is tied by a length of string to the floor of a closed automobile, which accelerates to the right with constant acceleration A. What will be the orientation of the string?

The fictitious gravitational field experienced in the accelerating car has magnitude A and points to the left. This must be added vectorially to the real

FIGURE 1.14 Point O is fixed in an inertial frame. Point P accelerates relative to O with constant acceleration **a**.

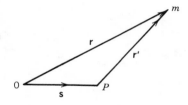

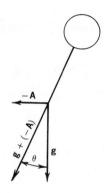

FIGURE 1.15 The string is parallel to the total effective gravitational field.

gravitational field, that has magnitude g and points downward. The total effective gravitational field points to the left of downward, by an angle $\theta = \arctan(g/A)$. The balloon string points in the effective "up" direction, so it points to the *right* of the upward vertical by the same angle.

This problem can also be solved by an analysis of the forces acting on the balloon (see Figure 1.16). Gravity acts downward with a force of magnitude mg. The string pulls at an angle θ to the vertical. Let the magnitude of the string force be f, so that

$$\mathbf{f}_s = f_s(-\cos\theta\,\hat{y} - \sin\theta\,\hat{x})$$

To calculate the force on the balloon due to the surrounding air, we use Archimedes' principle. Imagine that the helium-filled balloon is replaced by air enclosed within an imaginary surface of the same size and shape as the balloon. Let the mass of the air be m_a. The surrounding air exerts a force across the imaginary surface that serves to support the weight of the air and to give it an acceleration to the right of magnitude A. Thus, the force exerted by the surrounding air across the imaginary surface is

$$\mathbf{f}_a = m_a g\hat{y} + m_a A\hat{x}$$

and this is also the force that the surrounding air exerts on the balloon. Then the total force acting on the balloon is

$$\hat{x}(m_a A - f_s\sin\theta) + \hat{y}(m_a g - f_s\cos\theta - mg)$$

FIGURE 1.16 The forces acting on the balloon. The pressure gradient in the surrounding air causes the force that the air exerts on the balloon to vary over the surface of the balloon.

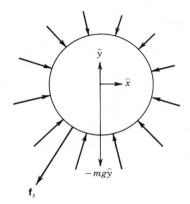

which must be set equal to m times the acceleration of the balloon $A\hat{x}$. This leads to

$$m_a A - f_s\sin\theta = mA$$
$$m_a g - f_s\cos\theta = mg$$

from which we calculate

$$f_s\sin\theta = (m_a - m)A$$
$$f_s\cos\theta = (m_a - m)g$$
$$\tan\theta = A/g$$

in agreement with the result that we reached much more quickly using the effective gravitational field equivalent to the accelerated reference frame.

PROBLEM 1.7.2 A small mass attached to a vertical spring undergoes simple harmonic oscillations in a laboratory at rest on the earth's surface (assumed to be an inertial frame). The measured oscillation frequency of the mass is ω. Now the system is mounted in an elevator accelerating upward with acceleration A. What will be the new oscillation frequency?

　　　If we work in a reference frame accelerating upward with acceleration A, the total effective gravitational field is downward and of magnitude $g + A$. Using this total effective gravitational field, we can analyze the motion of the mass using Newton's second law of motion. This leads to an oscillation frequency of $\omega = \sqrt{k/m}$ as before. Thus, the acceleration of the elevator has no effect on the oscillation frequency. However, the equilibrium length of the spring (see Section 1.2.3) would be increased from $l_0 + mg/k$ to $l_0 + m(g + A)/k$.

PROBLEM 1.7.3 A box of mass M is sliding down a plane inclined at an angle θ to the horizontal. The coefficient of friction between the box and the plane is μ. A pendulum bob of mass m is hung by a string from the top of the hollow cavity in the box and

FIGURE 1.17 (a) The box sliding down the plane; (b) free-body diagram showing the external forces acting on the box and its contents; (c) the components of the total effective gravitational field.

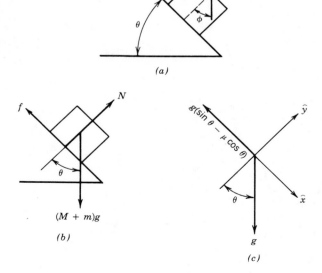

(a)

(b)

(c)

comes to equilibrium. Find the angle ϕ that the pendulum makes with the normal to the inclined plane.

First, we find the acceleration a of the box along the plane. Since there is no acceleration perpendicular to the plane, $N = (M + m)g \cos \theta$. The component of the net force down the plane is

$$(M + m)g \sin \theta - f = (M + m)g(\sin \theta - \mu \cos \theta)$$

so that a is given by

$$a = g(\sin \theta - \mu \cos \theta)$$

Now we go into a reference frame accelerating down the plane with the box. The total effective gravitational field is

$$g(-\hat{y} \cos \theta + \hat{x} \sin \theta) - \hat{x}g(\sin \theta - \mu \cos \theta) = g \cos \theta(\mu\hat{x} - \hat{y})$$

At equilibrium the string hangs in the direction of this effective gravitational field. The angle ϕ defined in Figure 1.17 is thus

$$\phi = \arctan \frac{(g_{\text{eff}})_x}{(-g_{\text{eff}})_y} = \arctan \mu$$

PROBLEM 1.7.4 A car is started from rest with one of its doors fully open (perpendicular to the body of the car). If the hinges of the door are toward the front of the car, the door will slam shut as the car picks up speed. The acceleration A of the car is constant, and the distance between the mass center of the door and the hinge axis is d. The radius of gyration of the door is r_0 $(\equiv \sqrt{I/m})$. Obtain a formula for the time needed for the door to close.

Go into a reference frame accelerating with the car. This produces a fictitious gravitational field of strength A, pointing toward the rear of the car. Thus, the closing door is analogous to a falling pendulum that has been released from rest in a horizontal position. This was Problem 1.4.5. The time required for the door to swing through an angle $\pi/2$ is

$$T = \sqrt{\frac{r_0^2}{2gd}} \int_0^{\pi/2} \frac{d\theta}{\sqrt{\sin \theta}} \quad \text{(cf., p. 18)}$$

1.7.2 Uniformly Rotating Reference Frames

Suppose that $\hat{x}$ and $\hat{y}$ are unit vectors attached to an inertial reference frame, and $\hat{x}'$ and $\hat{y}'$ are rotating with angular velocity $\omega\hat{z} = \boldsymbol{\omega}$ (see Figure 1.18). Thus,

$$\frac{d\hat{x}'}{dt} = \boldsymbol{\omega} \times \hat{x}', \qquad \frac{d\hat{y}'}{dt} = \boldsymbol{\omega} \times \hat{y}'$$

FIGURE 1.18 The $\hat{x}'$, $\hat{y}'$ axes rotate with angular speed ω relative to the fixed axes $\hat{x}$, $\hat{y}$.

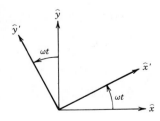

A vector **r** can be analyzed into its components (x, y) along $\hat{x}$, $\hat{y}$, or into its components x', y' along $\hat{x}'$, $\hat{y}'$,

$$\mathbf{r} = x\hat{x} + y\hat{y} = x'\hat{x}' + y'\hat{y}'$$

Then the velocity **v** is given by

$$\frac{d\mathbf{r}}{dt} = \frac{dx'}{dt}\hat{x}' + \frac{dy'}{dt}\hat{y}' + x'\frac{d\hat{x}'}{dt} + y'\frac{d\hat{y}'}{dt}$$

$$= \frac{dx'}{dt}\hat{x}' + \frac{dy'}{dt}\hat{y}' + \boldsymbol{\omega} \times (x'\hat{x}' + y'\hat{y}')$$

$$\frac{d\mathbf{r}}{dt} = \frac{dx'}{dt}\hat{x}' + \frac{dy'}{dt}\hat{y}' + \boldsymbol{\omega} \times \mathbf{r} \tag{1.28}$$

The first two terms on the right-hand side of (1.28) give the velocity of the endpoint of **r**, as seen by an observer rotating with $\hat{x}'$, $\hat{y}'$. This quantity is usually denoted by $d^*\mathbf{r}/dt$. Thus,

$$\frac{d\mathbf{r}}{dt} = \frac{d^*\mathbf{r}}{dt} + \boldsymbol{\omega} \times \mathbf{r} \tag{1.29}$$

Differentiating a second time leads to

$$\frac{d^2\mathbf{r}}{dt^2} = \frac{d}{dt}\left(\frac{d^*\mathbf{r}}{dt} + \boldsymbol{\omega} \times \mathbf{r}\right) = \frac{d^*}{dt}\left(\frac{d^*\mathbf{r}}{dt} + \boldsymbol{\omega} \times \mathbf{r}\right) + \boldsymbol{\omega} \times \left(\frac{d^*\mathbf{r}}{dt} + \boldsymbol{\omega} \times \mathbf{r}\right)$$

$$= \frac{d^{*2}\mathbf{r}}{dt} + 2\boldsymbol{\omega} \times \frac{d^*\mathbf{r}}{dt} + \boldsymbol{\omega} \times (\boldsymbol{\omega} \times \mathbf{r}) + \frac{d^*\boldsymbol{\omega}}{dt} \times \mathbf{r} \tag{1.30}$$

We now assume that $\boldsymbol{\omega}$ is constant, so $d^*\boldsymbol{\omega}/dt = 0$. Thus, Newton's second law for a particle of mass m acted upon by a real force **F** is expressed as

$$m\frac{d^{*2}\mathbf{r}}{dt^2} = \mathbf{F} - 2m\boldsymbol{\omega} \times \frac{d^*\mathbf{r}}{dt} - m\boldsymbol{\omega} \times (\boldsymbol{\omega} \times \mathbf{r}) \tag{1.31}$$

The two effective forces appearing on the right-hand side of (1.31) are called, respectively, the Coriolis force and the centrifugal force.

The calculation of the right-hand side of (1.31) is usually facilitated by choosing a set of unit vectors fixed in the rotating coordinate system. If $\boldsymbol{\omega}$ and **r** are expressed in terms of these vectors, the cross products in (1.31) can easily be worked out. This is illustrated in the examples that follow.

PROBLEM 1.7.5A A plumb bob is hung from a string tied to the top of the leaning tower of Pisa. At equilibrium, does the string point toward the center of the earth? If not, where does it point? Assume that the earth is a sphere of radius R.

Choose a coordinate system whose origin is at the center of the earth, with its $\hat{z}$ axis pointing to Pisa and its $\hat{x}$ axis in the plane defined by $\hat{z}$ and the rotation axis of the earth. The latitude of Pisa is θ. We see from Figure 1.19 that

$$\boldsymbol{\omega} = \omega(\hat{z}\sin\theta - \hat{x}\cos\theta)$$

If **r** locates a plumb bob in the vicinity of Pisa, then $\mathbf{r} = R\hat{z}$. Thus,

$$\boldsymbol{\omega} \times \mathbf{r} = \omega R(\hat{z}\sin\theta - \hat{x}\cos\theta) \times \hat{z} = \omega R\cos\theta\,\hat{y}$$

$$\boldsymbol{\omega} \times (\boldsymbol{\omega} \times \mathbf{r}) = \omega^2 R\cos\theta(\hat{z}\sin\theta - \hat{x}\cos\theta) \times \hat{y} = \omega^2 R\cos\theta(-\hat{x}\sin\theta - \hat{z}\cos\theta)$$

FIGURE 1.19 The rotating earth carries a coordinate system whose z axis points from the center to Pisa. The axis of the earth is in the $\hat{x} - \hat{z}$ plane.

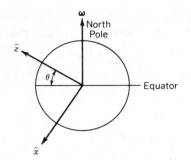

Since the bob is assumed to be stationary relative to the rotating earth, $d^*\mathbf{r}/dt = 0$. If we designate by $\mathbf{f}_s$ the force that the string exerts on the bob, the total effective force on the bob is

$$\mathbf{f}_s - mg\hat{z} + m\omega^2 R(\hat{z} \cos^2 \theta + \hat{x} \sin \theta \cos \theta)$$

At equilibrium this is zero, so that

$$-\mathbf{f}_s = -\hat{z}m(g - \omega^2 R \cos^2\theta) + \hat{x}\frac{m\omega^2}{2}R \sin 2\theta$$

Thus, the bob hangs south of vertical by an angle

$$\phi = \arctan\left(\frac{\frac{\omega^2}{2}R \sin 2\theta}{g - \omega^2 R \cos^2\theta}\right) \approx \arctan\left(\frac{\omega^2 R \sin 2\theta}{2g}\right)$$

$$\approx \arctan(.00173) \approx .00173 \text{ radians} = 0.1°$$

PROBLEM 1.7.5B Now suppose that the point of support of the string is traveling north at a constant speed of 60 mph. Even if the string is protected from air currents, the motion will result in an additional displacement. What is it?

The northbound motion implies that

$$\frac{d^*\mathbf{r}}{dt} = -v\hat{x} \qquad (v = 60 \text{ mph})$$

Thus, we must add the Coriolis force

$$-2m\boldsymbol{\omega} \times \frac{d^*\mathbf{r}}{dt} = -2m\omega(\hat{z} \sin \theta - \hat{x} \cos \theta) \times (-v\hat{x})$$

$$= 2m\omega v \sin \theta \hat{y}$$

and $-\mathbf{f}_s$ acquires a component in the $\hat{y}$ (eastward) direction. The angle ψ of eastward tilt of the string is approximately

$$\psi \approx \arctan\left(\frac{2v\omega \sin \theta}{g}\right) \approx \frac{2v\omega \sin \theta}{g} = .00028 \text{ radian}$$

$$\approx .016°$$

PROBLEM 1.7.6A At the Ford bridge, between Minneapolis and St. Paul, the Mississippi River flows due south at a speed of $v = 2$ m/s, and is 280 m wide. What is the difference in height of the water on the opposite banks of the river?

The main idea is to calculate the direction of the total effective vertical. The plane of the river surface will be perpendicular to this effective vertical. (Refer to Figure 1.19.) We have

$$\boldsymbol{\omega} = \omega(\hat{z}\sin\theta - \hat{x}\cos\theta)$$

$$\frac{d^*\mathbf{r}}{dt} = v\hat{x}$$

$$\mathbf{r} = R\hat{z}$$

The total effective force on a particle of mass m flowing with the river is

$$-mg\hat{z} - 2m\boldsymbol{\omega} \times \frac{d^*\mathbf{r}}{dt} - m\boldsymbol{\omega} \times (\boldsymbol{\omega} \times \mathbf{r})$$

$$= -mg\hat{z} - 2m\omega(\hat{z}\sin\theta - \hat{x}\cos\theta) \times v\hat{x} + m\omega^2 R(\hat{x}\sin\theta + \hat{z}\cos\theta)\cos\theta$$

$$= -\hat{z}m(g - \omega^2 R\cos^2\theta) + \hat{x}\frac{m\omega^2 R}{2}\sin 2\theta - \hat{y}2mv\omega\sin\theta$$

(cf. our work on p. 28). Here the downward direction of total effective gravity is tipped toward the west, by an angle of approximately

$$\arctan\left(\frac{2mv\omega\sin\theta}{m(g - \omega^2 R\cos^2\theta)}\right) \simeq \frac{2v\omega\sin\theta}{g}$$

If the width of the river is D, this tipping of the horizontal plane will cause the west bank of the river to be higher by

$$\frac{2v\omega\sin\theta}{g}D \simeq 5.9\text{ mm}$$

than the east bank ($\theta \simeq 45°$ N).

PROBLEM 1.7.6B Some distance downstream, the river flows east, with approximately the same width and speed. What is the effect that now dominates the difference between the water heights on the two banks?

Here we have

$$\frac{d^*\mathbf{r}}{dt} = v\hat{y}$$

$$-2m\boldsymbol{\omega} \times \frac{d^*\mathbf{r}}{dt} = -2mv\omega(\hat{z}\sin\theta - \hat{x}\cos\theta) \times \hat{y} = 2mv\omega(+\hat{x}\sin\theta + \hat{z}\cos\theta)$$

and the total effective gravitational field is

$$-\hat{z}m(g - \omega^2 R\cos^2\theta - 2v\omega\cos\theta) + \hat{x}m\left(\frac{\omega^2 R}{2}\sin 2\theta + 2v\omega\sin\theta\right)$$

Since $\omega R \gg 2$ m/s, the centrifugal term dominates in the coefficient of $\hat{x}$. Thus, the downward direction of total effective gravity is tipped toward the south by an angle of approximately $\omega^2 R\sin 2\theta/2g$, which causes the south bank of the river to be higher than the north bank by

$$\frac{\omega^2 R\sin 2\theta}{2g}D \simeq 0.48\text{ m}$$

PROBLEM 1.7.7   An object of mass 1 g is weighed on an accurate spring balance, first at Minneapolis (latitude 45° N) and then at the North Pole. Estimate the difference in the readings of the spring balance. Assume that the earth has a spherically symmetric mass distribution.

Let $\mathbf{F}_s$ be the force that the spring exerts on the mass. At latitude θ the equilibrium condition for the mass is

$$\mathbf{F}_s + (-mg\hat{z}) - m\boldsymbol{\omega} \times (\boldsymbol{\omega} \times \mathbf{R}) = 0$$

with $\boldsymbol{\omega} = \omega[\hat{z} \sin\theta - \hat{x} \cos\theta]$, $\mathbf{R} = R\hat{z}$. Working out the cross products we get

$$\mathbf{F}_s = (-mg + m\omega^2 R \cos^2\theta)\hat{z} + m\omega^2 R \cos\theta \sin\theta\,\hat{x}$$

and the magnitude of $\mathbf{F}_s$ is

$$F_s = m\sqrt{(-g + \omega^2 R \cos^2\theta)^2 + (\omega^2 R \cos\theta \sin\theta)^2}$$
$$= m\sqrt{(\omega^2 R \cos\theta)^2 + g^2 - 2g\omega^2 R \cos^2\theta}$$

At the North Pole (latitude 90°), this is $F_s(90°) = mg$. At Minneapolis (latitude 45° N) this is

$$F_s(45°\text{ N}) = m\sqrt{\left(\frac{\omega^2 R}{\sqrt{2}}\right)^2 + g^2 - g\omega^2 R} = mg\sqrt{1 - \frac{\omega^2 R}{g} + \left(\frac{\omega^2 R}{\sqrt{2}\,g}\right)^2}$$

$$\simeq mg\sqrt{1 - \frac{\omega^2 R}{g}} \simeq mg\left[1 - \frac{\omega^2 R}{2g}\right]$$

Thus,

$$F_s(90°) - F_s(45°\text{ N}) = mg\frac{\omega^2 R}{2g} \simeq 1.7 \text{ dynes}$$

PROBLEM 1.7.8   In the South Atlantic in 1914, during the battle of the Falkland Islands, the British naval gunners saw most of their shells fall almost 100 m to the left of their targets even though they had carefully adjusted their sights by practicing off the coast of England. What had they forgotten to allow for? Ignore aerodynamic effects.

The Coriolis acceleration of the shell is $-2\boldsymbol{\omega} \times d^*\mathbf{r}/dt$, where $d^*\mathbf{r}/dt$ is the velocity of the shell relative to the rotating earth. Suppose that the shell is fired almost horizontally, at an angle ϕ to the right of North (see Figure 1.20). Thus,

$$\mathbf{v}^* = v^*(\hat{n} \cos\phi + \hat{e} \sin\phi)$$
$$= v^*([\hat{z} \cos\theta - \hat{x} \sin\theta]\cos\phi + \hat{y} \sin\phi)$$
$$\boldsymbol{\omega} = \omega\hat{z}$$

and the Coriolis force is

$$\mathbf{F}_c = -2m\boldsymbol{\omega} \times \mathbf{v}^* = -2m\omega v^*(-\hat{y} \sin\theta \cos\phi - \hat{x} \sin\phi)$$
$$= 2m\omega v^*(\hat{x} \sin\phi + \hat{y} \sin\theta \cos\phi)$$

The $\hat{x}$ component of $\mathbf{F}_c$ is independent of latitude θ. In the northern hemisphere, $\theta > 0$ and $\sin\theta > 0$. If the shell is fired in a northerly direction, $\cos\phi > 0$ and $\mathbf{F}_c$ has an easterly component. If the shell is fired in a southerly direction, $\cos\phi < 0$ and $\mathbf{F}_c$ has a westerly component. In both cases $\mathbf{F}_c$ pushes the shell to the right. However, in the southern hemisphere $\theta < 0$ and $\sin\theta < 0$. In this case the above

FIGURE 1.20 **v*** is the velocity of the shell as measured
from the rotating Earth.

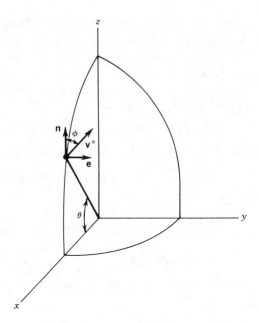

argument shows that $\mathbf{F}_c$ pushes the shell to the left. Thus, guns that are calibrated
in the northern hemisphere will fire shells that fall to the left of the target when
fired in the southern hemisphere.

PROBLEM 1.7.9 A bug of mass 1 g crawls out along a straight radial scratch on a phonograph
record rotating at $33\frac{1}{3}$ rpm (clockwise as viewed from above). The bug is 6 cm
from the center of the record and crawling at the rate of 1 cm/s.

(a) What is the net frictional force (magnitude and direction) on the bug?

Suppose that the bug is crawling along the $\hat{x}'$ axis, with speed v' relative to
that axis. Then

$$\mathbf{r} = v't\hat{x}', \qquad \frac{d^*\mathbf{r}}{dt} = v'\hat{x}', \qquad \frac{d^{*2}\mathbf{r}}{dt^2} = 0$$

The angular velocity of the record is $\boldsymbol{\omega} = -\omega\hat{z}$. Then the equation of motion of
the bug is

$$m\frac{d^{*2}\mathbf{r}}{dt^2} = 0 = \mathbf{F} - 2m\boldsymbol{\omega} \times \frac{d^*\mathbf{r}}{dt} - m\boldsymbol{\omega} \times (\boldsymbol{\omega} \times \mathbf{r})$$

$$\mathbf{F} = 2m\omega v'(-\hat{z} \times \hat{x}') + m\omega^2 v't\hat{z} \times (\hat{z} \times \hat{x}')$$

$$= -2m\omega v'\hat{y}' + m\omega^2 v't\hat{z} \times \hat{y}' = -2m\omega v'\hat{y}' - m\omega^2 v't\hat{x}'$$

If we use $m = 1$ g, and $\omega = 3.49$ rad/s $(= 33\frac{1}{3}$ rev/$m \times 2\pi$ rad/rev $\times$
$1m/60$ s), we get

$$\mathbf{F} = -6.98\hat{y}' - 73.08\hat{x}' \text{ dyne}$$

(b) What added power must the motor expend because of the bug?

The velocity of the bug is

$$\mathbf{v} = \frac{d^*\mathbf{r}}{dt} + \boldsymbol{\omega} \times \mathbf{r} = v'\hat{x}' - \omega v't\hat{z} \times \hat{x}' = v'\hat{x}' - \omega v't\hat{y}'$$

Thus, the rate at which work is being done on the bug is

$$\mathbf{F} \cdot \mathbf{v} = -\left(2m\omega v' \hat{y}' + m\omega^2 v't\hat{x}'\right) \cdot \left(v'\hat{x}' - \omega v't\hat{y}'\right)$$
$$= -m\omega^2 v'^2 t + 2m\omega^2 v'^2 t = m\omega^2 v'^2 t$$

PROBLEM 1.7.10 A two-dimensional harmonic oscillator consists of a particle of mass m confined to a horizontal frictionless plane, and subject to harmonic forces in the x and y directions. That is to say, if the Cartesian coordinates of the particle are x and y, the Cartesian components of the force on the particle are given by

$$F_x = -k_x x, \qquad F_y = -k_y y$$

Here k_x and k_y are constants, and are not necessarily equal.

(a) Write Newton's second law of motion for this system, and solve to find general expressions for x and y as functions of the time t and the constants of integration.

The equations of motion for $x(t)$ and $y(t)$ are

$$F_x = m\ddot{x} = -k_x x, \qquad F_y = m\ddot{y} = -k_y y$$

These are satisfied by simultaneous and independent harmonic oscillations in the x and y directions [cf. (1.15)]

$$x(t) = A \cos\left(\sqrt{\frac{k_x}{m}}\, t + \alpha\right), \qquad y(t) = B \cos\left(\sqrt{\frac{k_y}{m}}\, t + \beta\right)$$

(b) The entire apparatus described above is now mounted on a turntable with angular speed ω rad/s about the z axis. The center of attraction ($x = y = 0$) is directly on the axis of rotation. Write new equations of motion in terms of coordinates x and y measured with respect to coordinate axes rotating with the apparatus.

Let $\hat{x}$ and $\hat{y}$ be unit vectors in the plane of the turntable, rotating with the turntable. Then

$$\boldsymbol{\omega} = \omega\hat{z}, \qquad \mathbf{r} = x(t)\hat{x} + y(t)\hat{y}$$

$$\frac{d^*\mathbf{r}}{dt} = \dot{x}(t)\hat{x} + \dot{y}(t)\hat{y}, \qquad \frac{d^{*2}\mathbf{r}}{dt^2} = \ddot{x}(t)\hat{x} + \ddot{y}(t)\hat{y}$$

$$\mathbf{F} = -\left(k_x x\hat{x} + k_y y\hat{y}\right)$$

so the equation of motion

$$m\frac{d^{*2}\mathbf{r}}{dt^2} = \mathbf{F} - 2m\boldsymbol{\omega} \times \frac{d^*\mathbf{r}}{dt} - m\boldsymbol{\omega} \times (\boldsymbol{\omega} \times \mathbf{r})$$

becomes

$$m\left(\ddot{x}\hat{x} + \ddot{y}\hat{y}\right) = -\left(k_x x\hat{x} + k_y y\hat{y}\right) - 2m\omega\hat{z} \times \left(\dot{x}\hat{x} + \dot{y}\hat{y}\right)$$
$$- m\omega^2 \hat{z} \times \left(\hat{z} \times \left(x\hat{x} + y\hat{y}\right)\right)$$
$$= \hat{x}\left(-k_x x + 2m\omega\dot{y} + m\omega^2 x\right)$$
$$+ \hat{y}\left(-k_y \hat{y} - 2m\omega\dot{x} + m\omega^2 y\right)$$

and the coefficients of $\hat{x}$ and $\hat{y}$ give the pair of coupled equations

$$\ddot{x} = \left(\omega^2 - \omega_x^2\right)x + 2\omega\dot{y} \tag{1.32a}$$

$$\ddot{y} = \left(\omega^2 - \omega_y^2\right)y - 2\omega\dot{x} \tag{1.32b}$$

We have introduced the notation $\omega_x = \sqrt{k_x/m}$, $\omega_y = \sqrt{k_y/m}$ to simplify the formulas.

(c) Find the normal frequencies of the resulting oscillations of the particle. Let

$$x(t) = \xi e^{i\phi t}, \qquad y(t) = \eta e^{i\phi t}$$

Substitution in the equations of motion yields

$$\left(\omega^2 - \omega_x^2 + \phi^2\right)\xi + 2i\omega\phi\eta = 0$$

$$-2i\omega\phi\xi + \left(\omega^2 - \omega_y^2 + \phi^2\right)\eta = 0$$

For these homogeneous equations to have a nontrivial solution, we must have

$$\left(\omega^2 - \omega_x^2 + \phi^2\right)\left(\omega^2 - \omega_y^2 + \phi^2\right) - 4\omega^2\phi^2 = 0$$

This is a quadratic equation for ϕ^2, whose solutions are

$$\phi^2 = \frac{\omega_x^2 + \omega_y^2}{2} + \omega^2 \pm \sqrt{\left(\frac{\omega_x^2 + \omega_y^2}{2} + \omega^2\right)^2 - \left(\omega^2 - \omega_x^2\right)\left(\omega^2 - \omega_y^2\right)}$$

Note that if $\omega < \omega_x, \omega_y$ or $\omega > \omega_x, \omega_y$, then ϕ^2 is positive (ϕ is real) for both roots. However, if ω is between ω_x and ω_y, then ϕ^2 is negative (ϕ is imaginary) for one root. An imaginary value of ϕ corresponds to an exponentially growing or decaying solution, and we will not have stable oscillations about $x = y = 0$.

1.8 PRINCIPAL AXES

Suppose that a rigid body rotates with angular velocity $\boldsymbol{\omega}$. Then a point in the body located at position $\mathbf{r}$ relative to a point 0 on the axis of rotation has a velocity $\mathbf{v} = \boldsymbol{\omega} \times \mathbf{r}$, and the total angular momentum of the body relative to 0 is

$$\mathbf{L} = \int dm\, \mathbf{r} \times \mathbf{v} = \int dm\, \mathbf{r} \times (\boldsymbol{\omega} \times \mathbf{r}) = \int dm[(\mathbf{r} \cdot \mathbf{r})\boldsymbol{\omega} - \mathbf{r}(\mathbf{r} \cdot \boldsymbol{\omega})]$$

If we write the components of this equation along the Cartesian axes of a coordinate system centered at 0, we get

$$L_\alpha = \sum_\beta I_{\alpha\beta}\omega_\beta \tag{1.33a}$$

$$I_{\alpha\beta} = \int dm\left[r^2\delta_{\alpha\beta} - r_\alpha r_\beta\right] \tag{1.33b}$$

The set of nine numbers $I_{\alpha\beta}$ are the components of the moment-of-inertia tensor in the chosen coordinate system. It is usually convenient to choose this coordinate system to be fixed in the moving body, because then the $I_{\alpha\beta}$ do not change with time.

It can be shown[4] that it is always possible to find a set of three mutually perpendicular axes intersecting at 0 such that the components $I_{\alpha\beta}$ calculated with

---

[4]See, for example, Chapter 10 of *Mechanics* by K. R. Symon, Addison-Wesley Publishing Co., Inc., Reading, Mass., 1971.

respect to these axes form a diagonal matrix:

$$I_{\alpha\beta} = \delta_{\alpha\beta}I_\alpha = \delta_{\alpha\beta}\int dm\left[r^2 - r_\alpha^2\right] \qquad (1.34\text{a})$$

$$I = \begin{bmatrix} \int dm\left(y^2 + z^2\right) & 0 & 0 \\ 0 & \int dm\left(x^2 + z^2\right) & 0 \\ 0 & 0 & \int dm\left(x^2 + y^2\right) \end{bmatrix} \qquad (1.34\text{b})$$

If we substitute (1.34a) into (1.33a), we see that the components of **L** and $\boldsymbol{\omega}$ along these axes (called *principal axes*) satisfy the simpler relation

$$L_\alpha = I_\alpha \omega_\alpha \qquad (1.35\text{a})$$

or, more explicitly,

$$L_x = \omega_x \int dm\left(y^2 + z^2\right) \qquad (1.35\text{b})$$

$$L_y = \omega_y \int dm\left(x^2 + z^2\right) \qquad (1.35\text{c})$$

$$L_z = \omega_z \int dm\left(x^2 + y^2\right) \qquad (1.35\text{d})$$

It is clear from (1.33) and (1.35) that the directions of **L** and $\boldsymbol{\omega}$ need not be the same. However, if $\boldsymbol{\omega}$ lies along the $\hat{x}$ direction, then $\omega_y = \omega_z = 0$ and (1.35) yields

$$L_x = \omega \int dm\left(y^2 + z^2\right)$$

$$L_y = L_z = 0$$

Thus, if $\boldsymbol{\omega}$ lies along the $\hat{x}$ direction, then so does **L**. Analogous conclusions can be drawn if $\boldsymbol{\omega}$ lies along the $\hat{y}$ or $\hat{z}$ direction. *If $\boldsymbol{\omega}$ lies along a principal axis, **L** and $\boldsymbol{\omega}$ point in the same direction.* The diagonal components I_α calculated with respect to principal axes are called the principal moments of inertia. The symbol I used in Section 1.1 was really a principal moment of inertia corresponding to the axis of rotation, which was a principal axis in every example considered there.

The technique for finding the principal axes and principal moments of inertia of a given object is straightforward, but rather laborious. Fortunately, if the object has certain symmetries we can often find some of the principal axes by inspection.

1. If the object is symmetric with respect to reflection across a plane, any axis perpendicular to that plane is a principal axis.

2. If an object is symmetric with respect to any rotation about an axis, then that axis is a principal axis.

These statements can be verified by showing that $I_{\alpha\beta}$ is diagonal when calculated with respect to axes with the stated symmetry properties.

PROBLEM 1.8.1 Find the principal axes and principal moments of inertia, relative to the mass center, of a homogeneous block of mass M and sides a, b, and c.

From reflection symmetry with respect to planes through the center of the block and parallel to its faces, we conclude that the principal axes are perpendic-

FIGURE 1.21 The principal axes of the uniform rectangular block are perpendicular to the planes of reflection symmetry.

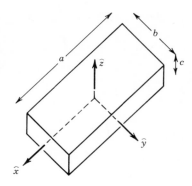

ular to the faces of the block. Then

$$I_{xx} = \int dm \left(y^2 + z^2 \right) = \frac{M}{abc} \int_{-c/2}^{c/2} dz \int_{-b/2}^{b/2} dy \int_{-a/2}^{a/2} dx \left(y^2 + z^2 \right) = \frac{1}{12} M(b^2 + c^2)$$

Similarly,

$$I_{yy} = \tfrac{1}{12} M(a^2 + c^2)$$

$$I_{zz} = \tfrac{1}{12} M(a^2 + b^2)$$

PROBLEM 1.8.2   Find the principal axes and principal moments of inertia, relative to the mass center, of a homogeneous cylinder of mass M, radius R, and length L.

The axial symmetry of the cylinder guarantees that the axis of the cylinder is one principal axis. Call it the $\hat{z}$ axis. Since the cylinder has reflection symmetry across any plane containing the $\hat{z}$ axis, we can choose any pair of orthogonal axes perpendicular to the $\hat{z}$ axis as the $\hat{x}$ and $\hat{y}$ principal axes. Then

$$I_{zz} = \int dm \left(x^2 + y^2 \right) = \frac{M}{\pi R^2 L} \int_{-L/2}^{L/2} dz \int_0^R dr \cdot 2\pi r \cdot r^2 = \frac{MR^2}{2}$$

$$I_{xx} = \int dm \left(y^2 + z^2 \right) = \frac{M}{\pi R^2 L} \int_{-L/2}^{L/2} dz \int_0^{2\pi} d\theta \int_0^R dr \cdot r(r^2 \sin^2\theta + z^2)$$

$$= \frac{ML^2}{12} + \frac{MR^2}{4} = I_{yy}$$

PROBLEM 1.8.3   Suppose that a uniform wheel of radius R, thickness d, and mass M is rotating at angular speed ω about an axis that passes through its mass center, but makes an angle θ with a line perpendicular to the wheel. Find the angular momentum of the wheel and the torque on the axis.

We use a coordinate system fixed in the wheel, coinciding with the principal axes. The vector $\boldsymbol{\omega}$ makes an angle θ with the $\hat{z}$ axis, and we choose our $\hat{x}$ axis so that $\boldsymbol{\omega}$ lies in the $\hat{x} - \hat{z}$ plane. Then

$$\omega_x = \omega \sin\theta, \qquad L_x = I_{xx}\omega_x = M\left(\frac{d^2}{12} + \frac{R^2}{4} \right) \sin\theta \cdot \omega$$

$$\omega_y = 0, \qquad L_y = 0$$

$$\omega_z = \omega \cos\theta, \qquad L_z = I_{zz}\omega_z = \frac{MR^2}{2} \cos\theta \cdot \omega$$

Thus, $\mathbf{L}$ lies in the plane of $\boldsymbol{\omega}$ and $\hat{z}$, and if $d < \sqrt{3}\,R$, it lies between $\boldsymbol{\omega}$ and $\hat{z}$.

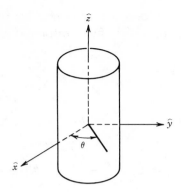

FIGURE 1.22 The principal $\hat{z}$ axis lies along the axis of the cylinder. The orientation about $\hat{z}$ of the principal $\hat{x}$ and $\hat{y}$ axes is arbitrary.

In the body-fixed principal axis system, ω and $I_{\alpha\beta}$ are constant, and so $\mathbf{L}$ is also constant. To calculate the torque we use (1.1b), but the time derivative there must be calculated relative to an inertial frame of reference. We can combine (1.1b) and (1.29) to write

$$\tau_{\text{ext}} = \frac{d\mathbf{L}}{dt} = \frac{d^*\mathbf{L}}{dt} + \omega \times \mathbf{L} \tag{1.36a}$$

Since $\mathbf{L}$ is stationary relative to our rotating coordinate system, $d^*\mathbf{L}/dt = 0$ and we get

$$\tau_{\text{ext}} = \omega \times \mathbf{L} = (\hat{x}\omega_x + \hat{z}\omega_z) \times (\hat{x}I_{xx}\omega_x + \hat{z}I_{zz}\omega_z)$$

$$= \hat{y}\omega_x\omega_z(I_{xx} - I_{zz})$$

$$= \hat{y}\frac{M\omega^2}{8} \sin 2\theta \left(\frac{d^2}{3} - R^2\right)$$

This torque must be supplied by the bearings that hold the axis in place.

If $\theta = 0$, $\tau_{\text{ext}} = 0$, and the wheel is said to be dynamically balanced. Note that if $d^2 = 3R^2$, the wheel is also dynamically balanced, independent of θ. In this case all three principal moments of inertia are equal, and any axis through the mass center is a principal axis. This is also the situation for the rectangular block of Problem 1.8.1, if $a = b = c$.

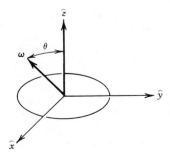

FIGURE 1.23 The $\hat{x}$ and $\hat{y}$ principal axes are fixed in the plane of the wheel, but ω is not perpendicular to this plane.

Euler's equations of motion are obtained by writing (1.36a) in terms of components of vectors along a body-fixed principal axis coordinate system:

$$\tau_x = I_{xx}\frac{d\omega_x}{dt} + (I_{zz} - I_{yy})\omega_y\omega_z \qquad (1.36b)$$

$$\tau_y = I_{yy}\frac{d\omega_y}{dt} + (I_{xx} - I_{zz})\omega_z\omega_x \qquad (1.36c)$$

$$\tau_z = I_{zz}\frac{d\omega_z}{dt} + (I_{yy} - I_{xx})\omega_x\omega_y \qquad (1.36d)$$

These equations are useful in the analysis of torque-free motion.

PROBLEM 1.8.4 Calculate the angular momentum of a thin rectangular plate with dimensions a, b, and mass m, rotating with constant angular velocity $\boldsymbol{\omega}$ about a vertical axis through a diagonal. Compute the magnitude of the torque (exerted by the bearings) that is required to sustain this motion.

We choose the origin of our principal axis system at the mass center. The symmetry of the plate implies that one principal axis will be perpendicular to the plate, and that the other two will be parallel to the edges of the plate. The components of $\boldsymbol{\omega}$ in this coordinate system are

$$\omega_y = 0, \qquad \omega_x = -\omega\frac{a}{\sqrt{a^2 + b^2}}, \qquad \omega_z = \omega\frac{b}{\sqrt{a^2 + b^2}}$$

FIGURE 1.24 (a) A thin rectangular plate rotating about a diagonal; (b) the $\hat{x}$ and $\hat{z}$ principal axes of the plate.

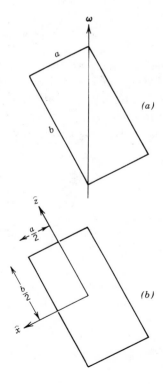

The components of the moment of inertia tensor are

$$I_{xx} = \int dm \left(y^2 + z^2 \right) = \frac{m}{ab} \int_{-a/2}^{a/2} dx \int_{-b/2}^{b/2} dz \cdot z^2 = \frac{m}{ab} \cdot a \cdot \frac{b^3}{12} = \frac{mb^2}{12}$$

$$I_{yy} = \frac{m(a^2 + b^2)}{12}, \qquad I_{zz} = \frac{ma^2}{12}$$

$$I_{xy} = -\int dm\, xy = 0 = I_{zy}, \qquad I_{xz} = -\int dm\, xz = -\frac{m}{ab} \int_{-a/2}^{a/2} x\, dx \int_{-b/2}^{b/2} z\, dz = 0$$

The components of the angular momentum are given by

$$L_\alpha = \sum_\beta I_{\alpha\beta} \omega_\beta$$

Written out in matrix form, this becomes

$$\begin{bmatrix} L_x \\ L_y \\ L_z \end{bmatrix} = m \begin{bmatrix} \dfrac{b^2}{12} & 0 & 0 \\ 0 & \dfrac{a^2 + b^2}{12} & 0 \\ 0 & 0 & \dfrac{a^2}{12} \end{bmatrix} \begin{bmatrix} -\omega \dfrac{a}{\sqrt{a^2 + b^2}} \\ 0 \\ \omega \dfrac{b}{\sqrt{a^2 + b^2}} \end{bmatrix} = \begin{bmatrix} -\dfrac{m\omega b^2 a}{12\sqrt{a^2 + b^2}} \\ 0 \\ \dfrac{m\omega a^2 b}{12\sqrt{a^2 + b^2}} \end{bmatrix}$$

Thus,

$$\mathbf{L} = \frac{m\omega ab}{12\sqrt{a^2 + b^2}} \left[-b\hat{x} + a\hat{z} \right]$$

Since **L** rotates with the plate, the torque that must be supplied by the bearings is

$$\boldsymbol{\tau} = \frac{d\mathbf{L}}{dt} = \boldsymbol{\omega} \times \mathbf{L} = \frac{m\omega^2 ab}{12(a^2 + b^2)} (-a\hat{x} + b\hat{z}) \times (-b\hat{x} + a\hat{z})$$

$$= \frac{m}{12}\omega^2 \frac{ab(a^2 - b^2)}{a^2 + b^2} \hat{y}$$

Note that this vanishes for $a = b$. In this case the matrix of I is

$$m \begin{bmatrix} \dfrac{a^2}{12} & 0 & 0 \\ 0 & \dfrac{a^2}{6} & 0 \\ 0 & 0 & \dfrac{a^2}{12} \end{bmatrix}$$

and it is easily verified that any line in the x–z plane through the mass center is a principal axis.

FIGURE 1.25 The light rod is pivoted at O.

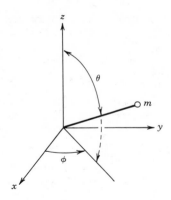

1.9 LAGRANGE'S AND HAMILTON'S EQUATIONS FOR A CONSERVATIVE SYSTEM

Let the generalized coordinates of a system with d degrees of freedom be labeled $q_1, q_2, \ldots, q_d$. The Lagrangian L is defined by

$$L \equiv T(q_1, \ldots, q_d, \dot{q}_1, \ldots, \dot{q}_d, t) - U(q_1, \ldots, q_d, t) \qquad (1.37a)$$

Here T and U are the kinetic and potential energies, respectively. The kinetic energy must be calculated relative to an inertial frame of reference. The generalized momentum p_i conjugate to the generalized coordinate q_i is defined by

$$p_i \equiv \frac{\partial L}{\partial \dot{q}_i} = \frac{\partial T}{\partial \dot{q}_i} \qquad (1.37b)$$

Lagrange's equations of motion are[5]

$$\frac{d}{dt} p_i = \frac{d}{dt}\left(\frac{\partial L}{\partial \dot{q}_i} \right) = \frac{\partial L}{\partial q_i} \qquad (1.37c)$$

These equations are equivalent to Newton's laws of motion, but are often easier to write down and solve.

PROBLEM 1.9.1   A bob of mass m is attached to a light rod of length l, which is pivoted at a fixed point 0. Find the motion of the bob.

The first step is the identification of the relevant degrees of freedom. In this problem they must specify the orientation of the rod, and we choose them to be the polar coordinates θ and ϕ (see Figure 1.25). Next we must express the kinetic and potential energies in terms of θ and ϕ and their time derivatives. Let us choose the condition of zero potential energy to be when the rod is horizontal

---

[5] For a derivation of the connection between Lagrange's equations and Newton's second law of motion, see, for example, Chapter 9 of *Mechanics* by K. R. Symon, Addison-Wesley Publishing Co., Inc., Reading, Mass., 1971. However, the most interesting approach to Lagrange's equations is from the calculus of variations (see, for example, Chapter 2 of *Classical Mechanics* by Herbert Goldstein, Addison-Wesley Publishing Co., Inc., Reading, Mass, 1980. The variational interpretation shows immediately why Lagrange's equations have the same form (1.37c) for any generalized coordinates. An excellent introduction to variational methods in physics is presented in Chapter 19, Volume II of *The Feynman Lectures on Physics*, Addison-Wesley Publishing Co., Inc., Reading, Mass., 1963.

($\theta = \pi/2$). Then

$$U(\theta, \phi) = mgl \cos \theta$$

$$T(\theta, \phi, \dot{\theta}, \dot{\phi}) = \tfrac{1}{2}mv^2 = \tfrac{1}{2}m\left[(l\dot{\theta})^2 + (l \sin \theta \dot{\phi})^2\right]$$

$$L(\theta, \phi, \dot{\theta}, \dot{\phi}) = \tfrac{1}{2}ml^2\left[\dot{\theta}^2 + \dot{\phi}^2 \sin^2 \theta\right] - mgl \cos \theta$$

$$p_\phi = \frac{\partial L}{\partial \dot{\phi}} = ml^2\dot{\phi} \sin^2 \theta, \qquad \frac{\partial L}{\partial \phi} = 0$$

Thus, Lagrange's equations (1.37c) are

$$\frac{d}{dt}\left(ml^2\dot{\theta}\right) = ml^2\left[\dot{\phi}^2 \sin \theta \cos \theta + \frac{g}{l} \sin \theta\right] \tag{1.38a}$$

$$\frac{d}{dt}\left(ml^2\dot{\phi} \sin^2 \theta\right) = 0 \tag{1.38b}$$

Equation (1.38b) implies that $p_\phi = ml^2\dot{\phi} \sin^2 \theta$ is a constant of the motion. In fact, it is the vertical component of the angular momentum of the bob, and its constancy is due to the absence of a component of torque about this vertical axis. We can express $\dot{\phi}$ in (1.38a) in terms of the constant p_ϕ, to get

$$\ddot{\theta} = \frac{p_\phi^2}{m^2 l^4} \frac{\cos \theta}{\sin^3\theta} + \frac{g}{l} \sin \theta \tag{1.38c}$$

whose solution can be expressed in terms of elliptic functions.

Often one wants to consider simple special cases, such as a motion of the pendulum with constant θ. If $\theta = \theta_0 =$ constant, then $\ddot{\theta} = 0$, and (1.38a) becomes

$$\left[\dot{\phi}^2\cos \theta_0 + \frac{g}{l}\right]\sin \theta_0 = 0$$

This equation can be satisfied in two ways:

1. $\theta_0 = 0$, and the pendulum is vertical;
2. $\theta_0 \neq 0$. Then $\dot{\phi}^2 \cos \theta_0 + g/l = 0$ and $\cos \theta_0 < 0$, so that $\theta_0 > \pi/2$, and the rod points below the horizontal plane. Furthermore, since $|\cos \theta_0| < 1$, such a solution is only possible when $\dot{\phi}^2 \geq g/l$.

Having found equilibrium values of $\dot{\phi}$ and θ, one can investigate solutions involving small deviations from these values. Equation (1.38c) has the form

$$\ddot{\theta} = f(\theta) \tag{1.39a}$$

and the equilibrium value of θ satisfies $f(\theta_0) = 0$. Now write $\theta(t) = \theta_0 + \psi(t)$ where we expect $\psi(t)$ to remain small. Then (1.39a) implies

$$\ddot{\theta} = \ddot{\psi} = f(\theta_0 + \psi) \simeq f(\theta_0) + \psi f'(\theta_0) + \dots$$
$$= \psi f'(\theta_0) + \dots \tag{1.39b}$$

If $f'(\theta_0) < 0$, we can find a harmonic solution of

$$\ddot{\psi} = \psi f'(\theta_0) \tag{1.39c}$$

where ψ remains at all times small enough to justify retaining only the terms in the expansion of $f(\theta_0 + \psi)$ shown in (1.39b). The circular frequency of the harmonic oscillations of ψ can be obtained from (1.39c),

$$\omega = \sqrt{-f'(\theta_0)} \tag{1.40}$$

If we apply this method to our spherical pendulum, we have

$$f(\theta) = \frac{p_\phi^2}{m^2 l^4} \frac{\cos\theta}{\sin^3\theta} + \frac{g}{l}\sin\theta$$

$$f'(\theta_0) = -\frac{p_\phi^2}{m^2 l^4} \frac{\sin^2\theta_0 + 3\cos^2\theta_0}{\sin^4\theta_0} + \frac{g}{l}\cos\theta_0 = -\dot\phi^2\left[1 + 3\cos^2\theta_0\right]$$

Thus, the oscillations about $\theta = \theta_0$ are stable, and occur with a circular frequency $\omega = \dot\phi\sqrt{1 + 3\cos^2\theta_0}$.

Usually, the most difficult step in the Lagrangian method is expressing the kinetic energy in terms of the generalized coordinates and their derivatives. The use of an orthonormal set of unit vectors often makes the process more systematic and efficient.

PROBLEM 1.9.2   Find the equations of motion for plane oscillations of the structure shown in Figure 1.26. The cord has length b and no mass. The stick has length a and mass M.

We choose to work with the generalized coordinates θ_1 and θ_2 shown in Figure 1.26. We also erect the set of unit vectors $\hat{x}$ and $\hat{y}$ at the point of support. First, we locate the mass center of the stick, since this will determine the

FIGURE 1.26 (a) The mass center of the system is at the midpoint of the stick; (b) locating the mass center relative to the point of support.

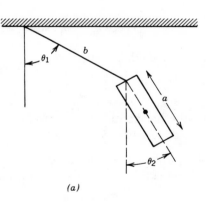

(a)

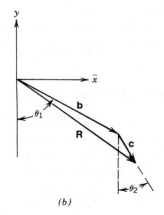

(b)

gravitational potential energy, and is also needed for the kinetic energy:

$$\mathbf{R} = \mathbf{b} + \mathbf{c} = b\left[\hat{x}\sin\theta_1 - \hat{y}\cos\theta_1\right] + \frac{a}{2}\left[\hat{x}\sin\theta_2 - \hat{y}\cos\theta_2\right]$$

$$= \hat{x}\left[b\sin\theta_1 + \frac{a}{2}\sin\theta_2\right] - \hat{y}\left[b\cos\theta_1 + \frac{a}{2}\cos\theta_2\right] \tag{1.41}$$

Thus, the potential energy is

$$U = MgR_y = -Mg\left[b\cos\theta_1 + \frac{a}{2}\cos\theta_2\right]$$

We use (1.26) to determine the kinetic energy of the stick. Here $\omega = \dot{\theta}_2$ and $I = \frac{1}{12}Ma^2$. Differentiating (1.41) gives

$$\dot{\mathbf{R}} = \hat{x}\left(b\dot{\theta}_1\cos\theta_1 + \frac{a}{2}\dot{\theta}_2\cos\theta_2\right) + \hat{y}\left(b\dot{\theta}_1\sin\theta_1 + \frac{a}{2}\dot{\theta}_2\sin\theta_2\right)$$

$$(\dot{\mathbf{R}})^2 = \left(b\dot{\theta}_1\cos\theta_1 + \frac{a}{2}\dot{\theta}_2\cos\theta_2\right)^2 + \left(b\dot{\theta}_1\sin\theta_1 + \frac{a}{2}\dot{\theta}_2\sin\theta_2\right)^2$$

$$= b^2\dot{\theta}_1^2 + \left(\frac{a}{2}\right)^2\dot{\theta}_2^2 + ab\dot{\theta}_1\dot{\theta}_2\cos(\theta_1 - \theta_2)$$

Thus,

$$T = \tfrac{1}{2}M\left[b^2\dot{\theta}_1^2 + \left(\frac{a}{2}\right)^2\dot{\theta}_2^2 + ab\dot{\theta}_1\dot{\theta}_2\cos(\theta_1 - \theta_2)\right] + \tfrac{1}{2}\left(\tfrac{1}{12}Ma^2\right)\dot{\theta}_2^2$$

and

$$L = T - U = \frac{1}{2}M\left[b^2\dot{\theta}_1^2 + \frac{a^2}{3}\dot{\theta}_2^2 + ab\dot{\theta}_1\dot{\theta}_2\cos(\theta_1 - \theta_2)\right]$$

$$+ Mg\left[b\cos\theta_1 + \frac{a}{2}\cos\theta_2\right]$$

The set of equations obtained by applying (1.37c) to this Lagrangian is

$$\frac{d}{dt}\left[b\dot{\theta}_1 + \frac{1}{2}a\dot{\theta}_2\cos(\theta_1 - \theta_2)\right] = -\frac{1}{2}a\dot{\theta}_1\dot{\theta}_2\sin(\theta_1 - \theta_2) - g\sin\theta_1$$

$$\tag{1.42a}$$

$$\frac{d}{dt}\left[\frac{1}{2}b\dot{\theta}_1\cos(\theta_1 - \theta_2) + \frac{1}{3}a\dot{\theta}_2\right] = \frac{1}{2}b\dot{\theta}_1\dot{\theta}_2\sin(\theta_1 - \theta_2) - \frac{g}{2}\sin\theta_2$$

$$\tag{1.42b}$$

We can simplify these equations if we restrict our attention to motions where θ_1, θ_2, and their derivatives are small. If we make the small-angle approximations

$$\sin\theta \simeq \theta, \qquad \cos\theta \simeq 1 - \theta^2/2$$

in (1.42) and keep only terms linear in θ_1, θ_2, and their derivatives, we get

$$b\ddot{\theta}_1 + \tfrac{1}{2}a\ddot{\theta}_2 = -g\theta_1 \tag{1.43a}$$

$$b\ddot{\theta}_1 + \tfrac{2}{3}a\ddot{\theta}_2 = -g\theta_2 \tag{1.43b}$$

To decouple these equations we look for the normal modes. These are solutions in which both θ_1 and θ_2 oscillate with the same frequency ω,

$$\theta_1 = Re\left(\xi_1 e^{i\omega t}\right)$$
$$\theta_2 = Re\left(\xi_2 e^{i\omega t}\right) \qquad (\xi_1, \xi_2 \text{ constant})$$

If these are substituted into (1.43), we get two homogeneous linear equations for ξ_1 and ξ_2,

$$\left(g - b\omega^2\right)\xi_1 - \tfrac{1}{2}a\omega^2\xi_2 = 0 \qquad (1.44a)$$

$$-b\omega^2\xi_1 + \left(g - \tfrac{2}{3}a\omega^2\right)\xi_2 = 0 \qquad (1.44b)$$

If these equations are to have a solution other than $\xi_1 = \xi_2 = 0$, the determinant formed from the coefficients of ξ_1 and ξ_2 must vanish,

$$\det\begin{bmatrix} g - b\omega^2 & -1/2a\omega^2 \\ -b\omega^2 & g - 2/3a\omega^2 \end{bmatrix} = \left(g - b\omega^2\right)\left(g - \tfrac{2}{3}a\omega^2\right) - \tfrac{1}{2}ab\omega^4 = 0$$

This yields a quadratic equation for ω^2, whose two solutions ω_+^2 and ω_-^2 are

$$\left(\omega_{\pm}\right)^2 = \frac{g}{ab}\left[(2a + 3b) \pm \sqrt{4a^2 + 6ab + 9b^2}\right] \qquad (1.45a)$$

ω_+ and ω_- are the normal frequencies of the system. If each is substituted into (1.44), we determine the ratios of ξ_1 and ξ_2, and thus of θ_1 and θ_2. The result of this analysis is a pair of solutions

$$\left(\theta_1\right)_{\pm} = \tfrac{1}{2}a\left(\omega_{\pm}\right)^2 \cos \omega_{\pm}t \qquad (1.45b)$$

$$\left(\theta_2\right)_{\pm} = \left[g - b\left(\omega_{\pm}\right)^2\right]\cos \omega_{\pm}t \qquad (1.45c)$$

The most general solution of (1.43) is a linear combination of the solutions (1.45),

$$\theta_i = A_+\left(\theta_i\right)_+ + A_-\left(\theta_i\right)_-, \qquad (i = 1, 2)$$

with the constants $A_{\pm}$ determined by the initial conditions.

PROBLEM 1.9.3   A pendulum bob of mass m is suspended by a string of length l from a point of support. The point of support moves back and forth along a horizontal x axis according to the equation

$$x = a \cos \omega t$$

Assume that the pendulum swings only in the vertical plane containing the x axis. Let the orientation of the pendulum be described by the angle θ with the downward vertical.

(a) Set up the Lagrangian function and write out Lagrange's equations.

Let $\mathbf{r}$ locate the bob relative to the origin:

$$\mathbf{r} = x\hat{x} + l\left(\hat{x} \sin\theta - \hat{y}\cos\theta\right)$$

$$= (x + l\sin\theta)\hat{x} - l\cos\theta\,\hat{y}; \quad \dot{\mathbf{r}} = \left(\dot{x} + l\dot{\theta}\cos\theta\right)\hat{x} + l\dot{\theta}\sin\theta\,\hat{y}$$

$$T = \tfrac{1}{2}m\dot{\mathbf{r}}^2 = \tfrac{1}{2}m\left[\left(\dot{x} + l\dot{\theta}\cos\theta\right)^2 + \left(l\dot{\theta}\sin\theta\right)^2\right] = \tfrac{1}{2}m\left[\dot{x}^2 + l^2\dot{\theta}^2 + 2\dot{x}\dot{\theta}l\cos\theta\right]$$

$$U = -mgl\cos\theta$$

$$L = T - U = \tfrac{1}{2}m\left(\dot{x}^2 + l^2\dot{\theta}^2 + 2\dot{x}\dot{\theta}l\cos\theta\right) + mgl\cos\theta$$

$$= \tfrac{1}{2}m\left(a^2\omega^2\sin^2\omega t + l^2\dot{\theta}^2 - 2a\omega l\sin\omega t \cdot \cos\theta \cdot \dot{\theta}\right) + mgl\cos\theta$$

$$\frac{\partial L}{\partial\dot{\theta}} = ml^2\dot{\theta} - ma\omega l\sin\omega t \cdot \cos\theta; \quad \frac{\partial L}{\partial\theta} = ma\omega l\sin\omega t \cdot \sin\theta \cdot \dot{\theta} - mgl\sin\theta$$

FIGURE 1.27 The pendulum swings in a vertical plane as its point of support oscillates back and forth in this plane.

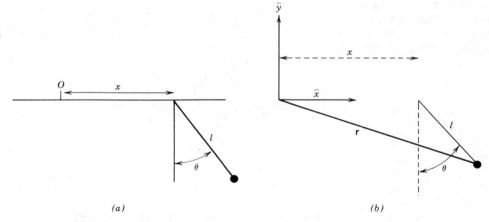

(a)   *(b)*

The equation of motion is

$$\frac{d}{dt}\left(ml^2\dot{\theta} - ma\omega l \sin \omega t \cdot \cos \theta\right) = ma\omega l \sin \omega t \cdot \sin \theta \cdot \dot{\theta} - mgl \sin \theta$$

(b) Show that for small values of θ the equation reduces to that of a forced harmonic oscillator, and find the corresponding steady-state motion.

For small θ, we neglect terms of second or higher order in θ and its derivatives,

$$ml^2\ddot{\theta} - ma\omega^2 l \cos \omega t = -mgl\theta$$

$$\ddot{\theta} + \frac{g}{l}\theta = \frac{a\omega^2}{l} \cos \omega t$$

This is the equation of motion of a forced oscillator. If we seek a solution $\theta(t) = A \cos \omega t$, we get

$$\left(\frac{g}{l} - \omega^2\right) A = \frac{a\omega^2}{l}$$

$$A = \frac{a\omega^2}{g - \omega^2 l}$$

This is the amplitude of the forced oscillations. It diverges as ω approaches the natural frequency of the free simple pendulum.

PROBLEM 1.9.4   A simple pendulum is hung from a block which is free to slide without friction on a horizontal surface. Two identical springs are stretched between the block and the walls (see Figure 1.28).

(a) Write the equations of motion of the system.

Let x be the displacement of m_1 to the right of its equilibrium position, and θ the angle of the pendulum to the right of the vertical. We put our origin at the equilibrium position of m_1. If $\mathbf{r}_1$ and $\mathbf{r}_2$ locate m_1 and m_2 relative to this origin,

$$\mathbf{r}_1 = x\hat{x}, \qquad \dot{\mathbf{r}}_1 = \dot{x}\hat{x}$$

$$\mathbf{r}_2 = x\hat{x} + l(\hat{x} \sin \theta - \hat{y} \cos \theta) = \hat{x}(x + l \sin \theta) - \hat{y}l \cos \theta$$

$$\dot{\mathbf{r}}_2 = \hat{x}(\dot{x} + l\dot{\theta} \cos \theta) + \hat{y}\dot{\theta}l \sin \theta$$

FIGURE 1.28 The mass m_1 slides on a smooth surface. The mass m_2 is at the end of a light pendulum pivoted at m_1.

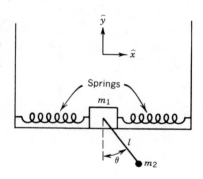

Thus, the kinetic and potential energies are

$$T = \tfrac{1}{2}m_1(\dot{\mathbf{r}}_1)^2 + \tfrac{1}{2}m_2(\dot{\mathbf{r}}_2)^2 = \tfrac{1}{2}m_1\dot{x}^2 + \tfrac{1}{2}m_2(\dot{x}^2 + l^2\dot\theta^2 + 2\dot{x}\dot\theta l\cos\theta)$$

$$U = 2\cdot\tfrac{1}{2}kx^2 + m_2 gl(1 - \cos\theta)$$

The Lagrangian is

$$L = T - U = \tfrac{1}{2}(m_1 + m_2)\dot{x}^2 + \tfrac{1}{2}m_2(l^2\dot\theta^2 + 2\dot{x}\dot\theta l\cos\theta)$$
$$- kx^2 + m_2 gl(\cos\theta - 1)$$

To get the equations of motion we calculate

$$\frac{\partial L}{\partial \dot{x}} = (m_1 + m_2)\dot{x} + m_2\dot\theta l\cos\theta, \qquad \frac{\partial L}{\partial \dot\theta} = m_2(l^2\dot\theta + \dot{x}l\cos\theta)$$

$$\frac{\partial L}{\partial x} = -2kx, \qquad \frac{\partial L}{\partial \theta} = -m_2\dot{x}\dot\theta l\sin\theta - m_2 gl\sin\theta$$

$$\frac{d}{dt}\left[(m_1 + m_2)\dot{x} + m_2\dot\theta l\cos\theta\right] = -2kx$$

$$\frac{d}{dt}(l^2\dot\theta + \dot{x}l\cos\theta) = -(\dot{x}\dot\theta + g)l\sin\theta$$

(b) Find the normal frequencies for small oscillations of the system.

We rewrite the equations of motion, retaining only terms linear in x, θ, and their derivatives.

$$(m_1 + m_2)\ddot{x} + m_2 l\ddot\theta = -2kx$$

$$l^2\ddot\theta + l\ddot{x} = -gl\theta$$

To find the normal frequencies we seek a solution of the form

$$x(t) = \xi e^{i\omega t}, \qquad \theta(t) = \eta e^{i\omega t}$$

Then we get

$$\left[2k - (m_1 + m_2)\omega^2\right]\xi - m_2 l\omega^2\eta = 0$$

$$-\omega^2\xi + \left[g - l\omega^2\right]\eta = 0$$

A nontrivial solution exists only when

$$\det\begin{bmatrix} 2k - (m_1 + m_2)\omega^2 & -m_2 l\omega^2 \\ -\omega^2 & g - l\omega^2 \end{bmatrix} = 0$$

$$\left[2k - (m_1 + m_2)\omega^2\right]\left[g - l\omega^2\right] - m_2 l\omega^4 = 0$$

The normal frequencies are thus

$$\omega_{\pm} = \frac{1}{\sqrt{2m_1 l}} \sqrt{(m_1 + m_2)g + 2kl \pm \sqrt{[(m_1 + m_2)g + 2kl]^2 - 8m_1 lkg}}$$

PROBLEM 1.9.5 A block of mass M, width $2l$, and height $2d$, rolls on a fixed cylinder of radius R. When the block is horizontal it is perfectly balanced on the cylinder. Find the equations of motion and determine whether the horizontal orientation represents stable or unstable equilibrium.

We choose as our generalized coordinate the angle that the block makes with the horizontal. If the block rolls on the cylinder through angle θ without sliding, the distance between the center of the bottom of the block and the point of contact of the block and cylinder is $a \cdot \theta$ (Figure 1.29b). Figure 1.29c shows how we can locate the mass center of the block relative to the center of the cylinder, as a sum of vectors parallel to the faces of the block. By inspection of Figure 1.29c we get

$$\mathbf{R} = (a + d)(\hat{x} \sin \theta + \hat{y} \cos \theta) + a\theta(-\hat{x} \cos \theta + \hat{y} \sin \theta)$$
$$= \hat{x}[(a + d)\sin \theta - a\theta \cos \theta] + \hat{y}[(a + d)\cos \theta + a\theta \sin \theta]$$

The potential and kinetic energies are

$$U = MgR_y = Mg[(a + d)\cos \theta + a\theta \sin \theta] \qquad (1.46a)$$

$$T = \frac{M}{2}\dot{R}^2 + \frac{1}{2}I\dot{\theta}^2 = \frac{M}{2}\dot{\theta}^2(d^2 + a^2\theta^2) + \frac{1}{2}\left[\frac{M}{12}(4l^2 + 4d^2)\right]\dot{\theta}^2$$

$$= \frac{1}{2}M\left(\frac{4}{3}d^2 + \frac{l^2}{3} + a^2\theta^2\right)\dot{\theta}^2 \qquad (1.46b)$$

from which the Lagrangian can be constructed. The question whether $\theta = 0$ is a position of stable equilibrium can be answered by expanding (1.46a) about $\theta = 0$,

$$U = Mg\left[(a + d)\left(1 - \frac{\theta^2}{2} + \cdots\right) + a\theta(\theta + \cdots)\right]$$

$$= Mg\left[a + d + \theta^2\left(\frac{a - d}{2}\right) + O(\theta^4)\right]$$

Thus, $\theta = 0$ is a local minimum of the potential energy if $a > d$. As $|\theta|$ increases from zero, the potential energy increases, so that the kinetic energy decreases. Eventually, the kinetic energy will vanish, which means that the excursion away from $\theta = 0$ has come to a stop, and then the block will begin its return toward $\theta = 0$. However, $\theta = 0$ is a local maximum of the potential energy if $d > a$. In this case, as $|\theta|$ increases from zero, the kinetic energy will increase, so that a small deviation from $\theta = 0$ will grow into a large one. Thus, $\theta = 0$ is a point of stable equilibrium if and only if $a > d$.

PROBLEM 1.9.6 A particle of mass m in the earth's gravitational field is constrained to move in a frictionless vertical plane which rotates about a vertical axis with constant angular velocity ω. Write the Lagrangian for this particle and find the equations of motion. Solve these equations if, at $t = 0$, the particle is at rest relative to the plane, at a distance x_0 from the axis of rotation and height h above the ground.

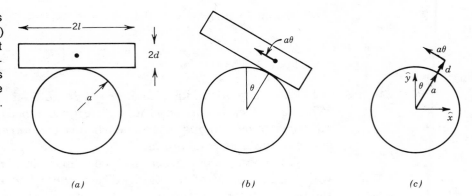

The statement of the problem suggests that we use Cartesian coordinates ρ, z in the rotating plane. The particle is located relative to a fixed origin on the axis by

$$\mathbf{r}(t) = \rho[\cos \omega t \cdot \hat{x} + \sin \omega t \cdot \hat{y}] + z\hat{z}$$

The particle velocity and kinetic energy are

$$\dot{\mathbf{r}}(t) = \rho\omega[-\hat{x}\sin \omega t + \hat{y}\cos \omega t] + \dot{\rho}[\hat{x}\cos \omega t + \hat{y}\sin \omega t] + \dot{z}\hat{z}$$

$$= \hat{x}[-\rho\omega\sin \omega t + \dot{\rho}\cos \omega t] + \hat{y}[\rho\omega\cos \omega t + \dot{\rho}\sin \omega t] + \dot{z}\hat{z}$$

$$T = \frac{1}{2}m\dot{\mathbf{r}}^2 = \frac{m}{2}[\rho^2\omega^2 + \dot{\rho}^2 + \dot{z}^2]$$

FIGURE 1.30 (a) Mass m slides in the vertical plane that rotates about a vertical axis with angular speed ω; (b) location of the particle in its plane at time t, $\hat{x}$, $\hat{y}$, and $\hat{z}$ are fixed.

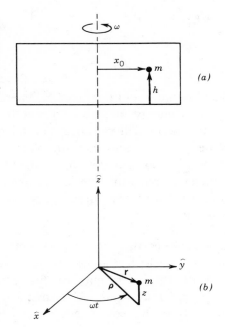

The potential energy is $U = mgz$, so the Lagrangian is

$$L = T - U = \frac{m}{2}\left[\rho^2\omega^2 + \dot\rho^2 + \dot z^2\right] - mgz$$

Then we calculate

$$\frac{\partial L}{\partial \dot\rho} = m\dot\rho, \qquad \frac{\partial L}{\partial \rho} = m\rho\omega^2, \qquad \frac{d}{dt}(m\dot\rho) = m\rho\omega^2$$

$$\frac{\partial L}{\partial \dot z} = m\dot z, \qquad \frac{\partial L}{\partial z} = -mg, \qquad \frac{d}{dt}(m\dot z) = -mg$$

Thus, the equations of motion can be written as

$$\ddot\rho = \omega^2\rho, \qquad \ddot z = -g$$

with the general solution

$$\rho(t) = \rho(0)\cosh\omega t + \frac{\dot\rho(0)}{\omega}\sinh\omega t$$

$$z(t) = z(0) + \dot z(0)t - \tfrac{1}{2}gt^2$$

Since

$$\rho(0) = x_0, \qquad z(0) = h, \qquad \dot\rho(0) = \dot z(0) = 0$$

we have

$$\rho(t) = x_0\cosh\omega t, \qquad z(t) = h - \tfrac{1}{2}gt^2$$

as the solution with the specified initial conditions.

PROBLEM 1.9.7 Consider a smooth wire helix with vertical axis and radius R such that $z = b\theta$. A glass bead slides down the helix under the influence of gravity.

(a) Find $z(t)$ by Lagrangian methods.

We will work with cylindrical coordinates $z, \theta, \rho = \sqrt{x^2 + y^2}$. The bead is located relative to the origin by

$$\mathbf{r} = z\hat z + R\hat\rho$$

where $\hat\rho$ is a unit vector in the x–y plane, directed away from the $\hat z$ axis,

$$\hat\rho = \hat x \cos\theta + \hat y \sin\theta$$

We also introduce a unit vector $\hat\theta$, perpendicular to $\hat z$ and ρ, defined by

$$\hat\theta = -\hat x \sin\theta + \hat y \cos\theta$$

FIGURE 1.31 The view of a portion of the helix as seen by someone looking down the $\hat z$ axis.

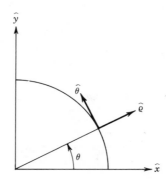

Since $\hat{x}$ and $\hat{y}$ are constant vectors, differentiation with respect to θ gives

$$\frac{d\hat{\rho}}{d\theta} = -\hat{x}\sin\theta + \hat{y}\cos\theta = \hat{\theta}$$

$$\frac{d\hat{\theta}}{d\theta} = -\hat{x}\cos\theta - y\sin\theta = -\hat{\rho}$$

The velocity of the bead is given by

$$\dot{\mathbf{r}} = \frac{d\mathbf{r}}{dt} = \dot{z}\hat{z} + R\frac{d\hat{\rho}}{dt} = \dot{z}\hat{z} + R\dot{\theta}\frac{d\hat{\rho}}{d\theta} = \dot{z}\hat{z} + R\dot{\theta}\hat{\theta}$$

$$= \dot{z}\left[\hat{z} + \left(\frac{R}{b}\right)\hat{\theta}\right]$$

where we have used the fact that $\theta = z/b$ along the helix. Since $\hat{z}$ and $\hat{\theta}$ are perpendicular unit vectors, we can write

$$T = \frac{1}{2}m\dot{\mathbf{r}}^2 = \frac{1}{2}m\dot{z}^2\left[1 + \left(\frac{R}{b}\right)^2\right]$$

The potential energy is given by $U = mgz$, so the Lagrangian is

$$L = \frac{m}{2}\left[1 + \left(\frac{R}{b}\right)^2\right]\dot{z}^2 - mgz$$

which leads to

$$\ddot{z} = -\frac{g}{1 + \left(\dfrac{R}{b}\right)^2}$$

as the equation of motion. The general solution is

$$z(t) = z(0) + \dot{z}(0)t - \frac{1}{2}\frac{g}{1 + \left(\dfrac{R}{b}\right)^2}t^2$$

(b) Find the restraining forces.

According to Newton, the total force on the bead is $m\,d^2\mathbf{r}/dt^2$. Thus, the force on the bead due to the helix alone, $\mathbf{F}_h$, is given by $\mathbf{F}_h = m\,d^2\mathbf{r}/dt^2 - (-mg\hat{z})$. We have already calculated $\dot{\mathbf{r}}$. We differentiate once more to get the acceleration,

$$\frac{d^2\mathbf{r}}{dt^2} = \frac{d\dot{\mathbf{r}}}{dt} = \ddot{z}\left(\hat{z} + \frac{R}{b}\hat{\theta}\right) + \dot{z}\frac{R}{b}\frac{d\hat{\theta}}{dt} = \ddot{z}\left(\hat{z} + \frac{R}{b}\hat{\theta}\right) - \dot{z}\dot{\theta}\frac{R}{b}\hat{\rho}$$

$$= \ddot{z}\left(\hat{z} + \frac{R}{b}\hat{\theta}\right) - (\dot{z})^2\frac{R}{b^2}\hat{\rho}$$

$$\mathbf{F}_h = m\ddot{z}\left(\hat{z} + \frac{R}{b}\hat{\theta}\right) - m\dot{z}^2\frac{R}{b^2}\hat{\rho} + mg\hat{z}$$

$$= mg\frac{(R/b)^2}{(R/b)^2 + 1}\hat{z} - mg\frac{R/b}{(R/b)^2 + 1}\hat{\theta} - m\left(\dot{z}(0) - \frac{gt}{(R/b)^2 + 1}\right)^2\frac{R}{b^2}\hat{\rho}$$

Let us check that this force is perpendicular to the helix, as it must be if the helix is frictionless. Along the helix,

$$d\mathbf{r} = dz \cdot \hat{z} + R\,d\hat{\rho} = dz\left[\hat{z} + R\frac{d\hat{\rho}}{d\theta}\frac{d\theta}{dz}\right]$$

$$= dz\left[\hat{z} + \frac{R}{b}\hat{\theta}\right]$$

We can now easily verify that $\mathbf{F}_h \cdot d\mathbf{r} = 0$ at all times.

Another formulation of the equations of motion is based on the *Hamiltonian*. For a conservative system with d degrees of freedom, the Hamiltonian is defined by

$$H(q_1, \ldots, q_d, p_1, \ldots, p_d, t) = \sum_i p_i \dot{q}_i - L \tag{1.47}$$

Note that the right-hand side of (1.47) is not the Hamiltonian until it has been expressed as a function of $q_1, \ldots, q_d, p_1, \ldots, p_d, t$. To do this, it is necessary to invert (1.37b) in order to obtain expressions for the $\dot{q}_i$ in terms of $q_1, \ldots, q_d, p_1, \ldots, p_d, t$. Hamilton's equations are derived by differentiating H of (1.47) partially with respect to p_i or q_i (keeping all the other ps and qs constant). The result is

$$\frac{\partial H}{\partial p_j} = \dot{q}_j \tag{1.48a}$$

$$\frac{\partial H}{\partial q_j} = -\dot{p}_j \tag{1.48b}$$

This set of $2d$ first-order equations has the same physical content as the set of d second-order equations (1.37c) that constitute Lagrange's equations.

The task of constructing the Hamiltonian is often made easier by the following result:

> If the kinetic energy is a homogeneous quadratic function of the generalized velocities, then the Hamiltonian is numerically equal to the total (kinetic + potential) energy.

To prove this we remark that if T is a homogeneous quadratic function of the $\dot{q}_i$, then multiplying every $\dot{q}_i$ by any number λ will have the effect of multiplying T by λ^2:

$$T(q_1, \ldots, q_d, \lambda\dot{q}_1, \ldots, \lambda\dot{q}_d, t) = \lambda^2 T(q_1, \ldots, q_d, \dot{q}_1, \ldots, \dot{q}_d, t)$$

If we differentiate this equation with respect to λ and then set $\lambda = 1$, we find

$$\sum_i \frac{\partial T}{\partial \dot{q}_i}\dot{q}_i = 2T = \sum_i p_i\dot{q}_i$$

In this case, (1.47) becomes

$$H = 2T - (T - U) = T + U = E$$

and the Hamiltonian and total energy are equal.

PROBLEM 1.9.8 Write Hamilton's equations for a one-dimensional harmonic oscillator.

Here we have only one degree of freedom, which we label $x(\equiv q_1)$. Since $T = 1/2 \cdot m\dot{x}^2$ is quadratic in $\dot{x}$, the Hamiltonian is simply

$$H = T + U = E$$

$$= \frac{p^2}{2m} + \frac{1}{2}m\omega^2 x^2 \tag{1.49}$$

Note that we have used $\dot{x} = p/m$ to remove from H any explicit dependence on $\dot{x}$. Hamilton's equations (1.48) are

$$\frac{\partial H}{\partial p} = \frac{p}{m} = \dot{x} \tag{1.50a}$$

$$\frac{\partial H}{\partial q} = m\omega^2 x = -\dot{p} \tag{1.50b}$$

Here (1.50a) just recovers the relationship between velocity and momentum, and (1.50b) is equivalent to Newton's second law of motion for this problem (1.15a).

Hamilton's equations (1.48) exhibit a high degree of symmetry between the generalized coordinates and momenta. It often turns out to be convenient to exploit this symmetry by regarding the $2d$ numbers $(q_1,\ldots,q_d; p_1,\ldots,p_d)$ as the coordinates of a single point, representing the instantaneous state of the system, in *phase space*. As time proceeds, the point traces out a path in phase space determined by its initial location and Hamilton's equations. For example, Figure 1.32a shows the phase-space trajectory for a one-dimensional harmonic oscillator. According to (1.49), the trajectory is an ellipse with semi-axes $\sqrt{2mE}$ and $\sqrt{2E/(m\omega^2)}$. Figure 1.32b shows a useful variation of this diagram, in which the axes are changed to $p/\sqrt{2m\omega}$ and $\sqrt{m\omega/2}\,x$. The trajectory is now a circle of radius $\sqrt{2E/\omega}$. It is easy to show that (1.50) implies that the point representing the instantaneous state of the oscillator moves around this circle with constant angular speed ω. The occurrence of trigonometric functions in the solution of the harmonic oscillator is associated with this circular motion in the phase space of Figure 1.32b.

FIGURE 1.32 (a) Phase-space diagram for a one-dimensional harmonic oscillator. The ellipse is determined by (1.49); (b) the representative point moves in a circle with constant angular speed ω.

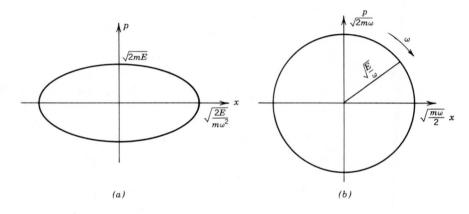

(a) (b)

PROBLEM 1.9.9 Write Hamilton's equations for the rotating two-dimensional oscillator of Problem 1.7.10, using the Cartesian coordinates of the particle relative to axes fixed in the rotating plane as generalized coordinates.

The kinetic energy is

$$T = \frac{1}{2}m(\mathbf{v})^2 = \frac{1}{2}m\left[\frac{d^*\mathbf{r}}{dt} + \omega\hat{z} \times \mathbf{r}\right]^2$$

$$= \frac{1}{2}m\left[\dot{x}\hat{x} + \dot{y}\hat{y} + \omega(x\hat{y} - y\hat{x})\right]^2$$

$$= \frac{1}{2}m\left[(\dot{x} - \omega y)^2 + (\dot{y} + \omega x)^2\right]$$

and the Lagrangian is

$$L = \tfrac{1}{2}m\left[(\dot{x} - \omega y)^2 + (\dot{y} + \omega x)^2\right] - \tfrac{1}{2}m\left[\omega_x^2 x^2 + \omega_y^2 y^2\right]$$

In this case the kinetic energy is *not* a homogeneous quadratic function of $\dot{x}$ and $\dot{y}$. To construct the Hamiltonian, we calculate

$$p_x = \frac{\partial L}{\partial \dot{x}} = m(\dot{x} - \omega y), \qquad p_y = \frac{\partial L}{\partial \dot{y}} = m(\dot{y} + \omega x)$$

$$H = p_x\dot{x} + p_y\dot{y} - L$$

$$= \frac{1}{2m}(p_x^2 + p_y^2) + \frac{1}{2}m(\omega_x^2 x^2 + \omega_y^2 y^2) - \omega(xp_y - yp_x)$$

Hamilton's equations (1.48) are

$$\frac{\partial H}{\partial p_x} = \frac{p_x}{m} + \omega y = \dot{x} \tag{1.51a}$$

$$\frac{\partial H}{\partial p_y} = \frac{p_y}{m} - \omega x = \dot{y} \tag{1.51b}$$

$$\frac{\partial H}{\partial x} = m\omega_x^2 x - \omega p_y = -\dot{p}_x \tag{1.51c}$$

$$\frac{\partial H}{\partial y} = m\omega_y^2 y + \omega p_x = -\dot{p}_y \tag{1.51d}$$

Equations (1.51a and 1.51b) give us no new information, but (1.51c and 1.51d) are equivalent to the equations of motion (1.32) we found by using Newton's second law in a rotating coordinate system.

In general, elementary problems in mechanics are just as easily solved with Lagrange's equations (1.37) as with Hamilton's equations (1.48). The principal reason for introducing Hamiltonian methods into the undergraduate curriculum is the important role they play in quantum mechanics. For example, the prescription for writing the Schrödinger equation (5.2.7) starts with the construction of the classical Hamiltonian (1.47). It is even possible to use Hamilton's equations in quantum mechanics, although the symbols H, q_j, p_j have to be interpreted as "operators" rather than as numerical quantities (see Section 5.2).

1.10 NORMAL MODE ANALYSIS OF SMALL OSCILLATIONS ABOUT EQUILIBRIUM

In many mechanical systems, the Lagrangian has the form

$$L(q_1, \ldots, q_n, \dot{q}_1, \ldots, \dot{q}_n) = \sum_{i=1}^{n} \frac{1}{2} m_i \dot{q}_i^2 - U(q_1, \ldots, q_n) \qquad (1.52)$$

Lagrange's equations of motion then take the form

$$m_i \ddot{q}_i = -\frac{\partial U}{\partial q_i}(q_1, \ldots, q_n) \qquad (i = 1, 2, \ldots, n) \qquad (1.53)$$

At equilibrium all the $\ddot{q}_i$ vanish, so the n equilibrium values of the q_i, which we designate by $\mathring{q}_i$, satisfy the n conditions

$$\frac{\partial U}{\partial q_i}\bigg|_{q_i = \mathring{q}_i} = 0 \qquad (i = 1, 2, \ldots, n) \qquad (1.54)$$

To study motions near equilibrium, we expand U in a Taylor series about $\mathring{q}_i$,

$$U(q_1, \ldots, q_n) = U(\mathring{q}_1, \ldots, \mathring{q}_n) + \sum_i (q_i - \mathring{q}_i) \frac{\partial U}{\partial q_i}\bigg|_{q = \mathring{q}}$$

$$+ \frac{1}{2} \sum_{i,j} (q_i - \mathring{q}_i)(q_j - \mathring{q}_j) \frac{\partial^2 U}{\partial q_i \partial q_j}\bigg|_{q = \mathring{q}}$$

$$\simeq U(\mathring{q}_1, \ldots, \mathring{q}_n) + \frac{1}{2} \sum_{ij} r_i a_{ij} r_j, \qquad + \ldots \qquad (1.55a)$$

with

$$r_i = q_i - \mathring{q}_i \qquad (1.55b)$$

$$a_{ij} = \frac{\partial^2 U}{\partial q_i \partial q_j}\bigg|_{q = \mathring{q}} = a_{ji} \qquad (1.55c)$$

The "harmonic" approximation (1.55) for U yields equations of motion that are coupled second-order linear differential equations in the deviations r_i,

$$m_i \ddot{r}_i = -\sum_j a_{ij} r_j$$

The method for solving such equations is a generalization of the method outlined in Section 1.9 for decoupling (1.43a) and (1.43b). The essential idea is to attempt to find solutions (called *normal modes*) in which all the r_i oscillate with the same frequency. For a system with n degrees of freedom there will be n linearly independent normal modes, and the most general motion of the system can be expressed as a superposition of them. In general, finding the normal modes involves diagonalizing the symmetric matrix

$$(b_{ij}) = \left(\frac{a_{ij}}{\sqrt{m_i m_j}} \right)$$

which can be a laborious procedure if n is greater than 2. However, in many cases the symmetries of the system allow us to write down at least some of the normal modes by inspection, and then to find the normal frequencies by means of a simple calculation.

The following is a convenient general procedure for finding the normal frequency once we have guessed the form of the normal mode:

1. Let ξ represent the amplitude of the mode ($\xi = 0$ corresponds to the equilibrium configuration).

2. Calculate the total potential energy of the system as a function of ξ. It will have the form

$$U(\xi) = \tfrac{1}{2}k_{\text{eff}}\xi^2 \qquad (1.56a)$$

where the "effective spring constant" k_{eff} depends on the restoring forces in the system when the components move in the chosen normal mode.

3. Calculate the total kinetic energy of the system as a function of $\dot{\xi}$. It will have the form

$$T(\xi) = \tfrac{1}{2}m_{\text{eff}}\dot{\xi}^2 \qquad (1.56b)$$

where the "effective mass" m_{eff} depends on the masses of the system and on the way they move in the chosen normal mode.

4. The frequency of the normal mode is then

$$\omega = \sqrt{k_{\text{eff}}/m_{\text{eff}}} \qquad (1.56c)$$

PROBLEM 1.10.1 Two identical masses m slide on a smooth surface, are connected to the walls by identical springs (spring constant k) and to each other by a spring with spring constant K. Find the two normal modes and normal frequencies.

Figure 1.33a shows the masses in equilibrium, and Figure 1.33b shows an out-of-phase oscillation in which the masses move by the same amount but in opposite directions. The symmetry of the system then dictates that the midpoint of the middle spring will remain at rest. The behavior of the left-hand mass is then the same as that of the mass illustrated in Figure 1.33c. Here the mass m is subjected to a harmonic restoring force of effective spring constant $k + 2K$, so the frequency with which it oscillates is $\sqrt{(k + 2K)/m}$. A similar observation is made about the right-hand mass, so that in the motion depicted in Figure 1.33b both masses oscillate with the same frequency, and this is indeed a normal mode of the system.

At equilibrium (Figure 1.33a), the tension is the same in all three springs. In the in-phase oscillation shown in Figure 1.33d, both masses have moved to the right of their equilibrium positions by the same amount. The spring K has the same length and, thus, the same tension, that it had in equilibrium. The net force on the left-hand mass is just the additional force provided by the left-hand spring due to the stretching of this spring. If the left-hand mass has moved x to the right, there is, therefore, a restoring force of magnitude kx. This mass will then execute an oscillation with frequency $\omega = \sqrt{k/m}$. Similarly, the net force on the right-hand mass will be the additional force exerted by the right-hand spring due to the compression of that spring, and this also produces a restoring force of magnitude kx. Thus, the two masses oscillate at frequency $\omega = \sqrt{k/m}$, maintaining the separation equal to their equilibrium separation.

PROBLEM 1.10.2 A model of the CO_2 molecule assumes that the atoms are aligned and connected by springs of equal spring constant k. Describe the normal modes of oscillation of

FIGURE 1.33 (a) The masses and springs in equilibrium; (b) out-of-phase oscillation of the masses (note that the center of the middle spring remains at rest); (c) if spring K is fixed at its center, each half has an effective spring constant of $2K$; (d) in-phase oscillation of the masses. Note that the separation of the masses is the same as in the equilibrium configuration.

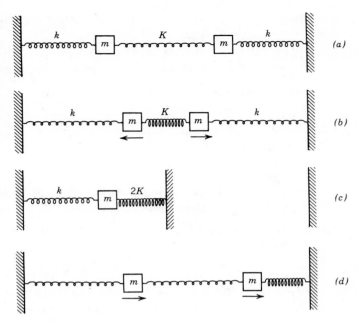

this system (along the line connecting the atoms) and find the frequencies in terms of k and the masses of the atoms.

Since there are no external forces acting on the molecule, the vibrational modes will leave the mass-center of the molecule at rest. One such mode would correspond to having the ^{12}C remain fixed, while the ^{16}O atoms oscillate $180°$ out of phase, corresponding to the pattern $\leftarrow x \rightarrow$. This oscillation frequency is $\omega = \sqrt{k/m_{(^{16}O)}}$. The other mode corresponds to the pattern $\rightarrow \leftarrow \rightarrow$, with the two ^{16}O atoms moving in phase, opposite to the motion of the ^{12}C. If each ^{16}O moves to the right by x, the C must move to the left by $2 \cdot 16/12 \cdot x$ in order to keep the mass-center fixed. Thus, the right-hand spring stretches by $x + (32/12) \cdot x$, while the left-hand spring compresses by $x + (32/12) \cdot x$.

^{16}O ^{12}C ^{16}O

k k

The potential energy associated with this displacement is

$$U = 2 \cdot \frac{1}{2} k \left(x + \frac{32}{12} x \right)^2 = \frac{1}{2} \cdot 2 \cdot \left(\frac{44}{12} \right)^2 k x^2$$

The kinetic energy is

$$T = 2 \cdot \frac{1}{2} m_{(^{16}O)} \dot{x}^2 + \frac{1}{2} m_{(^{12}C)} \left(\frac{32}{12} \dot{x} \right)^2 = \frac{1}{2} \left[2 \cdot \frac{44}{12} \cdot m_{(^{16}O)} \right] \dot{x}^2$$

FIGURE 1.34 (a) Two identical springs in series; (b) two identical springs in parallel.

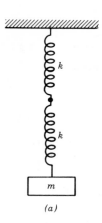

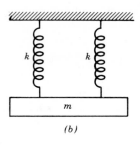

(a)

(b)

The frequency of this mode is thus

$$\omega = \sqrt{k_{\text{eff}}/m_{\text{eff}}} = \sqrt{\frac{2 \cdot \left(\frac{44}{12}\right)^2 k}{2 \cdot \frac{44}{12} m_{(^{16}O)}}} = \sqrt{\frac{44}{12} \frac{k}{m_{(^{16}O)}}}$$

PROBLEM 1.10.3   What is the frequency of oscillation for each of the systems shown in Figure 1.34?

(a) *Springs in series.* Let $x(t)$ be the vertical displacement from equilibrium. The kinetic energy is $T = (1/2)m\dot{x}^2$. Each spring stretches by $x/2$, so the potential energy of each spring is $(1/2)k \cdot (x/2)^2 = 1/8 \cdot kx^2$. Thus, the total potential energy is $U = 2 \times 1/8 \cdot kx^2 = (1/2)(k/2) \cdot x^2$. We have $m_{\text{eff}} = m$, $k_{\text{eff}} = (1/2)k$, so $\omega = \sqrt{k_{\text{eff}}/m_{\text{eff}}} = \sqrt{k/2m}$.

(b) *Springs in parallel.* Now a displacement x of the mass stretches each spring by x. Thus, $U = 2 \times (1/2)kx^2 = (1/2)(2k)x^2$ and $k_{\text{eff}} = 2k$. T is still $(1/2)m\dot{x}^2$, so $\omega = \sqrt{k_{\text{eff}}/m_{\text{eff}}} = \sqrt{2k/m}$.

PROBLEM 1.10.4   A nonviscous liquid of density ρ is in a U-tube of cross-section area A. The length of the fluid column is l. Find the frequency of small oscillations about the equilibrium position with equal levels in the two columns.

Let ξ be the distance that the left-hand level goes down, and, thus, the distance that the right-hand level goes up. The potential energy of the configuration shown in Figure 1.35 is the amount of work required to raise a column of

FIGURE 1.35 The dashed lines represent the equilibrium levels of the liquid.

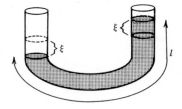

fluid of length ξ by a distance ξ. This is

$$U(\xi) = g \cdot \rho A \xi \cdot \xi = g\rho A \xi^2$$

Thus, $k_{\text{eff}} = 2g\rho A$. The kinetic energy is

$$T(\xi) = \tfrac{1}{2} m \dot{\xi}^2 = \tfrac{1}{2}(\rho A l) \dot{\xi}^2$$

and so $m_{\text{eff}} = \rho A l$. Thus, the frequency of small oscillations is

$$\omega = \sqrt{k_{\text{eff}}/m_{\text{eff}}} = \sqrt{\frac{2g\rho A}{\rho A l}} = \sqrt{\frac{2g}{l}}$$

PROBLEM 1.10.5 Three identical masses m are connected by three identical springs, of spring constant k (Figure 1.36a). Find the frequency of the "breathing mode" of the system, in which the masses move radially, in phase.

We choose to specify the amplitude of the mode by giving the distance ξ that each mass moves radially away from its equilibrium position. It is clear from Figure 1.36b that this results in a stretching (or compression) of each spring by an amount $2(\sqrt{3}/2)\xi = \sqrt{3}\,\xi$. Thus, the potential energy of each spring is $(1/2)k(\sqrt{3}\,\xi)^2$, and the total potential energy is

$$U(\xi) = 3 \cdot \tfrac{1}{2}k(\sqrt{3}\,\xi)^2 = \tfrac{1}{2}(9k)\xi^2 \tag{1.57a}$$

Comparison of (1.57a) and (1.56a) shows that we must take $k_{\text{eff}} = 9k$. The speed of each mass is $\dot{\xi}$, and its kinetic energy is $(1/2)m\dot{\xi}^2$. The total kinetic energy is

$$T(\xi) = 3 \cdot \tfrac{1}{2}m\dot{\xi}^2 = \tfrac{1}{2}(3m)\dot{\xi}^2 \tag{1.57b}$$

so that the effective mass in (1.56b) is given by $m_{\text{eff}} = 3m$. Finally, the frequency of the breathing mode is

$$\omega = \sqrt{k_{\text{eff}}/m_{\text{eff}}} = \sqrt{\frac{9k}{3m}} = \sqrt{\frac{3k}{m}}$$

PROBLEM 1.10.6 Find the period of small oscillations about equilibrium of a hemisphere of mass M and radius R, rocking without slipping on a horizontal surface.

The first step in the calculation of the potential and kinetic energy is the location of the center of mass. A straightforward integration shows that it is $(3/8)R$ below the center of the full sphere. We also need I_{CM}, the moment of inertia of the hemisphere about a horizontal axis through its mass-center [we know that this is a principal axis because it is perpendicular to a plane of reflection symmetry (see p. 35)]. We start with a full sphere of mass $2M$, whose moment of inertia about a horizontal axis through C is $I_C = (2/5)(2M)R^2$. Half of this is contributed by each hemisphere, so the moment of inertia of the hemisphere shown in Figure 1.37a about a horizontal axis through C is $(2/5)MR^2$. To get I_{CM}, we use the parallel axis theorem (see footnote p. 22):

$$I_C = \tfrac{2}{5}MR^2 = I_{\text{CM}} + M\left(\tfrac{3}{8}R\right)^2, \qquad I_{\text{CM}} = \tfrac{83}{320}MR^2$$

Figure 1.37c shows that the vector $\mathbf{R}$, which locates the mass-center relative to the original point of contact, is

$$\mathbf{R} = \hat{x}\left(R\theta - \tfrac{3}{8}R\sin\theta\right) + \hat{y}\left(R - \tfrac{3}{8}R\cos\theta\right) \tag{1.58}$$

Thus, the potential energy is

$$U(\theta) = Mg\left(R - \tfrac{3}{8}R\cos\theta\right) = \tfrac{5}{8}MgR + \tfrac{3}{16}MgR\theta^2 + \cdots,$$

FIGURE 1.36 (a) The three masses are connected symmetrically by three identical springs with spring constant k; (b) the inner triangle represents the equilibrium configuration. ξ is the distance each mass moves radially from its equilibrium position during a breathing mode.

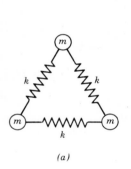
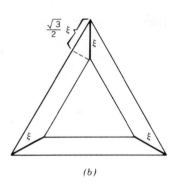

FIGURE 1.37 (a) The hemisphere in equilibrium. The center of mass is $3R/8$ below C, the center of the sphere. (b) The hemisphere has rolled through angle θ. (c) Locating the center of mass relative to the original point of contact.

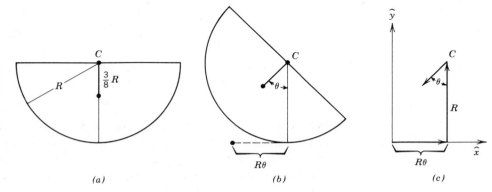

so that k_{eff} for small oscillations is $(3/8)MgR$. We use (1.26) to find the kinetic energy,

$$\dot{\mathbf{R}} = R\dot{\theta}\left[\hat{x}\left(1 - \tfrac{3}{8}\cos\theta\right) + \hat{y}\tfrac{3}{8}\sin\theta\right]$$

$$T = \tfrac{1}{2}MR^2\dot{\theta}^2\left(\tfrac{73}{64} - \tfrac{3}{4}\cos\theta\right) + \tfrac{1}{2}\left(\tfrac{83}{320}MR^2\right)\dot{\theta}^2 \approx \tfrac{1}{2}\left(\tfrac{13}{20}MR^2\right)\dot{\theta}^2$$

and $m_{\text{eff}} = (13/20)MR^2$. Finally, we find that the frequency of small oscillations about $\theta = 0$ is

$$\omega = \sqrt{k_{\text{eff}}/m_{\text{eff}}} = \sqrt{\frac{\tfrac{3}{8}MgR}{\tfrac{13}{20}MR^2}} = \sqrt{\frac{15}{26}\frac{g}{R}}$$

1.11 MOTION OF A PARTICLE IN A CENTRAL POTENTIAL

A particle of mass m moves in a fixed central force field given by (1.21). The torque on the particle, defined relative to the force center, is

$$\boldsymbol{\tau} = \mathbf{r} \times \mathbf{f} = g(r)\mathbf{r} \times \hat{r} = 0$$

so that (1.1b) implies that the angular momentum $\mathbf{l}$ of the particle is constant in time. Since $\mathbf{l} = m\mathbf{r} \times \mathbf{v}$, the vector $\mathbf{r}$ from the force center to the particle is always perpendicular to the fixed vector $\mathbf{l}$. Thus, the particle moves in a plane that contains the force center and is perpendicular to $\mathbf{l}$. Let us choose a coordinate system whose origin is at the force center, and whose x–y plane is the plane of the orbit. If we use polar coordinates r, θ to locate the particle in this plane, we have

$$T = \frac{m}{2}(\dot{r}^2 + r^2\dot{\theta}^2)$$

$$L = \frac{m}{2}(\dot{r}^2 + r^2\dot{\theta}^2) - U(r)$$

$$\frac{\partial L}{\partial \dot{r}} = m\dot{r} \qquad \frac{\partial L}{\partial \dot{\theta}} = mr^2\dot{\theta}$$

$$\frac{\partial L}{\partial r} = mr\dot{\theta}^2 - \frac{\partial U}{\partial r} \qquad \frac{\partial L}{\partial \theta} = 0$$

Lagrange's equations of motion are

$$\frac{d}{dt}(m\dot{r}) = mr\dot{\theta}^2 - \frac{\partial U}{\partial r} \tag{1.59a}$$

$$\frac{d}{dt}(mr^2\dot{\theta}) = 0 \tag{1.59b}$$

$$mr^2\dot{\theta} = l = \text{constant} \tag{1.60}$$

We use l to label the constant value $mr^2\dot{\theta}$, since it is easily verified that it is equal to the magnitude of the particle's angular momentum. The quantity $r^2\dot{\theta}$ is twice the rate at which **r** sweeps out area. Thus, (1.60) establishes Kepler's second law (the radius vector sweeps out equal areas in equal times), and shows that this law is true for any central force.

The total mechanical energy, which is also a constant of the motion, is

$$E = \frac{m}{2}(\dot{r}^2 + r^2\dot{\theta}^2) + U(r)$$

$$= \frac{m}{2}\dot{r}^2 + \frac{l^2}{2mr^2} + U(r) \tag{1.61}$$

This shows that we can formally consider the radial motion as if it were a one-dimensional problem, provided that we add to the true potential $U(r)$ an effective "centrifugal potential" $l^2/2mr^2$.

We can solve (1.61) for $\dot{r} = dr/dt$ and write the result as

$$dt = \frac{dr}{\pm\sqrt{\dfrac{2}{m}\left(E - U(r) - \dfrac{l^2}{2mr^2}\right)}}$$

If we integrate this equation, we get

$$\int_{t_0}^{t_1} dt = t_1 - t_0 = \int_{r_0}^{r_1} \frac{dr}{\pm\sqrt{\dfrac{2}{m}\left(E - U(r) - \dfrac{l^2}{2mr^2}\right)}} \tag{1.62a}$$

If we can evaluate this integral as a function of its upper limit we will have determined r as a function of t for an orbit of energy E and angular momentum l, subject to the initial condition $r(t_0) = r_0$. Once $r(t)$ is determined, (1.60) gives

$\dot{\theta}$ as a function of time from which we can obtain $\theta(t)$ by one more integration:

$$\theta(t) = \theta(t_0) + \int_{t_0}^{t} \frac{l}{m[r(t')]^2} \, dt' \qquad (1.62b)$$

1.12 MOTION OF A PARTICLE IN A 1/R POTENTIAL

We now apply the results of the last section to a central field whose radial dependence is

$$U(r) = K \cdot \frac{1}{r} \qquad (K = \text{constant}) \qquad (1.63)$$

If $K = -Gmm_2$, (1.63) describes the potential experienced by a particle of mass m moving in the gravitational field of a spherical object of mass m_2. If $K = G'q_1q_2$, (1.63) describes the Coulomb interaction between particles with charges q_1 and q_2; G' is a positive constant whose numerical value depends on the system of units being used. If the charges are opposite, q_1q_2 is negative and the force is attractive, whereas if the charges have the same sign the force is repulsive. In either case, if (1.63) is substituted into (1.62), the integrals can be done analytically, and one finds that the orbit is a conic section with the force center at one of the foci (Kepler's first law).

The qualitative features of the motion can be determined by plotting the "effective potential"

$$U_{\text{eff}}(r) = \frac{l^2}{2mr^2} + \frac{K}{r} \qquad (1.64)$$

as a function of r. Figure 1.38a shows the situation when K is negative, corresponding to the gravitational or Coulomb attraction.

The radial kinetic energy

$$\frac{m\dot{r}^2}{2} = E - U_{\text{eff}}(r) = E - \frac{l^2}{2mr^2} - \frac{K}{r} \qquad (1.65)$$

is zero when $\dot{r} = 0$, but is positive at all other times. Figure 1.38a shows that this implies that when $E = E_1 < 0$, the value of r is constrained by $r_1 \le r \le r_2$. Here r_1 and r_2 are the zeros of the right-hand side of (1.65),

$$\left.\begin{array}{c} r_1 \\ r_2 \end{array}\right\} = \frac{-K}{2|E|} \mp \sqrt{\left(\frac{K}{2|E|}\right)^2 - \frac{l^2}{2m|E|}} \qquad (1.66)$$

They are referred to as the "turning points." We see that when $E < 0$ the particle remains at all times within a finite distance from the force center. The orbit is an ellipse (Figure 1.39).

We can relate the semi-axes a and b to r_1 and r_2 by using the geometrical properties of the ellipse, which state that $r + s$ is constant as P moves around the ellipse. At points 1 and 2 in Figure 1.39, $\dot{r} = 0$, so these are the turning points. At point 2, $r = r_2 = a + d$, $s = r_1 = a - d$. Thus, $a = \frac{1}{2}(r_2 + r_1)$, $d = \frac{1}{2}(r_2 - r_1)$, and $r + s = 2a$. At point 3, $r = s = \frac{1}{2}(r + s) = a = \sqrt{b^2 + d^2}$. If

FIGURE 1.38 $U_{\text{eff}}(r)$ from (1.64) for attractive and repulsive $1/r$ potentials. Each horizontal line corresponds to motion at the specified energy. The radial kinetic energy $m\dot{v}^2/2$ is given by the vertical distance between the horizontal line and the $U_{\text{eff}}(r)$ curve below it. r_1, r_2, r_3, r_4 are turning points of the radial motion. (a) Attractive $1/r$ potential. K is negative. (b) Repulsive $1/r$ potential. K is positive.

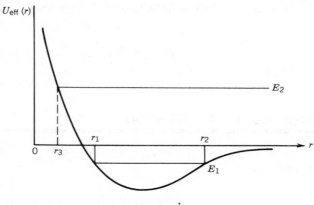

(a) Attractive $\frac{1}{r}$ potential;
K is negative.

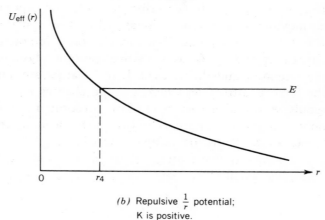

(b) Repulsive $\frac{1}{r}$ potential;
K is positive.

FIGURE 1.39 An elliptic orbit, with the attractive force center at one focus. The distance sum $r + s$ is constant as P moves around the ellipse. The points 1 and 2 are turning points of the radial motion.

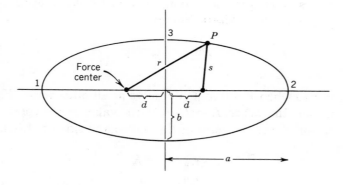

we use the expression (1.66) for r_1 and r_2, we get

$$a = \tfrac{1}{2}(r_1 + r_2) = \frac{-K}{2|E|} \tag{1.67a}$$

$$b = \sqrt{a^2 - d^2} = \sqrt{\left(\frac{r_2 + r_1}{2}\right) - \left(\frac{r_2 - r_1}{2}\right)^2} = \frac{l}{\sqrt{2m|E|}} \tag{1.67b}$$

The area within the elliptic orbit is $\pi ab = \pi |K| l / \sqrt{8m|E|^3}$. The time required for a complete excursion around the orbit (the period) is thus

$$\tau = \frac{\pi ab}{\frac{1}{2}r^2\dot{\theta}} = \frac{\pi|K|l}{\sqrt{8m|E|^3}} \bigg/ \frac{l}{2m} = \pi|K|\sqrt{\frac{m}{2|E|^3}} = 2\pi\sqrt{\frac{m}{|K|}}\,a^{3/2} \quad (1.68)$$

In the gravitational case, where $|K| = Gmm_2$, this becomes

$$\tau = \frac{2\pi}{\sqrt{Gm_2}}\,a^{3/2} \qquad\qquad (1.69)$$

This establishes that the square of the period of a planet is proportional to the cube of its semi-major axis, irrespective of its mass or energy or angular momentum (Kepler's third law).

If $|K|/2|E| = l/\sqrt{2m|E|}$, (1.66) and (1.67) imply that $r_1 = r_2$, $a = b$, and the orbit is circular. In this special case all the properties of the orbit can be derived most simply by equating the gravitational attraction to m times the centripetal acceleration in the circular orbit.

Reference to Figure 1.38 shows that if the energy E is positive, r has no maximum value. This is true both for attractive and repulsive $1/r$ potentials. The shape of the orbit is now a hyperbola, with the force center again at a focus.

An important application of positive-energy orbits is to the analysis of Coulomb scattering experiments. Initially, at $t = -\infty$, the particle is infinitely far from the scattering center, with a velocity $\mathbf{v}_0$. Since $U_{\text{eff}}(r) = 0$ at $r = \infty$, the total energy E is $mv_0^2/2$. The vector $\mathbf{v}_0$ and the scattering center determine a plane, and this is the plane of the orbit. Figure 1.40 illustrates typical orbits for attractive and repulsive potentials. The impact parameter b is defined to be the asymptotic distance between the orbit of the particle and a parallel line passing through the force center. Thus, at $t = -\infty$ the angular momentum of the particle is $l = mv_0 b$, and since the force field is central this must be the angular

FIGURE 1.40 Hyperbolic orbits for scattering by $1/r$ potentials. The impact parameter is b and the scattering angle is Θ. (a) Attractive $1/r$ potential. (b) Repulsive $1/r$ potential.

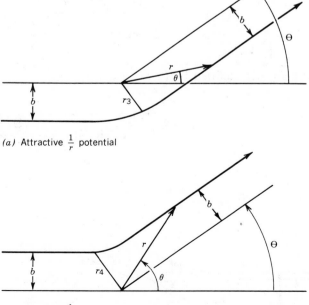

(a) Attractive $\frac{1}{r}$ potential

(b) Repulsive $\frac{1}{r}$ potential

momentum all along the orbit. In this way we see that specification of the impact parameter b and the asymptotic speed v_0 at $t = -\infty$ determine both the energy E and angular momentum l. We can then use (1.62a) and (1.62b) to obtain the time development of r and θ for this orbit. The minus sign on the square root in (1.62a) is used along the first half of the orbit (when $\dot{r} < 0$) and the plus sign along the second half. From the value of θ at $t = \infty$, we can calculate the *scattering angle* Θ. The scattering angle as a function of E and l (or b) is called the *deflection function*. For a $1/r$ potential it is given by

$$\Theta(E, b) = 2\arctan\left(\frac{K}{2Eb}\right) \tag{1.70}$$

We see that a large impact parameter corresponds to a distant orbit in which the particle is scattered very little, while a very small impact parameter corresponds to an almost head-on collision and a scattering angle near $180°$.

The differential cross-section $d\sigma/d\Omega$ is defined by

$$\frac{d\sigma}{d\Omega}(\Theta) = \frac{\left(\begin{array}{c}\text{rate at which particles are scattered}\\ \text{into unit solid angle about angle } \Theta\end{array}\right)}{\text{incident flux}} \tag{1.71}$$

Figure 1.41 shows how this can be calculated from the deflection function. Suppose that the incident flux (particles/unit area $\cdot$ unit time) is J. Then the rate at which particles are incident with impact parameters between b and $b + db$ is $2\pi b \, db \cdot J$. These are all scattered through angles between $\Theta(E, b)$ and $\Theta(E, b + db)$. The solid angle filled by these particles is

$$2\pi \sin\Theta(E, b) \cdot |d\Theta(E, b)| = 2\pi \sin\Theta(E, b) \cdot \left|\frac{d\Theta(E, b)}{db}\right| db$$

Thus, the number of particles per unit solid angle per unit time with scattering angle Θ is

$$\frac{2\pi b \cdot db \cdot J}{2\pi \sin\Theta(E, b) \left|\dfrac{d\Theta(E, b)}{db}\right| db}$$

and $d\sigma/d\Omega$, defined by (1.71) is

$$\frac{d\sigma}{d\Omega}(\Theta) = \frac{b}{\sin\Theta(E, b) \left|\dfrac{d\Theta(E, b)}{db}\right|} \tag{1.72}$$

FIGURE 1.41 Two orbits of the same energy with nearby impact parameters. The interaction is repulsive.

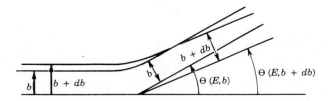

For the $1/r$ deflection function (1.70), this yields

$$\frac{d\sigma}{d\Omega} = \frac{|K/4E|^2}{\sin^4(\Theta/2)} \tag{1.73}$$

the Rutherford scattering formula.

PROBLEM 1.12.1 Show that circular motion in a central potential $V(r) = -k^2/r^n$ is unstable against radial disturbances for $n > 2$.

   We know that the spherical symmetry of the potential implies angular momentum conservation, which in turn implies motion in a plane. Choose the center of the coordinate system at the center of attraction. Then the Lagrangian is

$$L = \frac{m}{2}(\dot{r}^2 + r^2\dot{\theta}^2) + \frac{k^2}{r^n}$$

$$\frac{\partial L}{\partial \dot{r}} = m\dot{r} \qquad \frac{\partial L}{\partial r} = mr\dot{\theta}^2 - \frac{nk^2}{r^{n+1}}$$

$$\frac{\partial L}{\partial \dot{\theta}} = mr^2\dot{\theta}, \qquad \frac{\partial L}{\partial \theta} = 0, \qquad \text{so } mr^2\dot{\theta} = l = \text{constant}$$

The equation of motion for r is then

$$\frac{d}{dt}(m\dot{r}) = mr\dot{\theta}^2 - \frac{nk^2}{r^{n+1}} = \frac{l^2}{mr^3} - \frac{nk^2}{r^{n+1}}$$

Motion in a circle corresponds to $r = r_0$ ($=$ constant). Thus

$$\frac{l^2}{mr_0^3} - \frac{nk^2}{r_0^{n+1}} = 0, \qquad r_0^{n-2} = \frac{nmk^2}{l^2}$$

   Now consider small oscillations about circular motion. Let

$$r(t) = r_0 + \rho(t)$$

Then $\rho(t)$ satisfies the equation

$$m\ddot{\rho}(t) = \frac{l^2}{m(r_0 + \rho)^3} - \frac{nk^2}{(r_0 + \rho)^{n+1}}$$

$$\approx \left[\frac{l^2}{mr_0^3} - \frac{nk^2}{r_0^{n+1}}\right] + \rho\left[\frac{-3l^2}{mr_0^4} + \frac{n(n+1)k^2}{r_0^{n+2}}\right]$$

for sufficiently small ρ. The first bracketed term vanishes by virtue of the equilibrium conditions. If we replace r_0 in the second term by its expression in terms of n, m, k, l, we get

$$m\ddot{\rho} = \left(\frac{l^2}{m}\right)^{(n+2)/(n-2)} \frac{1}{(nk^2)^{4/(n-2)}}(n-2)\rho$$

Thus, if $n > 2$ the coefficient of ρ on the right-hand side is positive, and a small deviation from $\rho = 0$ will grow to become a large one. The circular motion is unstable against a radial disturbance.

PROBLEM 1.12.2 The orbital speed of the earth is 30 km/s. Neptune is 30 times as far from the sun as earth is. What is the orbital speed of Neptune? Assume that the orbits are circular.

Kepler's third law states that the periods τ and radii a are related by

$$\frac{\tau_e}{a_e^{3/2}} = \frac{\tau_n}{a_n^{3/2}}$$

Since the speed v is given by $v = 2\pi a/\tau$, we find that

$$\frac{V_n}{V_e} = \frac{a_n \tau_e}{a_e \tau_n} = \frac{a_n}{a_e}\left(\frac{a_e}{a_n}\right)^{3/2} = \sqrt{\frac{a_e}{a_n}} = \sqrt{\frac{1}{30}}$$

Thus, the orbital speed of Neptune is $30/\sqrt{30}$ km/s $= \sqrt{30}$ km/s.

PROBLEM 1.12.3 A "24-hour satellite" is to be put in orbit around the earth. This is a satellite that would remain overhead at the same spot on the equator for all 24 hours of every day if its orbit were in the equatorial plane. How high must such a satellite be placed?

If the satellite has angular speed ω in a circular orbit of radius r, its centripetal acceleration is $\omega^2 r$, so that the gravitational force of the earth on it must be

$$m\omega^2 r = \frac{mM_eG}{r^2}$$

Thus, $r = [M_eG/\omega^2]^{1/3}$. The gravitational attraction on a particle of mass m near the surface of the earth is

$$mg = m\frac{M_eG}{R_e^2}$$

so that $M_eG = gR_e^2$. Therefore, we can write the radius of the satellite orbit as

$$r = \left[\frac{gR_e^2}{\omega^2}\right]^{1/3}$$

If the satellite is to remain over the same point on the earth,

$$\omega = \frac{2\pi}{24}\frac{\text{rad}}{\text{h}}$$

Thus,

$$r = \left[\frac{9.8\text{m}/s^2\,(6.4\times10^6\text{m})^2}{\left(\dfrac{2\pi}{24}\dfrac{1}{3600}\dfrac{\text{rad}}{\text{s}}\right)^2}\right]^{1/3} = 4.23\times10^7\text{m}$$

The height above the surface of the earth is

$$h = r - R_e = 3.59\times10^7\text{m}$$

PROBLEM 1.12.4 A projectile starts from the earth's surface with its velocity vector inclined at an angle θ from the vertical direction. It moves freely under gravity. At large distance, the kinetic energy of the projectile approaches E, while its path approaches a straight line that is a distance b from the parallel through the earth's center. Ignore the rotation of the earth, and show that $b = \sqrt{(E + B)/E}\, R \sin\theta$, where B is a constant.

Energy and angular momentum are conserved. The total energy is given by $E = mv^2/2 - mM_eG/r$. Let v_∞ be the speed of the particle when $r = \infty$ and v_L is its speed at launch. Thus,

$$v_\infty = \sqrt{\frac{2E}{m}}\,, \qquad v_L = \sqrt{\frac{2E}{m} + \frac{2M_eG}{R}}\,, \qquad \frac{v_L}{v_\infty} = \sqrt{1 + \frac{mM_eG}{ER}}$$

At launch, the angular momentum of the particle relative to the center of the earth is $mv_L R \sin\theta$. When $r = \infty$, the angular momentum is $mv_\infty b$. Thus, angular momentum conservation yields

$$mv_L R \sin\theta = mv_\infty b$$

$$b = \frac{V_L}{V_\infty} R \sin\theta = \sqrt{1 + \frac{mM_eG}{ER}} \cdot R \sin\theta$$

which is the desired form, with $B = mM_eG/R$.

PROBLEM 1.12.5   A satellite is in an elliptical orbit about the earth, with an energy per unit mass given by $v^2/2 - GM_e/r = -GM_e/2a$, where $v = $ speed, $r = $ distance to the center of the earth, $2a = $ length of major axis, $M_e = $ mass of the earth, and $G = $ gravitational constant. The satellite receives an impulse that changes the magnitude of its speed by the amount Δv, but does not change the direction of the velocity.

(a) Show that the impulse changes the length of the major axis by $4va^2\,\Delta v/GM_e$.

The impulse changes the speed of the satellite without changing its location. Thus,

$$\Delta E = \Delta\left(\frac{mv^2}{2} - \frac{GmM_e}{r}\right) = mv\,\Delta v = +\frac{GmM_e}{2a^2} \cdot \Delta a$$

$$\Delta(2a) = \frac{4va^2 \cdot \Delta v}{GM_e} = \text{change in length of major axis}$$

(b) Show that the empty focus of the ellipse shifts by the same amount.

Rays from the foci to a point on the ellipse make equal angles θ with the tangent to the ellipse at that point. Since the impulse does not change the direction of the velocity, it does not change the direction of the tangent. Nor does it change the location of the earth, so θ remains unchanged. Thus, the new empty focus must be along the line from the satellite to the old empty focus.

FIGURE 1.42 The initial direction of the velocity makes an angle θ with the vertical direction.

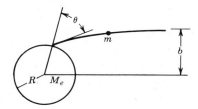

FIGURE 1.43 The shift of the empty focus when the velocity suddenly decreases in magnitude, but with no change in direction.

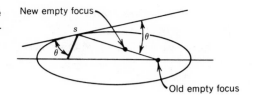

Since the sum of the distances from the satellite to the two foci equals $2a$, a sudden change in the value of $2a$ causes the empty focus to move a distance $2\,\Delta a$ along the line to the satellite.

PROBLEM 1.12.6 The radius of the earth's orbit is one "astronomical unit", 1 au. The earth's orbital speed is 30 km/s.

(a) Neglecting planetary perturbations, calculate the "escape speed" in km/sec for an object at 1 au. That is, what is the minimum speed that an object must have to escape from the solar system, starting at the earth's orbit.

An object of mass m at a distance R from the sun, with speed v, has total energy

$$E = \tfrac{1}{2}mv^2 - \frac{mM_sG}{R}$$

To be able to reach $R = \infty$ with $v \geq 0$, this energy E must be ≥ 0. Thus, the escape speed v_e at solar distance R is given by

$$v_e^2 = \frac{2M_sG}{R}$$

To calculate this quantity, we use the information given about the earth's orbital speed. If we assume that the orbit is a circle and relate the centripetal acceleration of the earth to its gravitational attraction to the sun, we get

$$\frac{mv_0^2}{R} = \frac{mM_sG}{R^2}, \qquad v_0^2 = \frac{M_sG}{R}$$

where m is the mass of the earth and v_0 is its orbital speed. Thus, we have

$$v_e^2 = 2v_0^2 \qquad v_e = \sqrt{2}\,v_0 = \sqrt{2} \times 30 \ \frac{\text{km}}{\text{s}} = 42.4 \ \frac{\text{km}}{\text{s}}$$

(b) An asteroid crosses the earth's orbit with velocity components v(radial) $= -30$ km/s (toward the sun), and v(tangential) $= 30$ km/s (perpendicular to the solar direction). Describe the asteroid's orbit. Calculate its perihelion (closest distance to the sun).

FIGURE 1.44 When the asteroid crosses the orbit of the earth, its velocity has components 30 km / s in the radial and tangential directions.

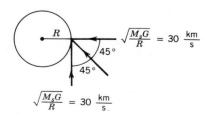

The speed of the asteroid is

$$\sqrt{(30)^2 + (30)^2} \ \frac{\text{km}}{\text{s}} = 2\sqrt{30} \ \frac{\text{km}}{\text{s}}$$

which we have seen to be the escape speed. The asteroid has just enough energy to escape to infinity, which means that its orbit is a parabola (boundary between an ellipse, for $E < 0$, and a hyperbola, for $E > 0$). To get the perihelion distance, we write the energy in terms of r and $\dot{r}$,

$$E = \tfrac{1}{2}m\dot{r}^2 + \frac{l^2}{2mr^2} - \frac{mM_sG}{r}$$

We know $E = 0$, and the perihelion is given by $\dot{r} = 0$. Thus,

$$r = \frac{l^2}{2m^2M_sG}$$

The angular momentum l is given by

$$l = mR\sqrt{\frac{M_sG}{R}} = m\sqrt{M_sGR}$$

Thus,

$$r = \frac{m^2M_sGR}{2m^2M_sG} = \frac{R}{2} = \tfrac{1}{2} \ \text{au}$$

1.13 RELATIVE AND CENTER-OF-MASS COORDINATES

We now consider the important case of two particles interacting with each other, with no external forces. We assume that the potential energy is a function only of the relative positions of the particles, $U = U(\mathbf{r}_2 - \mathbf{r}_1)$. An example would be the system consisting of the sun and one of its planets. Since U depends only on $\mathbf{r}_2 - \mathbf{r}_1$, we choose to work in relative and center-of-mass coordinates

$$\mathbf{r} = \mathbf{r}_2 - \mathbf{r}_1$$
$$\mathbf{R} = \frac{m_1\mathbf{r}_1 + m_2\mathbf{r}_2}{m_1 + m_2}$$

The inverse transformation is

$$\mathbf{r}_1 = \mathbf{R} - \frac{m_2}{m_1 + m_2}\mathbf{r}$$

$$\mathbf{r}_2 = \mathbf{R} + \frac{m_1}{m_1 + m_2}\mathbf{r}$$

The kinetic energy can then be expressed in terms of $\dot{\mathbf{R}}$ and $\dot{\mathbf{r}}$,

$$T = \frac{m_1}{2}\dot{\mathbf{r}}_1^2 + \frac{m_2}{2}\dot{\mathbf{r}}_2^2$$

$$= \frac{m_1}{2}\left(\dot{\mathbf{R}} - \frac{m_2}{m_1 + m_2}\dot{\mathbf{r}}\right)^2 + \frac{m_2}{2}\left(\dot{\mathbf{R}} + \frac{m_1}{m_1 + m_2}\dot{\mathbf{r}}\right)^2$$

$$= \frac{m_1 + m_2}{2}(\dot{\mathbf{R}})^2 + \tfrac{1}{2}\frac{m_1m_2}{m_1 + m_2}(\dot{\mathbf{r}})^2$$

Thus, if we define the *total mass*, M, and the *reduced mass*, μ, by

$$M \equiv m_1 + m_2 \tag{1.74a}$$

$$\mu \equiv \frac{m_1 m_2}{m_1 + m_2} \tag{1.74b}$$

we can write the Lagrangian of the system in the form

$$L(\mathbf{R}, \mathbf{r}, \dot{\mathbf{R}}, \dot{\mathbf{r}}) = \tfrac{1}{2} M (\dot{\mathbf{R}})^2 + \tfrac{1}{2} \mu (\dot{\mathbf{r}})^2 - U(\mathbf{r}) \tag{1.74c}$$

Notice that U does not depend on $\mathbf{R}$. Lagrange's equations for the components of $\mathbf{R}$ and $\mathbf{r}$ are then

$$\frac{d}{dt}(M\dot{X}) = \frac{d}{dt}(M\dot{Y}) = \frac{d}{dt}(M\dot{Z}) = 0 \tag{1.75a}$$

$$\frac{d}{dt}(\mu\dot{x}) = -\frac{\partial U}{\partial x}(x, y, z) \tag{1.75b}$$

$$\frac{d}{dt}(\mu\dot{y}) = -\frac{\partial U}{\partial y}(x, y, z) \tag{1.75c}$$

$$\frac{d}{dt}(\mu\dot{z}) = -\frac{\partial U}{\partial z}(x, y, z) \tag{1.75d}$$

We see from (1.75) that the motions of $\mathbf{R}$ and $\mathbf{r}$ separate completely. The two particles move in such a way that their mass center moves with constant velocity. Furthermore, the relative motion of the two particles, determined by (1.75b, c, d) is formally identical to the motion of a single particle of mass μ in the potential $U(\mathbf{r})$. Thus, the analysis given in Section 1.12 of the motion of a single particle in a fixed k/r potential can immediately be applied to the relative motion of a planet and the sun. If $\mathbf{r}$ locates the planet relative to the sun, an observer on the sun would conclude that the planet moves in an elliptical orbit with the sun at one focus, such that $\mathbf{r}$ sweeps out equal areas in equal times. However, Kepler's third law is no longer valid, since the period of relative motion is given by (1.68) with the particle mass m replaced by the reduced mass μ of (1.74b). But $|K|$ is still $G m\, m_2$. Thus, (1.69) must be replaced by

$$\tau = \frac{2\pi}{\sqrt{G(m_1 + m_2)}} a^{3/2}$$

and the period of a planet depends on its mass as well as on its semi-major axis. However, since the mass of the sun is so much greater than the mass of any of the planets, $m_1 + m_2$ changes by a small fraction of itself from one planet to another, and τ is very nearly proportional to $a^{3/2}$.

The momenta conjugate to $\mathbf{R}$ and $\mathbf{r}$ are obtained by differentiating (1.74c) with respect to $\dot{\mathbf{R}}$ and $\dot{\mathbf{r}}$,

$$\mathbf{P_R} = \frac{\partial L}{\partial \dot{\mathbf{R}}} = M\dot{\mathbf{R}} = m_1 \dot{\mathbf{r}} + m_2 \dot{\mathbf{r}}_2 = \mathbf{p}_1 + \mathbf{p}_2 \tag{1.76a}$$

$$\mathbf{P_r} = \frac{\partial L}{\partial \dot{\mathbf{r}}} = \mu\dot{\mathbf{r}} = \frac{m_1 m_2}{m_1 + m_2}(\dot{\mathbf{r}}_2 - \dot{\mathbf{r}}_1) = \frac{m_1 \mathbf{p}_2 - m_2 \mathbf{p}_1}{m_1 + m_2} \tag{1.76b}$$

It can easily be verified that $\mathbf{p_r}$ and $-\mathbf{p_r}$ are, respectively, the momenta of particles 2 and 1, as measured by an observer in the center-of-mass system.

FIGURE 1.45 The string is in equilibrium along the x axis, and moves in the $x-y$ plane.

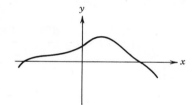

1.14 MECHANICAL WAVES

1.14.1 Transverse Waves on Stretched Strings

At equilibrium, the string shown in Figure 1.45 is stretched along the x axis with tension K. We consider small *transverse* motions of the string in the x–y plane. Let $y(x, t)$ be the vertical displacement at time t of the point on the string a distance x to the right of the origin. Thus, $y(x, t)$ gives the shape of the string at time t, and $\partial y(x, t)/\partial t$ gives the (vertical) speed at time t of the point on the string labeled by x. We also assume that the slope of the string remains small, so that $|\partial y(x, t)/\partial x| \ll 1$. Figure 1.46 shows that in this case the vertical force exerted by the string to the right of x on the string to the left of x is approximately equal to $K\partial y(x, t)/\partial x$.

Newton's second law, applied to the segment of string between $x + h/2$, and $x - h/2$, of mass ρh, gives

$$K\frac{\dfrac{\partial y}{\partial x}\left(x + \dfrac{h}{2}, t\right) - \dfrac{\partial y}{\partial x}\left(x - \dfrac{h}{2}, t\right)}{h} \simeq \rho\frac{\partial^2 y(x, t)}{\partial t^2} \tag{1.77}$$

Here ρ is the linear density of the string, assumed to be constant. We have neglected the gravitational force acting on the string; this is justified as long as we consider motions in which $|\partial^2 y(x, t)/\partial t^2| \gg g$. If we take the limit of (1.77) as $h \to 0$, we get

$$K\frac{\partial^2 y}{\partial x^2}(x, t) = \rho\frac{\partial^2 y(x, t)}{\partial t^2}$$

FIGURE 1.46 The arrow represents the force **K** exerted on the string to the left of x by the string to the right of x. **K** points along the direction of the string. Its vertical component is $K\sin\theta$. For $\theta \ll 1$, this is approximately $K\tan\theta = K\partial y/\partial x$.

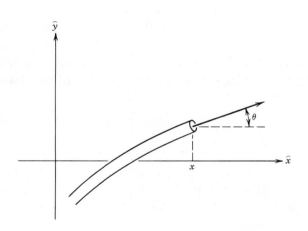

or, equivalently,

$$\frac{\partial^2 y}{\partial x^2}(x, t) = \frac{1}{c^2}\frac{\partial^2 y(x, t)}{\partial t^2} \tag{1.78a}$$

with

$$c = \sqrt{K/\rho} \tag{1.78b}$$

Any $y(x, t)$ that satisfies the wave equation (1.78a) describes a possible motion of the string.

Let $f(r)$ and $g(s)$ be any continuously differentiable functions of their independent variables r and s. Then it is easy to verify that $y(x, t)$ defined by

$$y(x, t) = f(r) + g(s) \tag{1.79a}$$

with

$$r = x - ct \tag{1.79b}$$

$$s = x + ct \tag{1.79c}$$

satisfies (1.78a). Here $f(r) \equiv f(x - ct)$ describes a shape moving to the right with speed c, while $g(s) \equiv g(x + ct)$ describes a shape moving to the left with speed c. Any solution of (1.78a) can be written in the general form (1.79a). These are called *transverse* waves, because the direction of propagation of the waves is transverse (perpendicular) to the direction of motion of the particles of the string.

PROBLEM 1.14.1 At $t = 0$, the string shown in Figure 1.45 has shape $\alpha(x)$, and the point on the string labeled by x has vertical speed $\beta(x)$. Find the shape of the string at any future time.

At $t = 0$, $r = s = x$, and (1.79a) becomes

$$y(x, 0) = \alpha(x) = f(x) + g(x) \tag{1.80a}$$

The vertical speed of the string at $t = 0$ is

$$\frac{\partial y(x, t)}{\partial t}\bigg|_{t=0} = \beta(x) = \left[\frac{\partial r}{\partial t}f'(r) + \frac{\partial s}{\partial t}g'(s)\right]_{t=0}$$

$$= -cf'(x) + cg'(x)$$

Integrating this equation yields

$$\frac{1}{c}\int_{x_1}^{x}\beta(x')\,dx' = -f(x) + g(x) \tag{1.80b}$$

The lower integration limit x_1 is a constant, which will drop out of the final result. We now can solve (1.80a) and (1.80b) simultaneously to get

$$f(x) = \frac{1}{2}\left[\alpha(x) - \frac{1}{c}\int_{x_1}^{x}\beta(x')\,dx'\right]$$

$$g(x) = \frac{1}{2}\left[\alpha(x) + \frac{1}{c}\int_{x_1}^{x}\beta(x')\,dx'\right]$$

so that the subsequent motion of the string is given by

$$y(x, t) = f(x - ct) + g(x + ct)$$

$$= \frac{1}{2}\left[\alpha(x - ct) + \alpha(x + ct) + \frac{1}{c}\int_{x_1}^{x+ct} \beta(x')\,dx' - \frac{1}{c}\int_{x_1}^{x-ct} \beta(x')\,dx \right]$$

$$= \frac{1}{2}\left[\alpha(x - ct) + \alpha(x + ct) + \frac{1}{c}\int_{x-ct}^{x+ct} \beta(x')\,dx' \right] \tag{1.80c}$$

1.14.2 Energy Flux

The rate at which energy flows to the right along the string, past the point x, is equal to the rate at which the string to the left of x does work on the string to the right of x. This is given by

$$S(x, t) = -K\frac{\partial y(x, t)}{\partial x}\frac{\partial y(x, t)}{\partial t} \qquad \text{(energy flux)} \tag{1.81}$$

For the general solution (1.79a) we have

$$\frac{\partial y(x, t)}{\partial x} = \frac{\partial r}{\partial x}f'(r) + \frac{\partial s}{\partial x}g'(s) = f'(r) + g'(s)$$

$$\frac{\partial y(x, t)}{\partial t} = \frac{\partial r}{\partial t}f'(r) + \frac{\partial s}{\partial t}g'(s) = c[-f'(r) + g'(s)]$$

and the energy flux is

$$S(x, t) = Kc\left\{ [f'(x - ct)]^2 - [g'(x + ct)]^2 \right\} \tag{1.82}$$

Thus, we can regard the motions $f(x - ct)$ and $g(x + ct)$ as carrying energy along the string, to the right and left, respectively, with speed c.

1.14.3 Harmonic Waves

An important special case of shapes $f(r)$ and $g(s)$ is provided by harmonic wave trains:

$$\left.\begin{array}{c} f(r) \\ g(s) \end{array}\right\} = Re\left[\alpha e^{\frac{2\pi i}{\lambda}(x \mp ct)} \right] = Re\left[\alpha e^{i(kx \mp \omega t)} \right] \tag{1.83}$$

The parameters introduced here to describe the wave are

$$\alpha \equiv \text{amplitude (perhaps complex)} \tag{1.84a}$$

$$\lambda \equiv \text{wavelength} \tag{1.84b}$$

$$\frac{c}{\lambda} = \nu \equiv \text{frequency} \tag{1.84c}$$

$$\frac{2\pi c}{\lambda} = \omega \equiv \text{circular frequency} \tag{1.84d}$$

$$\frac{2\pi}{\lambda} = k \equiv \text{propagation vector} \tag{1.84e}$$

From (1.84c), (1.84d), and (1.84e) we have the useful relations

$$\lambda \nu = c \qquad (1.85a)$$

$$\omega = ck \qquad (1.85b)$$

If we use (1.83) in (1.82) and time-average the flux over an integral number of cycles, we get

$$\overline{S(x)} = \pm \frac{1}{2} K k \omega |\alpha|^2 = \pm \frac{1}{2} K \frac{\omega^2}{c} |\alpha|^2 \qquad (1.86)$$

PROBLEM 1.14.2   A point mass m is attached to the string of Figure 1.47 at $x = 0$. Suppose that a harmonic wave of circular frequency ω is incident on this mass from the right. Calculate the fraction of the incident energy that is reflected (the reflectance) and the fraction transmitted (the transmittance).

The experimental conditions are described by a solution

$$y(x, t) = Re\left[\beta e^{i(kx+\omega t)}\right], \qquad x \le 0 \qquad (1.87a)$$

$$= Re\left[e^{i(kx+\omega t)} + \alpha e^{i(-kx+\omega t)}\right], \qquad x \ge 0 \qquad (1.87b)$$

The energy fluxes for $x \lessgtr 0$ are

$$\overline{S(x)} = -\frac{1}{2} K \frac{\omega^2}{c} |\beta|^2, \qquad x < 0 \qquad (1.88a)$$

$$= \frac{1}{2} K \frac{\omega^2}{c} \left[-1 + |\alpha|^2\right], \qquad x > 0 \qquad (1.88b)$$

Thus, $|\alpha|^2$ and $|\beta|^2$ are the fractions of the incident flux reflected and transmitted, respectively.

To calculate α and β, we impose the following two conditions:

1. Since the string is unbroken at $x = 0$, $y(x, t)$ is continuous at $x = 0$. This is satisfied by (1.87) if

$$\beta = 1 + \alpha \qquad (1.89a)$$

2. The net vertical force exerted by the string on the mass m produces its acceleration, $\partial^2 y(0, t)/\partial t^2$. Thus,

$$m \frac{\partial^2 y(0, t)}{\partial t^2} = K\left[\frac{\partial y(x, t)}{\partial x}\bigg|_{x=0^+} - \frac{\partial y(x, t)}{\partial x}\bigg|_{x=0^-}\right] \qquad (1.89b)$$

This condition is satisfied by (1.87) if

$$-m\omega^2 \beta = ikK(1 - \alpha - \beta) \qquad (1.89c)$$

FIGURE 1.47 A point mass is attached to the string at $x = 0$. The incident amplitude is 1, the reflected amplitude is α and the transmitted amplitude is β.

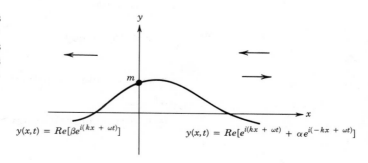

$$y(x, t) = Re[\beta e^{i(kx + \omega t)}]$$

$$y(x, t) = Re[e^{i(kx + \omega t)} + \alpha e^{i(-kx + \omega t)}]$$

Solving (1.89a) and (1.89c) simultaneously gives

$$\alpha = \frac{1}{-1 + i\frac{2kK}{m\omega^2}}$$

$$\beta = \frac{i2Kk/m\omega^2}{-1 + i\frac{2kK}{m\omega^2}}$$

and the reflected and transmitted fluxes are

$$\text{reflectance} = |\alpha|^2 = \frac{1}{1 + \left(2kK/m\omega^2\right)^2} \qquad (1.90a)$$

$$\text{transmittance} = |\beta|^2 = \frac{\left(2kK/m\omega^2\right)^2}{1 + \left(2kK/m\omega^2\right)^2} \qquad (1.90b)$$

Note that the sum of the reflectance and transmittance equals unity. This is an expression of the conservation of energy. Using (1.78b), (1.85b), and (1.84e), we have

$$\frac{2kK}{m\omega^2} = \frac{2k\rho c^2}{m\omega^2} = \frac{2\rho}{m} \cdot \frac{1}{k} = \frac{\lambda}{m\pi}$$

$$= \frac{1}{\pi} \frac{\text{mass of a wavelength of string}}{\text{particle mass}} \qquad (1.91)$$

The ratio (1.91) gives us a criterion for judging whether a mass m is large enough to reflect a substantial fraction of the incident energy.

PROBLEM 1.14.3   Two strings with linear mass densities ρ_1 and ρ_2 are connected at $x = 0$ and pulled with tension K. A wave of circular frequency ω is incident from the right. What fraction of the incident energy is reflected from the discontinuity in ρ?

Here the experimental conditions are described by a solution

$$y(x, t) = Re\left[\beta e^{i(k_2 x + \omega t)}\right], \qquad x \leq 0 \qquad (1.92a)$$

$$y(x, t) = Re\left[e^{i(k_1 x + \omega t)} + \alpha e^{i(-k_1 x + \omega t)}\right], \qquad x \geq 0 \qquad (1.92b)$$

The tension K and circular frequency ω are common to both strings, so k_1 and k_2 are given by (1.78b) and (1.85b) to be

$$k_1 = \frac{\omega}{c_1} = \omega\sqrt{\frac{\rho_1}{K}} \qquad (1.93a)$$

$$k_2 = \frac{\omega}{c_2} = \omega\sqrt{\frac{\rho_2}{K}} \qquad (1.93b)$$

FIGURE 1.48 The two strings are joined at $x = 0$. The tension has the same value K in both strings, but their linear mass densities are different.

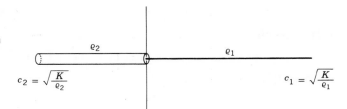

$$c_2 = \sqrt{\frac{K}{\rho_2}} \qquad\qquad c_1 = \sqrt{\frac{K}{\rho_1}}$$

The fluxes for $x \gtrless 0$ are

$$\overline{S(x)} = -\frac{1}{2}\frac{K\omega^2}{c_2}|\beta|^2, \qquad x < 0$$

$$= \frac{1}{2}K\frac{\omega^2}{c_1}[-1 + |\alpha|^2], \qquad x > 0$$

so that

$$\text{reflectance} = |\alpha|^2 \tag{1.94a}$$

$$\text{transmittance} = \frac{c_1}{c_2}|\beta|^2 = \sqrt{\frac{\rho_2}{\rho_1}}\,|\beta|^2 \tag{1.94b}$$

As before, $y(x, t)$ is continuous at $x = 0$. In addition, $\partial y(x, t)/\partial x$ is continuous at $x = 0$, if there is no point mass at the junction of the two strings [cf. (1.89b) with $m = 0$]. These conditions are satisfied by (1.92), if

$$\beta = 1 + \alpha, \qquad k_2\beta = k_1(1 - \alpha)$$

whose solution is

$$\alpha = \frac{k_1 - k_2}{k_1 + k_2} = \frac{\sqrt{\rho_1} - \sqrt{\rho_2}}{\sqrt{\rho_1} + \sqrt{\rho_2}} \tag{1.95a}$$

$$\beta = \frac{2k_1}{k_1 + k_2} = \frac{2\sqrt{\rho_1}}{\sqrt{\rho_1} + \sqrt{\rho_2}} \tag{1.95b}$$

Thus,

$$\text{reflectance} = \left[\frac{\sqrt{\rho_1} - \sqrt{\rho_2}}{\sqrt{\rho_1} + \sqrt{\rho_2}}\right]^2 \tag{1.96a}$$

$$\text{transmittance} = 4\sqrt{\frac{\rho_2}{\rho_1}}\frac{\rho_1}{\left[\sqrt{\rho_1} + \sqrt{\rho_2}\right]^2} = \frac{4\sqrt{\rho_1\rho_2}}{\left[\sqrt{\rho_1} + \sqrt{\rho_2}\right]^2} \tag{1.96b}$$

It is readily verified that the sum of the reflectance and transmittance is unity, as required by energy conservation. If $\rho_1 = \rho_2$, so that there is no discontinuity at $x = 0$, (1.96) gives the expected result that all the energy is transmitted. It follows from (1.95a) that if $\rho_1 > \rho_2$, then α is positive. Thus, if the incident wave is along the denser string, the reflected wave is in phase with it. Conversely, if the incident wave is along the less dense string, the incident and reflected waves are out of phase. In both cases, β of (1.95b) is positive, so the incident and transmitted waves are always in phase.

1.14.4 Standing Waves

If the boundary conditions on the string are

$$y\left(-\tfrac{L}{2}, t\right) = y\left(\tfrac{L}{2}, t\right) = 0 \qquad \text{(all } t\text{)} \tag{1.97}$$

it is convenient to work with solutions of the form

$$y(x, t) = X(x)T(t) \tag{1.98}$$

Then (1.97) is satisfied, if

$$X\left(-\tfrac{L}{2}\right) = X\left(+\tfrac{L}{2}\right) = 0 \tag{1.99}$$

If we substitute (1.98) into the wave equation (1.78), we get

$$X''(x)T(t) = \frac{1}{c^2}X(x)T''(t)$$

$$\frac{X''(x)}{X(x)} = \frac{1}{c^2}\frac{T''(t)}{T(t)} \tag{1.100}$$

The right-hand side of (1.100) is, by construction, independent of x. If $X''(x)/X(x)$ is independent of x, it must be constant,

$$X''(x) = -k^2 X(x) \qquad (k \text{ constant}[6]) \tag{1.101}$$

The general solution of (1.101) is

$$X(x) = Ae^{ikx} + Be^{-ikx} \tag{1.102}$$

If this is to vanish at $x = \pm L/2$, then

$$Ae^{ikL/2} = -Be^{-ikL/2} \tag{1.103a}$$

$$Ae^{-ikL/2} = -Be^{ikL/2} \tag{1.103b}$$

If we multiply these equations we get $A^2 = B^2$, so that either $A = B$ or $A = -B$. If $A = B$, then (1.103) implies that

$$\left.\begin{array}{l} e^{ikL} = -1, \qquad kL = (2n+1)\pi \\[2mm] X(x) = 2A\cos(2n+1)\dfrac{\pi x}{L} \end{array}\right\} \quad n = 0,1,2,\ldots \tag{1.104a}$$

Alternatively, if $A = -B$, we get

$$\left.\begin{array}{l} e^{ikL} = 1, \qquad kL = 2n\pi \\[2mm] X(x) = 2iA\sin\dfrac{2n\pi x}{L} \end{array}\right\} \quad n = 1,2,\ldots \tag{1.104b}$$

Associated with each acceptable value of k, we get an equation for $T(t)$,

$$T''(t) = -k^2 c^2 T(t) = -\omega^2 T(t)$$

with general solution

$$T(t) = a_n \cos n\omega t + b_n \sin n\omega t$$

Because (1.78a) is a linear partial differential equation, a sum of solutions is also a solution. The general solution of (1.78a), subject to the boundary conditions (1.97), is then

$$y(x,t) = \sum_{n=0}^{\infty} \cos(2n+1)\frac{\pi x}{L}\left[a_n\cos(2n+1)\frac{\pi ct}{L} + b_n\sin(2n+1)\frac{\pi ct}{L}\right]$$

$$+ \sum_{n=1}^{\infty} \sin\left(2n\frac{\pi x}{L}\right)\left[c_n\cos\frac{2n\pi ct}{L} + d_n\sin\frac{2n\pi ct}{L}\right] \tag{1.105}$$

PROBLEM 1.14.4 Suppose that $\alpha(x)$ and $\beta(x)$ in Problem 1.14.1 satisfy

$$\alpha\left(-\tfrac{L}{2}\right) = \beta\left(-\tfrac{L}{2}\right) = \alpha\left(\tfrac{L}{2}\right) = \beta\left(\tfrac{L}{2}\right) = 0$$

---

[6]Writing the constant in the form $-k^2$ is done for later convenience. It will soon be shown that k, introduced in this way, must be real.

Find the shape of the string at any future time.

At $t = 0$, we have

$$y(x, 0) = \alpha(x) = \sum_{n=0}^{\infty} a_n \cos(2n + 1)\frac{\pi x}{L} + \sum_{n=1}^{\infty} c_n \sin\frac{2n\pi x}{L}$$

$$\frac{\partial y(x, 0)}{\partial t} = \beta(x) = \frac{\pi c}{L}\left[\sum_{n=0}^{\infty} b_n(2n + 1)\cos(2n + 1)\frac{\pi x}{L} + \sum_{n=1}^{\infty} d_n(2n)\sin\frac{2n\pi x}{L}\right]$$

Thus,

$$a_n = \frac{2}{L}\int_{-L/2}^{L/2} \alpha(x')\cos(2n + 1)\frac{\pi x'}{L}\, dx' \tag{1.106a}$$

$$c_n = \frac{2}{L}\int_{-L/2}^{L/2} \alpha(x')\sin\frac{2n\pi x'}{L}\, dx' \tag{1.106b}$$

$$b_n = \frac{2\pi c}{2n + 1}\int_{-L/2}^{L/2} \beta(x')\cos(2n + 1)\frac{\pi x'}{L}\, dx' \tag{1.106c}$$

$$d_n = \frac{2\pi c}{2n}\int_{-L/2}^{L/2} \beta(x')\sin\frac{2n\pi x'}{L}\, dx' \tag{1.106d}$$

These equations determine the Fourier components a_n, b_n, c_n, d_n, which are to be used in the harmonic expansion (1.105).

PROBLEM 1.14.5   Suppose that a point mass m is fixed to the midpoint ($x = 0$) of the string we have just been considering (i.e., $y(L/2, t) = y(-L/2, t) = 0$). Find the possible standing wave solutions.

We have seen in (1.89b) that the presence of m will cause a discontinuity in $\partial y/\partial x|_{x=0}$. Thus, we seek a solution of the form

$$y(x, t) = \sin\left(k\left(\tfrac{L}{2} - x\right)\right)\sin \omega t \qquad x \geq 0 \tag{1.107a}$$

$$= \sin\left(k\left(\tfrac{L}{2} + x\right)\right)\sin \omega t \qquad x \leq 0 \tag{1.107b}$$

This solution satisfies the boundary conditions at $x = \pm L/2$, and it satisfies the wave equation if $\omega = kc$. To find the allowed values of k (and, hence, of ω), we impose the condition (1.89b):

$$-m\omega^2 \sin\frac{kL}{2} = -2Kk\cos\frac{kL}{2}$$

$$\cot\frac{kL}{2} = \frac{m\omega^2}{2Kk} = \frac{mc^2}{2K}k = \frac{m}{\rho L}\left(\frac{kL}{2}\right) \tag{1.108}$$

This is a transcendental equation, whose solutions give the allowed values of k. Note that if $m \ll \rho L$ (the mass of the string), we get solutions of (1.108) near

$$\cot\frac{kL}{2} \approx 0$$

$$\frac{kL}{2} \approx (2n + 1)\pi \qquad (n = 0, 1, 2, \dots)$$

which corresponds to the sequence of symmetric solutions (1.94a) of the unweighted string. On the other hand, if $m \gg \rho L$, we get solutions of (1.108) near

$$\cot \frac{kL}{2} \approx \infty$$

$$\frac{kL}{2} \approx n\pi \qquad (n = 1, 2, \ldots)$$

which forms the entire set of solutions (1.104a and 1.104b) for a string of length $L/2$.

If we try to find an odd solution analogous to (1.107),

$$y(x, t) = \sin\left(k\left(\tfrac{L}{2} - x\right)\right)\sin \omega t \qquad x \geq 0 \qquad (1.109a)$$

$$= -\sin\left(k\left(\tfrac{L}{2} + x\right)\right)\sin \omega t \qquad x \leq 0 \qquad (1.109b)$$

continuity at the origin requires that

$$\sin \frac{kL}{2} = 0, \qquad kL = 2n\pi \qquad (n = 1, 2, \ldots)$$

If this condition is satisfied by k, then (1.89b) is automatically satisfied by (1.109). The solution we have reached is precisely the same as the odd solution (1.104) for the unweighted string. Since this odd solution has a node at $x = 0$, attaching the mass m to this point has no effect on the motion.

1.14.5 Longitudinal Waves in Fluids

We consider a region of space filled with a continuous fluid. Let $\rho(\mathbf{r}, t)$ and $\mathbf{v}(\mathbf{r}, t)$ be, respectively, the mass density and velocity of the fluid at point $\mathbf{r}$ at time t. If no new mass is created or destroyed within the fluid (there are no sources or sinks), then ρ and $\mathbf{v}$ are related by[7]

$$\frac{\partial \rho}{\partial t}(\mathbf{r}, t) + \mathrm{div}[\rho(\mathbf{r}, t)\mathbf{v}(\mathbf{r}, t)] = 0 \qquad (1.110)$$

For a perfect fluid (one that can support no transverse stresses), Newton's second law of motion is expressed by the three equations

$$\rho \left[\frac{\partial v_\alpha}{\partial t} + \mathbf{v} \cdot \nabla v_\alpha \right] = f_\alpha(\mathbf{r}, t) - \frac{\partial}{\partial r_\alpha} p(\mathbf{r}, t) \qquad (\alpha = x, y, z) \quad (1.111)$$

Here p is the pressure within the fluid, and $\mathbf{f}$ represents an external force (per unit volume) that acts from a distance on the matter within the fluid, such as the gravitational force of the earth. Equations (1.111) are called Euler's equations. Because they are nonlinear in the components of $\mathbf{v}$, they are very difficult to solve, except in some very special cases.

Let us neglect any external force $\mathbf{f}$, and suppose that the pressure at any point depends only on the fluid density at that point. Then (1.111) becomes

$$\rho \left[\frac{\partial v_\alpha}{\partial t} + \mathbf{v} \cdot \nabla v_\alpha \right] = -\frac{dp(\rho)}{d\rho} \frac{\partial \rho}{\partial r_\alpha} \qquad (\alpha = x, y, z) \qquad (1.112)$$

---

[7]The derivation of (1.110) and (1.111) involves the application of Gauss' theorem to the conservation of matter and of momentum. See, for example, Chapter 8 of *Mechanics* by K. R. Symon, Addison-Wesley Publishing Co., Inc., Reading, Mass., 1971.

If $p(\rho)$ is a given function (the *equation of state* of the fluid), then (1.110) and (1.112) represent a set of four equations for the four functions v_x, v_y, v_z, and ρ. A trivial solution is a uniform fluid at rest ($v_x = v_y = v_z = 0$, $\rho(\mathbf{r}, t) = \rho_0$). We now consider small deviations from this uniform fluid,

$$\mathbf{v}(\mathbf{r}, t) = \mathbf{u}(\mathbf{r}, t) \tag{1.113a}$$

$$\rho(\mathbf{r}, t) = \rho_0 + \phi(\mathbf{r}, t) \tag{1.113b}$$

If we substitute these expressions for $\mathbf{v}$ and ρ into (1.110) and (1.112), and keep only terms of first order in the small quantities $\mathbf{u}$ and ϕ, we get

$$\frac{\partial \phi}{\partial t} + \rho_0 \operatorname{div} \mathbf{u} = 0 \tag{1.114a}$$

$$\rho_0 \frac{\partial \mathbf{u}}{\partial t} = - \left. \frac{dp}{d\rho} \right|_{\rho_0} \nabla \phi \tag{1.114b}$$

We can get an equation for ϕ alone by subtracting the divergence of (1.114b) from the time derivative of (1.114a):

$$\operatorname{div} \nabla \phi = \nabla^2 \phi = \frac{\partial^2 \phi}{\partial x^2} + \frac{\partial^2 \phi}{\partial y^2} + \frac{\partial^2 \phi}{\partial z^2} = \frac{1}{c^2} \frac{\partial^2 \phi}{\partial t^2} \tag{1.115a}$$

with

$$c = \sqrt{\left. \frac{dp(\rho)}{d\rho} \right|_{\rho = \rho_0}} \tag{1.115b}$$

Thus, we see that the small pressure fluctuations ϕ satisfy the wave equation (1.115a), with a wave propagation speed given by (1.115b). This only makes sense if $dp/d\rho|_{\rho_0} > 0$. This condition guarantees the "springiness" of the fluid (i.e., a density fluctuation will result in a restoring force that tends to reduce the fluctuation). For example, if an increase in density in a small region causes the pressure to increase there, then the resultant pressure gradient would drive fluid away from this region of high density. Conversely, if $dp/d\rho < 0$, the pressure would be unstable with respect to small deviations away from the uniform solution.

An explicit wavelike solution of (1.115a) is

$$\phi(\mathbf{r}, t) = Re\left[\phi_0 e^{i(\mathbf{k} \cdot \mathbf{r} - \omega t)} \right] \tag{1.116}$$

It is easily verified that (1.116) satisfies (1.115a) if $\omega = kc$. This function describes a wave moving with speed c in the direction of $\hat{k}$, with wave length $2\pi/k$, and frequency $\nu = \omega/2\pi = c/\lambda$. The amplitude of the wave is ϕ_0. The quantity $\mathbf{k} \cdot \mathbf{r} - \omega t$ is called the *phase* of the wave and (1.116) is called a *plane* wave because this phase has the same value for all $\mathbf{r}$ on a plane perpendicular to $\mathbf{k}$.

By eliminating ϕ between (1.114a) and (1.114b), we get

$$\nabla (\operatorname{div} \mathbf{u}) = \frac{1}{c^2} \frac{\partial^2 \mathbf{u}}{\partial t^2} \tag{1.117}$$

If we attempt to get a plane wave solution of (1.117)

$$\mathbf{u}(\mathbf{r}, t) = Re\left[\left(\mathbf{u}_0 e^{i(\mathbf{k} \cdot \mathbf{r} - \omega t)} \right) \right] \tag{1.118a}$$

we find that we can satisfy (1.117) provided that

$$(\mathbf{k} \cdot \mathbf{u}_0)\mathbf{k} = \frac{\omega^2}{c^2}\mathbf{u}_0 \tag{1.118b}$$

This implies that $\mathbf{k}$ and $\mathbf{u}_0$ are parallel. Since $\mathbf{u}_0$ in (1.118a) gives the direction of the fluid velocity, (1.118b) tells us that we have here a wave in which the fluid particles move in the direction of wave propagation. This is a *longitudinal* wave, as opposed to the transverse waves on strings considered above.

PROBLEM 1.14.6 Set up the equations required to calculate the frequencies of standing waves within a rigid closed surface S.

In analogy to (1.98), we seek a solution of the form

$$\phi(x, y, z, t) = \psi(x, y, z)T(t) \tag{1.119}$$

We substitute this in (1.115) and get

$$T(t)\nabla^2\psi(x, y, z) = \frac{1}{c^2}\psi(x, y, z)T''(t)$$

$$\frac{1}{\psi(x, y, z)}\nabla^2\psi(x, y, z) = \frac{1}{c^2T(t)}T''(t) = -k^2 \quad \text{(a constant)}$$

Thus, ψ and T satisfy

$$\nabla^2\psi(x, y, z) + k^2\psi(x, y, z) = 0 \tag{1.120a}$$

$$T''(t) + \omega^2 T(t) = 0 \tag{1.120b}$$

with $\omega = kc$. Equation (1.120a) is called the *Helmholtz* equation. We must supplement it by specifying the boundary conditions that ψ must satisfy. The problem specifies that the fluid is surrounded by a rigid closed surface S. Thus, the fluid velocity $\mathbf{u}$ at the surface must be tangential to the surface,

$$\mathbf{u} \cdot d\mathbf{S} = 0 \quad \text{(on the surrounding surface } S)$$

According to (1.114b) this implies that

$$\nabla\phi \cdot d\mathbf{S} = 0 \quad \text{(on the surrounding surface } S)$$

and this is possible only if

$$\nabla\psi(x, y, z) \cdot d\mathbf{S} = 0 \quad \text{(on the surrounding surface } S) \tag{1.121}$$

Equations (1.120a) and (1.121) define an eigenvalue problem whose solution depends only on the shape of S. The solution to the problem is a set of eigenvalues k_n, and eigenfunctions ψ_n. Once this problem is solved, the frequencies of the associated standing waves are given by $\omega_n = ck_n$, $\nu_n = \omega_n/2\pi$.

PROBLEM 1.14.7 Find the frequencies of the standing waves inside a rigid rectangular box, whose sides have lengths l_x, l_y, l_z.

The shape of the box suggests that we attempt to separate the solution ψ in Cartesian coordinates:

$$\psi(x, y, z) = A(x)B(y)C(z) \tag{1.122}$$

If we substitute this in (1.115a), we get

$$A''(x) + k_x^2 A(x) = 0 \tag{1.123a}$$

$$B''(y) + k_y^2 B(y) = 0 \tag{1.123b}$$

$$C''(z) + k_z^2 C(z) = 0 \tag{1.123c}$$

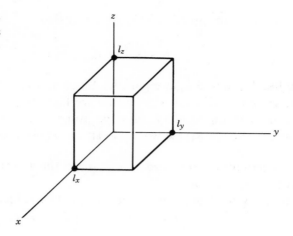

with

$$k^2 = k_x^2 + k_y^2 + k_z^2 \tag{1.123d}$$

The allowed values of k_x, k_y, k_z are obtained by imposing the boundary condition (1.121):

$$A'(0) = A'(l_x) = 0 \tag{1.124a}$$

$$B'(0) = B'(l_y) = 0 \tag{1.124b}$$

$$C'(0) = C'(l_z) = 0 \tag{1.124c}$$

(cf. Figure 1.49). The solutions of (1.123c) that satisfy (1.124c) are

$$C(z) = \cos k_z z \tag{1.125a}$$

with

$$k_z = \frac{n_z \pi}{l_z} \qquad (n_z = 0, 1, 2, \ldots) \tag{1.125b}$$

and similarly for $A(x)$ and $B(y)$. Thus, the eigenvalues and eigenfunctions of (1.120a) and (1.121) are

$$k_{n_x n_y n_z} = \pi \sqrt{\left(\frac{n_x}{l_x}\right)^2 + \left(\frac{n_y}{l_y}\right)^2 + \left(\frac{n_z}{l_z}\right)^2} \qquad (n_x, n_y, n_z = 0, 1, \ldots) \tag{1.126a}$$

$$\psi_{n_x n_y n_z} = \cos\left(\frac{n_x \pi}{l_x} x\right) \cos\left(\frac{n_y \pi}{l_y} y\right) \cos\left(\frac{n_z \pi}{l_z} z\right) \tag{1.126b}$$

and the associated frequencies are $\nu_{n_x n_y n_z} = (c/2\pi) k_{n_x n_y n_z}$.

PROBLEM 1.14.8 Repeat the previous problem, but for a box whose top is open to the atmosphere. Assume that the steady-state pressure inside the box is also atmospheric.

The only change associated with the open top is that the pressure at $z = l_z$ must be atmospheric at all times. This means that the density fluctuations $\phi(x, y, z, t)$ must vanish at $z = l_z$. Thus, the boundary condition (1.124c) is to be replaced by

$$C'(0) = 0, \qquad C(l_z) = 0 \tag{1.124d}$$

The solutions of (1.123c), subject to (1.124d), are

$$C(z) = \cos k_z z \tag{1.127a}$$

with

$$k_z = \frac{(n_z + 1/2)}{l_z} \pi \qquad (n_z = 0, 1, 2, \dots) \tag{1.127b}$$

The open top has no effect on the (x, y) motions, so the standing wave frequencies are $\nu_{n_x n_y n_z} = (c/2\pi)k_{n_x n_y n_z}$ with

$$k_{n_x n_y n_z} = \pi \sqrt{\left(\frac{n_x}{l_x}\right)^2 + \left(\frac{n_y}{l_y}\right)^2 + \left(\frac{n_z + 1/2}{l_z}\right)^2} \tag{1.127c}$$

with n_x, n_y, n_z taking on the same sequence of values as in (1.126).

REVIEW PROBLEMS

1.1. Three equal masses are attached to the ends and midpoint of a light uniform rod, which lies on a smooth horizontal plane. One of the end masses is struck by an impulsive force perpendicular to the direction of the rod. Find the ratios of velocities of the three masses immediately after the impulse.

Answer: $5 : 2 : -1$.

1.2. A particle of mass m moves on the inside surface of a smooth cone whose axis is vertical and whose half-angle is α. Find the period of small oscillations about a horizontal circular orbit a distance h above the vertex.

Answer: $\dfrac{2\pi}{\cos \alpha} \sqrt{\dfrac{h}{3g}}$.

1.3. A sphere of radius a is balanced on top of a fixed sphere of radius b, and then given an infinitesimal sideways push so that it rolls off. Find the angle between the vertical and a line between the centers at the instant the moving sphere separates from the fixed one.

Answer: $\arccos\left(\dfrac{10}{17}\right)$.

1.4. Find the period of small oscillations of a simple pendulum whose bob swings near the surface of the earth, but whose pivot is located infinitely far away from the earth.

Answer: $2\pi \sqrt{\dfrac{R_e}{g}}$.

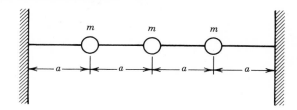

1.5. Express the rotational kinetic energy of a rigid body in terms of its angular velocity and its moment of inertia tensor, and show that for a given angular speed the kinetic energy is maximum if the direction of the angular velocity coincides with the principal axis with largest principal moment of inertia.

Answer: $T = \dfrac{1}{2} \displaystyle\sum_{\alpha, \beta} \omega_\alpha I_{\alpha\beta} \omega_\beta$.

1.6. The diagram shows a string under constant tension T, connecting three equal masses m separated by equal distances a. Find the normal frequencies of small transverse oscillations in a plane. Neglect gravity.

Answer: $\dfrac{1}{2\pi} \sqrt{\dfrac{2T}{ma}} \qquad \dfrac{1}{2\pi} \sqrt{\dfrac{2 \pm \sqrt{2}}{ma} T}$.

1.7. A string of length L, linear mass density ρ, with tension T is attached at one end to a wall. The other end is attached to a ring of mass m which can slide without friction along a vertical rod. Find the equation whose solutions gives the frequencies of standing waves. Neglect gravity. Check your answer in the limits $m \ll M$ and $m \gg M$ ($M = \rho L =$ mass of string)

$$\text{Answer: } \tan(kL) = \frac{M}{mkL} \qquad \omega = k\sqrt{\frac{T}{\rho}}.$$

1.8. A thin rod of mass M and length L is hinged at the bottom, and almost balanced vertically. It starts to fall. Find expressions for the following quantities as functions of r (the distance along the rod from the hinge) and θ (the angle between the rod and the vertical):

(a) The tangential component of the force that the part of the rod below r exerts on the part of the rod above r.

$$\text{Answer: } F_t = \frac{Mg(L - r)(3r - L)}{4L^2}\sin\theta.$$

(b) Suppose that the rod will break if the tangential stress exceeds a limiting value S. What condition (or conditions) must be satisfied by M, L, and S to guarantee that the stick will fall without breaking?

$$\text{Answer: } Mg < 12S.$$

1.9. A shot is fired vertically upwards with speed V_0, from a point on the rotating earth with latitude λ. Where does it land? Neglect air resistance, and all terms higher than first order in the angular velocity of the earth.

$$\text{Answer: } \frac{4}{3}\omega V_0^3 \frac{\cos\lambda}{g^2} \quad \text{west of initial position.}$$

1.10. Protons of energy E are scattered by a spherical target nucleus of charge Z and radius R. If the distance between the proton and the center of the target nucleus exceeds R, it feels only the Coulomb repulsion. If the proton comes within R of the center of the target nucleus, it is absorbed. Find an expression for $d\sigma/d\Omega$ as a function of the scattering angle θ. Assume $E > Ze^2/R$.

$$\text{Answer: } \frac{d\sigma}{d\Omega}(\theta) = \text{Rutherford scattering}$$
$$\text{formula for } \theta < \theta_0$$
$$= 0 \text{ for } \theta > \theta_0$$

$$\text{where } \theta_0 = 2\arctan\left(\frac{Ze^2}{2ER\sqrt{1 - Ze^2/R}}\right).$$

What would happen if E were $< Ze^2/R$?

$$\text{Answer: Rutherford scattering for all } \theta.$$

1.11. A particle is acted upon by an attractive central force of magnitude kr^n.

(a) For what values of n can there be stable circular orbits concentric with the center of interaction?

$$\text{Answer: } n > -3.$$

(b) What is the frequency of small radial oscillations about the circular orbits?

$$\text{Answer: } \frac{1}{2\pi}\sqrt{\frac{3 + n}{m}kr_0^{n-1}}.$$

(c) For what values of n will this radial oscillation frequency be an integral multiple of the orbital frequency? What does this imply about the change of orbit with time?

$$\text{Answer: } n = p^2 - 3, \qquad p = 1, 2, 3, \ldots.$$

1.12. The speed of a wave propagating in a dispersive medium depends upon the wavelength, so that $v = v(\lambda)$. Consider a "wave packet" consisting of the superposition of two waves with nearly equal wavelengths (λ and $\lambda + d\lambda$). Find an expression for the speed of the point where these two waves are in phase with each other. This is called the "group velocity."

$$\text{Answer: } v_g = v - \lambda\frac{dv}{d\lambda}.$$

1.13. Derive the parallel axis theorem for the moment of inertia tensor

$$I_{\alpha\beta} = I_{\alpha\beta}^{CM} + M\left(R^2\delta_{\alpha\beta} - R_\alpha R_\beta\right)$$

Here $I_{\alpha\beta}$ and $I_{\alpha\beta}^{CM}$ refer to two coordinate systems with parallel axes, whose origins are separated by vector $\mathbf{R}$.

1.14. Consider the uniform thin bent plate of mass M illustrated in the drawing.

(a) Locate the mass center.

$$\text{Answer: } \frac{1}{2}\frac{b^2\hat{y} + c^2\hat{z}}{b + c}.$$

(b) Calculate the elements of the moment of inertia tensor defined relative to axes parallel and perpendicular to the plate with origin at the mass center. You will probably find it easier to make use of the parallel axis theorem (Problem 1.13).

Answer:

$$
\begin{bmatrix}
\dfrac{b^3 + a^3}{3(b + c)} - \dfrac{b^4 + c^4}{4(b + c)^2} & 0 & 0 \\[3mm]
0 & \dfrac{a^2}{12} + \dfrac{c^3}{3(b + c)} - \dfrac{c^4}{4(b + c)^2} & \dfrac{b^2 c^2}{4(b + c)^2} \\[3mm]
0 & \dfrac{b^2 c^2}{4(b + c)^2} & \dfrac{a^2}{12} + \dfrac{b^3}{3(b + c)} - \dfrac{b^4}{4(b + c)^2}
\end{bmatrix}.
$$

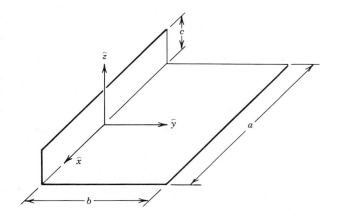

(c) Suppose the plate is rotating with constant angular speed about an axis parallel to the $\hat{z}$ axis, passing through the mass center. What is the angular momentum?

Answer: $\mathbf{L} = m\omega \left[\dfrac{b^2 c^2}{4(b + c)^2} \hat{y} \right.$

$$\left. + \left(\dfrac{a^2}{12} + \dfrac{b^3}{3(b + c)} - \dfrac{b^4}{4(b + c)^2} \right) \hat{z} \right].$$

(d) What is the torque acting at the instant shown in the drawing.

Answer: $- M\omega^2 \dfrac{b^2 c^2}{4(b + c)^2} \hat{x}.$

1.15. Particles move in a central potential with radial dependence given by A/r^n. For a given asymptotic speed, the impact parameter, b, and the distance of closest approach to the origin, a, are related by the formula $a^3 = \text{constant}/(a^2 - b^2)$. Find n.

Answer: $n = 5.$

1.16. An incompressible ideal fluid is put in a bucket which rotates with constant angular velocity ω about a vertical axis. The fluid eventually achieves a condition of rest relative to the rotating bucket. Calculate the shape of the surface of the fluid.

Answer: $z = \dfrac{\omega^2}{2g}(x^2 + y^2) + C.$

1.17. A charge q with mass m is midway between two fixed charges Q with the same sign as q. Calculate the frequency of small oscillations of q along the line of the three charges.

Answer: $\dfrac{1}{2\pi} \sqrt{\dfrac{32 q Q}{m L^3}}.$

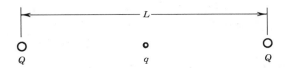

1.18. A string of length L fixed at both ends has a fundamental frequency ν_0. The string is excited with finite amplitude at a point $L/4$ from one end while an observer touches the string lightly at its midpoint.

(a) In this situation, what is the lowest frequency that can be excited?

Answer: $2\nu_0.$

(b) What is the frequency of the next resonance in this same situation?

Answer: $6\nu_0.$

1.19. (a) The propagation of a shock wave is determined by the total energy in the shock and the density ρ of the medium. Use dimensional analysis to determine how the distance $R(t)$ traveled by the shock from a point source in time t depends on the relevant parameters.

Answer: $R(t) \propto \left(\dfrac{E t^2}{\rho} \right)^{1/5}$

(b) The frequency of a deep-water gravity wave is given by

$$\omega = \sqrt{2\pi} \, \rho^a g^b \lambda^c.$$

where ρ is the water mass density, g is the gravitational acceleration, λ is the wavelength. What are the values of the exponents a, b, c? What is the ratio of wave velocities V_{group}/V_{phase}?

Answer: $a = 0,$ $\qquad b = \frac{1}{2} = -c$

$V_{group}/V_{phase} = \frac{1}{2}.$

CHAPTER 2

SPECIAL RELATIVITY

In relativity theory, we focus our attention on *events*, which are occurrences at particular points in space at particular instants in time. The time evolution of a system is associated with sequences of events. To understand, or predict, these sequences, we need equations of motion derived from physical laws. However, the equations of motion are not expressed in terms of the events themselves. They are expressed in terms of the numerical *coordinates* of the events, as measured with a particular set of coordinate axes and a particular clock (a coordinate *system*). These coordinates depend on the relationship between the events and the coordinate system and, therefore, different observers using different coordinate systems would use different coordinates to label the same events. Correspondingly, the equations of motion obeyed by these coordinates can be expected to vary from one coordinate system to another.

According to Einstein's principle of relativity, there exists a class of coordinate systems (called *Lorentz frames*) that have the following remarkable property: although observers in different Lorentz frames disagree about the numerical coordinates that label the same events, *they all agree on the form of the equations of motion*. In this sense, the Lorentz frames are physically equivalent, since the fundamental laws of physics, as expressed by the equations of motion, are the same in all Lorentz frames.

Two Lorentz frames can differ from each other in any of the following ways:

1. Their coordinate axes are displaced (but at rest) relative to each other, but their clocks are the same.

2. They have the same coordinate axes, but their clocks have a fixed time difference.

3. They have the same clock, but their coordinate axes differ by a fixed rotation or an inversion or both.

4. One set of coordinate axes moves relative to the other with constant velocity.

5. Any combination of the above.

Einstein also postulated that the speed of electromagnetic waves in vacuum ($c = 3 \times 10^{10}$ cm/s) is the same in all Lorentz frames. It is consistent with our prerelativistic ideas that observers whose coordinate systems are at rest relative to each other (as in 1, 2, 3 above) should agree on the speed of any phenomenon they watch. However, it is more difficult to understand how observers in relative motion (as in 4 above) could both watch the same electromagnetic wave, and agree about its speed. This is accomplished if we use the *Lorentz transformation* to relate the coordinates of the same event as seen by observers in Lorentz frames in relative motion. The main theme of this chapter is the way the Lorentz transformation determines all the essential kinematic results of special relativity. The dynamical part of relativity comes from the assumption that the energy and momentum also undergo a Lorentz transformation when we change from one Lorentz frame to another.

2.1 LORENTZ TRANSFORMATIONS IN ONE SPATIAL DIMENSION[1]

Moe and Joe are two observers in Lorentz frames, whose x axes are collinear but slide relative to each other with constant speed v. The origins of the two axes coincide at a time that both Moe and Joe agree to call the zeros of their time scales. Now Moe and Joe observe the *same* event. Moe says that it occurs at location x and at time t; Joe says it occurs at location x' and at time t'. According to the theory of special relativity, the Lorentz transformation relating these pairs of coordinates is

$$x = \gamma\left(x' + \frac{v}{c}ct' \right) \tag{2.1a}$$

$$ct = \gamma\left(ct' + \frac{v}{c}x' \right) \tag{2.1b}$$

$$y = y', \; z = z'$$

The constant γ used here is defined by

$$\gamma \equiv \frac{1}{\sqrt{1 - (v/c)^2}} \geq 1 \tag{2.2}$$

The inverse of the transformation (2.1a, b) is

$$x' = \gamma\left(x - \frac{v}{c}ct \right) \tag{2.1c}$$

$$ct' = \gamma\left(ct - \frac{v}{c}x \right) \tag{2.1d}$$

According to (2.1c), events at Joe's origin (where $x' = 0$) appear to Moe to satisfy the relation $x = vt$. This is consistent with the statement that Joe moves with speed v relative to Moe. Conversely, (2.1a) tells us that Moe moves with speed $-v$ relative to Joe.

Now consider two events, which we label a and b. It is easily verified that (2.1) and (2.2) imply that

$$\left(ct_b - ct_a \right)^2 - \left(x_b - x_a \right)^2 = \left(ct_b' - ct_a' \right)^2 - \left(x_b' - x_a' \right)^2 \tag{2.3}$$

Thus, although Moe and Joe disagree about the distance between the events, and about the time interval between them, they agree about the value of $(c\,\Delta t)^2 - (\Delta x)^2$. This is called the (square of) the *invariant interval* between the two events.

---

[1] A good introduction to the principles of special relativity and to the experimental evidence that supports them is presented in *Special Relativity* by A. P. French, W. W. Norton and Company, Inc., New York, 1968. This book also contains a useful bibliography of other texts in this field.

If (2.3) is positive, the interval is said to be time-like, since there is then a Lorentz frame in which the two events occur at the same place and differ only in time. If (2.3) is negative, the interval is said to be space-like, since there then exists a Lorentz frame in which the two events occur at the same time and differ only in place.

2.2 TIME DILATION

Let events a and b occur at the same place in Moe's Lorentz frame, so that $x_a = x_b$ (and assume that $t_b > t_a$). Thus, the interval between these two events is time-like. According to (2.1d), Joe says that the time interval between these two events is

$$t_b' - t_a' = \gamma \left[(t_b - t_a) - \frac{v}{c^2}(x_b - x_a) \right] = \gamma(t_b - t_a) \tag{2.4}$$

Thus, the time interval between the events is *shortest* in the Lorentz frame in which the events occur at the *same* place.

PROBLEM 2.2.1 A particle travels a distance L relative to the laboratory during its half-life τ (measured in its own rest frame). How fast is it moving relative to the laboratory?

The time interval elapsed, as measured in the laboratory, is $\gamma\tau$. Thus, the speed v as measured in the laboratory is

$$v = \frac{L}{\gamma\tau} = \sqrt{1 - \left(\frac{v}{c}\right)^2} \, \frac{L}{\tau}$$

Solving for v we get

$$\frac{v}{c} = \frac{L/c\tau}{\sqrt{1 + (L/c\tau)^2}} \tag{2.5}$$

Note that L/τ can be arbitrarily large, but v will not exceed c.

An example of this phenomenon is provided by a μ-meson which is formed by a cosmic-ray interaction at the top of the atmosphere, and then travels about 10 km to reach the ground during its 2×10^{-6} second half-life. In this case $L/c\tau = 100/6$, and (2.5) gives

$$\frac{v}{c} = \frac{(100/6)}{\sqrt{1 + (100/6)^2}} = .998$$

PROBLEM 2.2.2 According to an observer on the earth, a certain star is l light-years away. A spaceship travels from the earth to the star at a uniform speed and takes l years to get there by the pilot's measure of time. What is the speed of the spaceship relative to the earth?

This is another application of (2.5). A distance of "l light-years" really means that $L = cl$. The ship travels this distance in a time $\tau = l$ measured on the ship. Thus, $L/c\tau = cl/cl = 1$, and (2.5) implies that $v/c = 1/\sqrt{2}$.

2.3 LORENTZ CONTRACTION

Consider a point fixed on Joe's x axis ($x' = L$). According to (2.1a and b), Moe says that it has coordinates

$$x = \gamma L + \gamma v t'$$

$$t = \gamma t' + \gamma \frac{v}{c^2} L$$

If we eliminate t', we get Moe's version of the motion of Joe's fixed point:

$$x = vt + \frac{L}{\gamma}$$

This implies that if Joe says that the distance between two fixed points is ΔL, Moe will say that the distance between these two points is $\Delta L/\gamma$, which is $\leq L$. Thus, an object is *longest* in the Lorentz frame in which it is at rest.

PROBLEM 2.3.1 Atoms are arranged in a cubic lattice with a number density of n_0 atoms/cm^3 when the lattice is at rest. What number density would be observed if the lattice were moving with speed v?

   The Lorentz contraction of the lattice along its direction of motion reduces one linear dimension by a factor $1/\gamma$. It does not affect the two perpendicular dimensions. Thus, the number density of the moving lattice is

$$n = \gamma n_0 = \frac{n_0}{\sqrt{1 - \dfrac{v^2}{c^2}}}$$

2.4 LORENTZ TRANSFORMATION OF VELOCITY AND ACCELERATION

Moe and Joe from Section 2.1, who are in Lorentz frames with constant relative velocity, both watch the motion of the same particle. Moe summarizes his observations with a function $x(t)$, from which he calculates by differentiation the particle velocity $dx(t)/dt$ and acceleration $d^2x(t)/dt^2$. Correspondingly, Joe uses $x'(t')$, $dx'(t')/dt'$ and $d^2x'(t')/dt'^2$ for his version of the position, velocity, and acceleration of the particle. We want to relate Moe's and Joe's measurements of velocity and acceleration.

   We start by differentiating (2.1a) with respect to t':

$$\frac{dx(t)}{dt'} = \frac{d}{dt'}\gamma[x' + vt'] = \gamma\left[\frac{dx'(t')}{dt'} + v\right] \tag{2.6a}$$

We can get another expression for dx/dt' by using (2.1b),

$$\frac{dx(t)}{dt'} = \frac{dt}{dt'}\frac{dx(t)}{dt} = \gamma\left(1 + \frac{v}{c^2}\frac{dx'(t')}{dt'}\right)\frac{dx(t)}{dt} \tag{2.6b}$$

Comparing (2.6a) and (2.6b) leads to

$$\frac{dx'(t')}{dt'} + v = \left(1 + \frac{v}{c^2}\frac{dx'(t')}{dt'}\right)\frac{dx(t)}{dt} \tag{2.7a}$$

$$\frac{dx}{dt} = \frac{\dfrac{dx'(t')}{dt'} + v}{1 + \dfrac{v}{c^2}\dfrac{dx'(t')}{dt'}} \tag{2.7b}$$

Equation (2.7b) expresses the particle velocity measured by Moe in terms of the particle velocity measured by Joe, when Joe moves in the $+\hat{x}$ direction relative to

Moe with constant speed v. Note that if $dx'(t')/dt' = c$, then (2.7b) yields

$$\frac{dx(t)}{dt} = \frac{c + v}{1 + v/c} = c$$

Thus, an object moving with speed c relative to Joe also moves with speed c relative to Moe, irrespective of Moe's and Joe's relative speed. This confirms that we have not strayed from Einstein's second principle of relativity. Furthermore, it is not difficult to show that if $|dx'/dt'| < c$ and $v < c$, then $|dx/dt|$ given by (2.7b) is also less than c. An object moving more slowly than c in one Lorentz frame, moves more slowly than c in all Lorentz frames.

To relate Moe's and Joe's versions of the particle acceleration, we differentiate (2.7a), again with respect to t':

$$\frac{d^2x'(t')}{dt'^2} = \frac{v}{c^2}\frac{d^2x'(t')}{dt'^2}\frac{dx(t)}{dt} + \gamma\left(1 + \frac{v}{c^2}\frac{dx'}{dt'}\right)^2\frac{d^2x(t)}{dt^2}$$

If we use (2.7b) to eliminate dx/dt, this becomes

$$\frac{d^2x(t)}{dt^2} = \frac{\left(1 - \dfrac{v^2}{c^2}\right)^{3/2}}{\left(1 + \dfrac{v}{c^2}\dfrac{dx'(t')}{dt'}\right)^3}\frac{d^2x'(t')}{dt'^2} \tag{2.8}$$

This relation is true for any two Lorentz frames. Now let us specify that Joe's Lorentz frame is the one in which the particle is instantaneously at rest, so that $dx'/dt' = 0$. Then (2.7b) and (2.8) imply that

$$\frac{dx(t)}{dt} = v \tag{2.9a}$$

$$\frac{d^2x(t)}{dt^2} = \left.\left(1 - \frac{v^2}{c^2}\right)^{3/2}\frac{d^2x'(t')}{dt'^2}\right]_r$$

$$= \left.\left(1 - \frac{1}{c^2}\left(\frac{dx}{dt}\right)^2\right)^{3/2}\frac{d^2x'(t')}{dt'^2}\right]_r \tag{2.9b}$$

We have used the symbol $]_r$ to indicate that the acceleration is measured in the Lorentz frame in which the particle is instantaneously at rest.

An interesting application of (2.9b) is the situation in which the particle's acceleration relative to its instantaneous rest frame has a constant value g. Then (2.9b) can be written

$$\frac{d\left(\dfrac{dx}{dt}\right)}{\left[1 - \dfrac{1}{c^2}\left(\dfrac{dx}{dt}\right)^2\right]^{3/2}} = g\,dt$$

which can be integrated to yield

$$\frac{dx(t)}{dt} = \frac{g(t - t_0)}{\sqrt{1 + \left[\dfrac{g(t - t_0)}{c}\right]^2}} \tag{2.10a}$$

Here t_0 is the time at which the speed of the particle is zero. One more integration gives

$$x(t) = \frac{c^2}{g}\left(\sqrt{1 + \left[\frac{g(t - t_0)}{c}\right]^2} - 1 \right) + x_0 \qquad (2.10b)$$

with x_0 the position of the particle at time t_0. The "world-line" of the particle is a hyperbola. When $t - t_0 \ll c/g$, $x(t)$ is approximately equal to the nonrelativistic constant-acceleration expression, $(1/2)g(t - t_0)^2$. On the other hand, when $t - t_0 \gg c/g$, $x(t)$ goes asymptotically to the ultrarelativistic limit, $x = c(t - t_0)$.

PROBLEM 2.4.1   A fast luxury liner makes excursions to a star a distance L from the earth, maintaining normal gravity on board by accelerating halfway out and decelerating the remainder of the trip.

(a) How long will the one-way trip take according to an earthbound observer?

The ship starts from rest at the earth ($x_0 = t_0 = 0$), and travels a distance $L/2$ in time $T/2$ maintaining constant acceleration g with respect to its instantaneous rest frame. Thus, (2.10b) yields

$$\frac{L}{2} = \frac{c^2}{g}\left(\sqrt{1 + \left(\frac{gT}{2c}\right)^2} - 1 \right)$$

$$T = \frac{2c}{g}\sqrt{\left(1 + \frac{gL}{2c^2}\right)^2 - 1} \qquad (2.11a)$$

(b) How long will the one-way trip take according to a passenger on the ship?

Let τ represent the time measured on the ship (its "proper" time). According to (2.4), when the ship has speed v relative to the earth,

$$d\tau = \frac{dt}{\gamma} = \sqrt{1 - \left(\frac{v}{c}\right)^2}\, dt$$

If we use (2.10a) to express v ($= dx/dt$) in terms of t, we get

$$\tau(t) = \int_0^t dt'\sqrt{1 - \left(\frac{v(t')}{c}\right)^2} = \int_0^t \frac{dt'}{\sqrt{1 + \left(\frac{gt'}{c}\right)^2}}$$

$$\tau(t) = \frac{c}{g}\ln\left[\frac{gt}{c} + \sqrt{1 + \left(\frac{gt}{c}\right)^2}\right]$$

This applies for the first half of the outward journey, when the acceleration is $+g$. If we substitute $T/2$ from (2.11a), we get the proper time of half the journey, and from this we calculate the total proper time to be

$$2\tau\left(\frac{T}{2}\right) = \frac{2c}{g}\ln\left[1 + \frac{gL}{2c^2} + \sqrt{\left(1 + \frac{gL}{2c^2}\right)^2 - 1}\right] \qquad (2.11b)$$

For a long journey with $gL/2c^2 \gg 1$, the ratio of (2.11b) to (2.11a) is

$(\ln[\,gL/2c^2\,])/[\,gL/2c^2\,]$. Thus, the passenger ages much less during the trip than does his friend who watched him from the earth.

2.5 MOMENTUM AND ENERGY

A particle moves along the x axis. Moe says that is has momentum p and energy E; Joe says it has momentum p' and energy E'. According to special relativity, the relationship between the pairs $(\,p,\,E/c\,)$ and $(\,p',\,E'/c\,)$ is exactly the same in form as the relationship (2.1) between the pairs $(x,\,ct\,)$ and $(x',\,ct'\,)$.

$$p = \gamma\left(\,p' + \frac{v}{c}\frac{E'}{c}\,\right) \tag{2.12a}$$

$$\frac{E}{c} = \gamma\left(\frac{E'}{c} + \frac{v}{c}p'\,\right) \tag{2.12b}$$

$$p' = \gamma\left(\,p - \frac{v}{c}\frac{E}{c}\,\right) \tag{2.12c}$$

$$\frac{E'}{c} = \gamma\left(\frac{E}{c} - \frac{v}{c}p\,\right) \tag{2.12d}$$

The Lorentz-invariant combination of E and p is

$$(E/c)^2 - p^2 = (E'/c)^2 - (p')^2 \tag{2.13}$$

Now let Joe's Lorentz frame be the one in which the particle is at rest (at least momentarily). Then $p' = 0$. The rest mass of the particle, m, is defined by

$$E' = mc^2 \tag{2.14}$$

where E' is the energy of the particle in the frame in which $p' = 0$ (the rest frame). Then (2.14) and (2.13) yield

$$(E/c)^2 - p^2 = \left(\frac{mc^2}{c}\right)^2$$

$$E = \sqrt{c^2p^2 + m^2c^4} \tag{2.15a}$$

$$p = \gamma\frac{v}{c}\frac{mc^2}{c} = \gamma mv \tag{2.15b}$$

$$E = \gamma mc^2 \tag{2.15c}$$

In particular, if $m = 0$ (photon, neutrino), (2.15a) becomes $E = cp$.

The kinetic energy of a particle is defined as the difference between its total energy E and its rest energy mc^2,

$$T = E - mc^2 = (\gamma - 1)mc^2 = \left(\frac{v^2}{2c^2} + \frac{3}{8}\frac{v^4}{c^4} + \cdots\right)mc^2$$

$$= \frac{1}{2}mv^2 + \frac{3}{8}mv^2\left(\frac{v}{c}\right)^2 + \cdots \tag{2.16}$$

A comparison of (2.7b), (2.15), and (2.16) with the corresponding nonrelativistic expressions shows that relativistic effects are unimportant when $v \ll c$, $\gamma \simeq 1$, $T \ll mc^2$.

The three-dimensional generalization of (2.7) is

$$(E/c)^2 - \left(p_x^2 + p_y^2 + p_z^2 \right) = (E'/c)^2 - \left(p_x'^2 + p_y'^2 + p_z'^2 \right) \qquad (2.17)$$

The four quantities $(E/c, p_x, p_y, p_z)$ are the components of the 4-momentum vector, whose Lorentz-invariant length is given by (2.17). It is often convenient to use (2.15) and (2.16) to express all the components of the 4-momentum vector in terms of T or $\mathbf{p}$,

$$(E/c, \mathbf{p}) = \left(\sqrt{p^2 + m^2 c^2}, \mathbf{p} \right) = \left(\frac{T}{c} + mc, \hat{p} \sqrt{\left(\frac{T}{c} \right)^2 + 2Tm} \right)$$

$$= \left(\frac{T}{c}, \frac{T}{c} \hat{p} \right) \qquad \text{if } m = 0 \qquad (2.18)$$

In Section 2.7 we will see that the 4-momentum vector is a useful tool for solving collision problems.

PROBLEM 2.5.1 The Cambridge electron accelerator produces electrons with a kinetic energy of 8 GeV in the laboratory. Calculate the difference between c, the speed of light, and v the speed of the emerging electrons as observed in the laboratory system. How would this result be modified if the electrons were to be observed in a coordinate system moving with a speed of 2 m/s toward the emerging electron beam?

In problems of this sort, one can use (2.16) to express γ in terms of the kinetic energy T, and then calculate v from γ:

$$T = mc^2(\gamma - 1)$$

$$8 \times 10^9 \, \text{eV} = .51 \times 10^6 \, \text{eV} \cdot (\gamma - 1)$$

$$\gamma = 1 + \frac{8}{.51} \times 10^3 = 1.57 \times 10^4$$

$$\frac{v}{c} = \sqrt{1 - \frac{1}{\gamma^2}} \approx 1 - \frac{1}{2\gamma^2} \qquad \text{for } \gamma \gg 1$$

$$v \simeq c - \frac{c}{2\gamma^2}, \qquad c - v \simeq \frac{c}{2\gamma^2} = .61 \, \text{m/s}$$

Here v is the electron speed relative to the laboratory. An observer moving toward the emerging beam with a speed of w m/s relative to the laboratory would observe an electron speed of

$$v' = \frac{v + w}{1 + \dfrac{wv}{c^2}}$$

Thus,

$$c - v' = c - \frac{v + w}{1 + \dfrac{vw}{c^2}} = \frac{(c - v)\left(1 - \dfrac{w}{c} \right)}{1 + \dfrac{vw}{c^2}} \simeq (c - v)\left(1 - \frac{w}{c}\left\{ 1 + \frac{v}{c} \right\} \right)$$

Since $v/c \simeq 1$ and $w/c \approx 10^{-8}$, it is clear that $c - v'$ and $c - v$ are very nearly equal.

PROBLEM 2.5.2 What is the momentum (in units of MeV/c) of an electron accelerated through a potential difference of 2×10^6 volts?

The kinetic energy is 2×10^6 eV, so the total energy E is 2×10^6 eV + $.51 \times 10^6$ eV = 2.51 MeV. Then we have

$$E^2 = c^2 p^2 + m^2 c^4$$

$$cp = \sqrt{E^2 - (mc^2)^2} = \sqrt{(2.51)^2 - (.51)^2} \text{ MeV} = 2.46 \text{ MeV}$$

$$p = 2.46 \text{ MeV}/c$$

PROBLEM 2.5.3 An electron is moving so fast that its relativistic mass is 1000 times its rest mass. What is its kinetic energy in eV?

$$E = \gamma mc^2 = 1000 mc^2, \quad \text{so } \gamma = 1000$$

$$T = .51 \times 10^6 \text{ eV} \cdot (\gamma - 1) = .51 \times 10^6 \text{ eV} \cdot (999)$$

$$= .51 \times 10^9 \text{ eV}$$

PROBLEM 2.5.4 Suppose that 1 g of antimatter hits the earth and annihilates. How much matter could you lift to a height of 1 km with the energy generated in the process?

Altogether 2 g of matter are converted into energy. This yields an amount of energy equal to $mc^2 = 2 \times (3 \times 10^{10})^2$ g cm^2/s^2 = 18×10^{20} erg. If we set this equal to the potential energy of a mass M raised to height h above the earth, we get

$$Mgh = 18 \times 10^{20} \text{ erg}$$

$$M = \frac{18 \times 10^{20} \text{ erg}}{981 \dfrac{\text{cm}}{\text{s}^2} \times h \text{ cm}} = \frac{18 \times 10^{20}}{981 \times 10^5} \text{ g} = 1.8 \times 10^{13} \text{g}$$

PROBLEM 2.5.5 It has been observed that a few cosmic ray particles have a measured energy of approximately 1 j (10^7erg or about 10^{19} eV). If a proton ($mc^2 = 10^9$ eV) has this energy, how long would it take this proton to cross our galaxy (diameter = 10^5 light-years) as measured on a clock carried with the proton? Express your answer in seconds.

Again we calculate γ from the kinetic energy:

$$\gamma - 1 = \frac{T}{mc^2} = \frac{10^{19} \text{ eV}}{10^9 \text{ eV}} = 10^{10}$$

A γ-value as large as this implies that v is nearly equal to c. Now consider two events, the arrival of the proton at one edge of the galaxy, and its departure out through the opposite edge. Since an observer fixed in the galaxy says that the proton moves with a speed nearly equal to c, he will say that the time interval between the events is 10^5 years. An observer moving with the proton says the time interval is shorter by a factor γ,

$$\frac{1}{\gamma} \times 10^5 \text{ yr} = 10^{-10} \times 10^5 \text{ yr} = 10^{-5} \text{ yr}$$

$$= 10^{-5} \times 3 \times 10^7 \text{ s} = 300 \text{ s}$$

PROBLEM 2.5.6 The half-life of a free neutron is about 10 min. Estimate the kinetic energy of a neutron that travels from here to Pluto (distance = 5×10^9 km) during its half-life. Give your answer in eV.

The neutron travels at a speed close to c relative to an earthbound observer. Thus, it appears to the earthbound observer to require a time of (5×10^9 km/3×10^5 km/s) = $5/3 \times 10^4$ s to complete the trip. If an observer moving with the neutron says that it requires only 10 min (600 s), it must be that

$$\gamma = \frac{5/3 \times 10^4}{600} = \frac{50{,}000}{1800}$$

which is large compared to one, confirming our assumption that $v \approx c$. Thus, the kinetic energy of the neutron is

$$T = mc^2(\gamma - 1) \approx 10^9 \text{ eV} \left(\frac{500}{18} - 1 \right) \simeq 2.7 \times 10^{10} \text{ eV}$$

PROBLEM 2.5.7 A particle as observed in a certain reference frame has a total energy of 13 GeV and a momentum of 5 GeV/c and a lifetime of 10^{-6} s.

(a) What is its rest mass in u (1 u = 931.5 MeV/c^2)?

$$E^2 = c^2 p^2 + m^2 c^4$$

$$mc^2 = \sqrt{E^2 - c^2 p^2} = \sqrt{13^2 - 5^2} = 12 \text{ GeV} = 12{,}000 \text{ MeV}$$

Thus,

$$m = 12{,}000 \text{ MeV}/c^2 = 12.88 \ u$$

(b) What is its total energy in a frame in which its momentum is equal to 12 GeV/c?

$$E = \sqrt{(cp)^2 + (mc^2)^2} = \sqrt{12^2 + 12^2} \text{ GeV}$$

$$= 12\sqrt{2} \text{ GeV}$$

(c) What is its lifetime in the new reference frame?

In the original frame

$$\gamma = \frac{E}{mc^2} = \frac{13 \text{ GeV}}{12 \text{ GeV}} = \frac{13}{12}$$

Since its lifetime measured in this reference frame is 10^{-6} s, the lifetime of the particle in its rest frame is $12/13 \times 10^{-6}$ s. In the new reference frame $E = 12\sqrt{2}$ GeV so

$$\gamma = \frac{12\sqrt{2} \text{ GeV}}{mc^2} = \sqrt{2}$$

and the particle lifetime measured there is

$$\sqrt{2} \times \frac{12}{13} \times 10^{-6} \text{ s} = 1.31 \times 10^{-6} \text{ s}$$

2.6 THE DOPPLER SHIFT

Suppose that Moe has at his origin a device that sends electromagnetic signals to the right at constant intervals Δ. The nth signal moves with x, t values related by

$$x = c(t - n\Delta) \tag{2.19}$$

The (x', t') values that Joe attributes to this signal can be obtained by using (2.1) and (2.19),

$$\gamma(x' + vt') = c\left[\gamma\left(t' + \frac{v}{c^2}x'\right) - n\Delta\right]$$

These signals will reach Joe's origin ($x' = 0$) when

$$\gamma vt' = c(\gamma t' - n\Delta)$$

$$t' = \frac{n\Delta c}{\gamma(c - v)} = n\Delta\sqrt{\frac{c + v}{c - v}}$$

Thus, if the signals are sent out by Moe separated by intervals Δ, they will be received by Joe separated by intervals $\Delta\sqrt{(c + v)/(c - v)}$. If Moe's signals have frequency f ($= 1/\Delta$), Joe receives them with frequency f', where

$$f' = \sqrt{\frac{c - v}{c + v}}\, f \tag{2.20}$$

If $v > 0$, then $f' < f$. This describes the situation in which the receiver recedes from the sender, so each successive pulse must travel farther before it is received. If the receiver approaches the sender, the same formula (2.20) can be used, but with a negative value of v.

PROBLEM 2.6.1 Bill is moving at a constant velocity (of 80% of the velocity of light) with respect to Herman. As Bill and Herman pass each other, their clocks both read zero. One year later by his clock, Herman sends a light pulse to Bill. At what time by his (Bill's) clock does Bill receive the light pulse?

   If Herman sends out light pulses with interval Δ, Bill will receive them with interval

$$\Delta\sqrt{\frac{c + v}{c - v}} = \Delta\sqrt{\frac{1 + v/c}{1 - v/c}} = \Delta\sqrt{\frac{1.8}{0.2}} = 3\Delta$$

Thus, Bill's clock reads 3 years when he receives Herman's signal. Note that Bill is moving away from Herman, so the interval is increased by their relative motion.

PROBLEM 2.6.2 Observer A is on a rocket moving to the right at half the speed of light relative to observer C. Observer B is on a rocket moving to the left at half the speed of light relative to observer C. Observer B is approaching observer A.

   (a) What is the speed of B as seen by A?

   The formula (2.7b) for the addition of velocities gives

$$V_{B\text{ rel to }A} = \frac{V_{B\text{ rel to }C} + V_{C\text{ rel to }A}}{1 + \dfrac{V_{BC}V_{CA}}{c^2}} = \frac{-\dfrac{c}{2} - \dfrac{c}{2}}{1 + \dfrac{1}{c^2}\left(-\dfrac{c}{2}\right)^2}$$

$$= -\frac{4}{5}c$$

FIGURE 2.1 *A* moves to the right and *B* moves to the left relative to *C*, each with a speed of *c* / 2.

(b) Suppose that A sends a steady stream of photons towards B, at the rate of 100/s from A's standpoint. If these all strike B's rocket, what will B say is the rate at which they arrive?

We use (2.20), with $v = -(4/5)c$:

$$f_B = \sqrt{\frac{c + (4/5)c}{c - (4/5)c}} f_A$$

$$= \sqrt{\frac{9}{1}} \times 100/\text{s} = 300/\text{s}$$

(c) According to observer C, at what rate will the photons strike B's rocket?

The time dilation formula (2.4) implies that C will see a longer interval between arrivals at B than B does, longer by a factor of

$$\gamma = \frac{1}{\sqrt{1 - \left(\dfrac{.5c}{c}\right)^2}} = \sqrt{4/3}$$

Thus, C says that the arrival frequency is

$$\sqrt{3/4} \times 300/\text{s} = 150\sqrt{3}\,/\text{s}$$

PROBLEM 2.6.3   Three Lorentz frames have velocities v_i ($i = 1, 2, 3$) and relative velocities v_{ij} ($\equiv (v_i - v_j)/(1 - (v_i v_j/c^2))$). Their relative *rapidities* are defined by

$$Y_{ij} \equiv \ln \sqrt{\frac{c + v_{ij}}{c - v_{ij}}} \tag{2.21a}$$

Show that these Y_{ij} satisfy the following addition theorem:

$$Y_{31} = Y_{32} + Y_{21} \tag{2.21b}$$

Let observer 1 send out a signal with frequency f_1, which passes observer 2 on its way to observer 3. Observer 3 will receive it with frequency

$$f_1 \sqrt{\frac{c - v_{31}}{c + v_{31}}}$$

if he regards it as coming directly from observer 1, and with frequency

$$f_1 \sqrt{\frac{c - v_{21}}{c + v_{21}}} \cdot \sqrt{\frac{c - v_{32}}{c + v_{32}}}$$

if he regards it as coming from observer 2. Thus,

$$\sqrt{\frac{c - v_{31}}{c + v_{31}}} = \sqrt{\frac{c - v_{32}}{c + v_{32}}} \cdot \sqrt{\frac{c - v_{21}}{c + v_{21}}}$$

The logarithm of this expression yields (2.21b).

It is easy to verify that the definition (2.21a) implies that $\cosh(Y_{ij}) = \gamma$, $\sinh(Y_{ij}) = \gamma v_{ij}/c$. Thus, the coefficients in the Lorentz transformation (2.1) are very simply expressed in terms of the relative rapidity.

2.7 COLLISIONS

In a collision process, we start with two particles (call them 1 and 2) converging on each other from a large initial separation. Let E_i and $\mathbf{p}_i$ $(i = 1, 2)$ be their energies and momenta before they come close enough to interact. After the collision, particles will move away from the region of interaction. Call them $3, 4, \ldots$ (although some or all of the final particles may be identical with the incoming particles 1 and 2), and let E_i and $\mathbf{p}_i$ $(i = 3, 4, \ldots)$ be their energies and momenta after they are too far apart to have any further interaction with each other. Conservation of total energy and momentum implies that

$$E_1 + E_2 = E_3 + E_4 + \cdots \tag{2.22a}$$

$$\mathbf{p}_1 + \mathbf{p}_2 = \mathbf{p}_3 + \mathbf{p}_4 + \cdots \tag{2.22b}$$

The total energy can be written as a sum of rest energy and kinetic energy [cf. (2.16)],

$$E_1 + E_2 = T_{\text{initial}} + m_1 c^2 + m_2 c^2 \tag{2.23a}$$

$$E_3 + E_4 + \cdots = T_{\text{final}} + m_3 c^2 + m_4 c^2 + \cdots \tag{2.23b}$$

The *Q-value* of the reaction is defined by

$$Q = (m_1 + m_2)c^2 - (m_3 + m_4 + \cdots)c^2 = T_{\text{final}} - T_{\text{initial}} \tag{2.24}$$

If Q is positive the reaction is said to be exothermic; if it is negative the reaction is said to be endothermic. An example of an endothermic reaction is proton-anti-proton production in a proton-proton collision,

$$p + p \rightarrow p + p + p + \bar{p} \tag{2.25}$$

The Q-value for this reaction is $-(m_p + m_{\bar{p}})c^2 = -2 \times 938$ MeV.

For a reaction to occur we must be able to satisfy energy and momentum conservation, with both T_{initial} and T_{final} positive. The minimum value of T_{initial} for which this is possible is called the *threshold* energy of the reaction. Suppose we observe the collision in the center-of-mass system. This is the Lorentz frame in which the total momentum is zero,

$$(\mathbf{p}_1 + \mathbf{p}_2)_{\text{CM}} = 0 = (\mathbf{p}_3 + \mathbf{p}_4 + \cdots)_{\text{CM}}$$

The lowest energy consistent with $\mathbf{p}_3 + \mathbf{p}_4 + \cdots = 0$ is $(m_3 + m_4 + \cdots)c^2$, and arises when particles $3, 4, \ldots$ are at rest. Thus, the total energy in the center-of-mass system, $(E_{\text{tot}})_{\text{CM}}$, must satisfy

$$(E_{\text{tot}})_{\text{CM}} \geq (m_3 + m_4 + \cdots)c^2 \tag{2.26a}$$

A similar argument implies that

$$(E_{\text{tot}})_{\text{CM}} \geq (m_1 + m_2)c^2 \tag{2.26b}$$

However, we know that this is always true since the reaction is initiated by particles 1 and 2 in relative motion, so $T_{\text{initial}} > 0$. If $Q > 0$, so that $m_1 + m_2 > m_3 + m_4 + \cdots$, then (2.26b), which we know to be true, implies (2.26a). However, if $Q < 0$, then (2.26a) is not necessarily implied by (2.26b). The *threshold energy* is the minimum value of T_{initial} for which (2.26a) is satisfied, and it is obtained by using the equals sign in (2.26a). If T_{initial} is smaller than this, it is impossible to satisfy energy and momentum conservation, with both T_{initial} and T_{final} positive.

The simplest way to calculate $(E_{\text{total}})_{\text{CM}}$ is to use the fact that $(E_{\text{total}}/c)^2 - (\mathbf{p}_{\text{total}} \cdot \mathbf{p}_{\text{total}})$ has the same value in all Lorentz frames [cf. (2.17)]. Its value in the center-of-mass system, where $\mathbf{p}_{\text{total}} = 0$, is just $((E_{\text{total}})_{\text{CM}}/c)^2$. Thus,

$$\left[\left(\frac{E_{\text{tot}}}{c}\right)^2 - \mathbf{p}_{\text{tot}} \cdot \mathbf{p}_{\text{tot}}\right]_{\text{any Lorentz frame}} = \left(\frac{(E_{\text{tot}})_{\text{CM}}}{c}\right)^2 \qquad (2.27)$$

Now assume that in the laboratory frame, particle 2 (the target) is at rest and particle 1 (the projectile) has kinetic energy T_1. Then

$$E_2 = m_2 c^2, \qquad \mathbf{p}_2 = 0$$

$$E_1 = T_1 + m_1 c^2, \qquad \mathbf{p}_1 = \hat{p}_1 \sqrt{\left(\frac{T_1}{c}\right)^2 + 2T_1 m_1}$$

[cf. (2.18)]. Thus, in this particular Lorentz frame,

$$E_{\text{tot}} = T_1 + (m_1 + m_2) c^2$$

$$\mathbf{p}_{\text{tot}} = \hat{p}_1 \sqrt{\left(\frac{T_1}{c}\right)^2 + 2T_1 m_1}$$

If we use these expressions in the left-hand side of (2.27), we get

$$(E_{\text{tot}})_{\text{CM}} = \sqrt{\left[(m_1 + m_2) c^2\right]^2 + 2T_1 m_2 c^2} \qquad (2.28)$$

The threshold condition for an endothermic reaction $[(E_{\text{tot}})_{\text{CM}} = (m_3 + m_4 + \cdots)c^2]$ is then

$$(T_1)_{\text{threshold}} = \frac{(m_3 + m_4 + \cdots)^2 - (m_1 + m_2)^2}{2m_2} c^2 \qquad (2.29\text{a})$$

$$= -Q \cdot \frac{m_1 + m_2 + m_3 + m_4 + \cdots}{2m_2} \qquad (2.29\text{b})$$

For the $p\bar{p}$ production reaction (2.25), this gives a threshold energy of

$$2m_p c^2 \cdot \frac{6m_p}{2m_p} \simeq 6m_p c^2 \simeq 6 \text{ GeV}$$

PROBLEM 2.7.1 A π-meson of mass M_π comes to rest and disintegrates into a μ-meson of mass M_μ and a neutrino of mass zero. Find the kinetic energy of the μ-meson.

Initially we have a π-meson at rest. Thus,

$$E_{\text{tot}} = M_\pi c^2, \qquad \mathbf{p}_{\text{tot}} = 0$$

Finally we have a μ and a neutrino. Let T_μ be the μ-kinetic energy,

$$E_\mu = T_\mu + M_\mu c^2, \qquad \mathbf{p}_\mu = \hat{p}_\mu \sqrt{\left(\frac{T_\mu}{c}\right)^2 + 2M_\mu T_\mu}$$

Since $\mathbf{p}_{\text{tot}} = 0$, $\mathbf{p}_\nu = -\mathbf{p}_\mu$. Furthermore $E_\nu = cp_\nu$ since $M_\nu = 0$. Thus,

$$E_{\text{tot}} = M_\pi c^2 = \left(T_\mu + M_\mu c^2\right) + c\sqrt{\left(\frac{T_\mu}{c}\right)^2 + 2M_\mu T_\mu}$$

$$\left[(M_\pi + M_\mu) c^2 - T_\mu\right]^2 = T_\mu^2 + 2M_\mu T_\mu c^2$$

from which we get

$$T_\mu = \frac{\left(M_\pi - M_\mu\right)^2 c^2}{2M_\pi}$$

PROBLEM 2.7.2 The proton ($M_p c^2 = 938$ MeV) has an excited state called the Δ ($M_\Delta c^2 = 1236$ MeV). The Δ can be produced in the reaction $\gamma + p \to \Delta$. What is the minimum energy of the γ-ray beam needed for the production of Δ's in a liquid hydrogen target?

Since we end up with a single particle, the Δ, it will be at rest in the center-of-mass system. The threshold condition is thus $(E_{tot})_{CM} = M_\Delta c^2$. We can assume that the target proton is at rest in the laboratory. Let T_γ be the laboratory energy of the γ. Thus,

$$E_p = M_p c^2, \mathbf{p}_p = 0; \quad E_\gamma = T_\gamma, \mathbf{p}_\gamma = \frac{T_\gamma}{c}\hat{p}_\gamma$$

$$E_{total} = T_\gamma + M_p c^2, \qquad \mathbf{p}_{total} = \frac{T_\gamma}{c}\hat{p}_\gamma \qquad \text{(in the laboratory)}$$

$$\left[\frac{(E_{total})_{CM}}{c}\right]^2 = (M_\Delta c)^2$$

$$= \left(\frac{E_{total}}{c}\right)^2 - \mathbf{p}_{total} \cdot \mathbf{p}_{total} = \left(\frac{T_\gamma}{c} + M_p c\right)^2 - \left(\frac{T_\gamma}{c}\right)^2$$

$$= M_p^2 c^2 + 2M_p T_\gamma$$

Thus, the threshold condition is

$$T_\gamma = \frac{(M_\Delta c)^2 - (M_p c)^2}{2M_p} = \frac{(1236)^2 - (938)^2}{2 \times 938} = 345 \text{ MeV}$$

We have essentially rederived (2.29a), with $m_1 = 0$, $m_2 = M_p$ and $m_3 + m_4 + \cdots = M_\Delta$.

PROBLEM 2.7.3 The main accelerator at the Fermi National Laboratory can produce protons of total energy 400 GeV. It has been proposed to build a smaller accelerator nearby to produce protons of total energy 25 GeV that can collide head-on with the protons from the main accelerator. If the reaction

$$p + p \to p + p + W$$

is studied, what is the largest mass particle that can be produced with these machines? Assume that $M_p c^2 = 1$ GeV.

For a given value of $(E_{total})_{CM}$, the largest mass W that can be produced is $(E_{total})_{CM} - 2 \cdot M_p c^2$, corresponding to $(T_{final})_{CM} = 0$. To calculate $(E_{total})_{CM}$, we go to the initial situation in the laboratory frame, where we have

$$E_1 = 400 \text{ GeV}, \qquad \mathbf{p}_1 = \frac{1}{c}\sqrt{(400 \text{ GeV})^2 - \left(M_p c^2\right)^2}\,\hat{n}$$

$$= \sqrt{160{,}000 - 1}\;\frac{\text{GeV}}{c}\,\hat{n}$$

$$E_2 = 25 \text{ GeV}, \qquad \mathbf{p}_2 = -\frac{1}{c}\sqrt{(25 \text{ GeV})^2 - \left(M_p c^2\right)^2}\,\hat{n} = -\sqrt{625 - 1}\;\frac{\text{GeV}}{c}\,\hat{n}$$

$$E_{\text{total}} = 425 \text{ GeV}, \qquad \mathbf{p}_{\text{total}} = \left(\sqrt{159{,}999} - \sqrt{624}\right)\frac{\text{GeV}}{c}\,\hat{n}$$

$$\left[\frac{(E_{\text{total}})_{\text{CM}}}{c}\right]^2 = \left[\left(\frac{425}{c}\right)^2 - \left(\sqrt{159{,}999} - \sqrt{624}\right)^2\right]\left(\frac{\text{GeV}}{c}\right)^2$$

$$(E_{\text{total}})_{\text{CM}} = 200 \text{ GeV}$$

and the most massive W that could be produced has a rest energy of 200 GeV $- 2 \times 1$ GeV $= 198$ GeV.

PROBLEM 2.7.4   The apparatus for a Compton scattering experiment is arranged so that the scattered photon and the recoil electron are detected only if their paths are at right angles to one another. Show that under these conditions the energy of the scattered photon equals $m_e c^2$, where m_e is the rest mass of the electron.

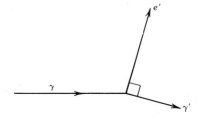

    This problem can be solved by using conservation of energy in the laboratory frame. The evaluation of $[(E_{\text{total}})_{\text{CM}}]^2$ after the scattering involves

$$(\mathbf{p}_e' + \mathbf{p}_\gamma')^2 = p_e'^2 + p_\gamma'^2 + 2\mathbf{p}_e' \cdot \mathbf{p}_\gamma'$$

which equals $p_e'^2 + p_\gamma'^2$, since we are told that $\mathbf{p}_e'$ and $\mathbf{p}_\gamma'$ are perpendicular. If T_e' and E_γ' are the final electron kinetic and photon energies, one finds that

$$\frac{T_e'}{m_e c^2} E_\gamma' = T_e'$$

so that $E_\gamma' = m_e c^2$, independent of the final kinetic energy of the electron.

REVIEW PROBLEMS

2.1. A particle of rest mass M decays at rest into two particles whose rest masses are m_1 and m_2. Calculate the kinetic energy of particle 1 as seen in the rest frame of particle 2.

Answer: $\dfrac{M^2 - (m_1 + m_2)^2}{2m_2}\, c^2$.

2.2. A π^0 meson (rest energy = 135.0 MeV) decays in flight into two γ rays, which emerge at right angles to each other as seen in the laboratory. What is the smallest π^0 laboratory kinetic energy for which this is possible?

Answer: $M_{\pi^0} c^2 \times (\sqrt{2} - 1)$.

2.3. Through what voltage difference should an electron be accelerated to give it a speed of 80% of the speed of light?

Answer: $.34 \times 10^6$ volts.

2.4. The BEVALAC at Lawrence Berkeley Laboratory can accelerate a nucleus with A nucleons to a kinetic energy of about $2A \times 10^9$ eV. What is the speed of such a nucleus?

Answer: $.948c$.

2.5. Section 2.1 referred to Lorentz transformations in one spatial dimension. To generalize to 3 dimensions we first resolve an arbitrary vector **r** into components parallel to and perpendicular to **v**:

$$\mathbf{r} = \mathbf{r}_{\parallel} + \mathbf{r}_{\perp}$$

$$\mathbf{r}_{\parallel} = (\mathbf{r} \cdot \hat{v})\hat{v}, \qquad \mathbf{r}_{\perp} = \mathbf{r} - (\mathbf{r} \cdot \hat{v})\hat{v}$$

Then the generalization of (1a) and (1b) is

$$r_{\parallel} = \gamma\left(r'_{\parallel} + \frac{v}{c}ct' \right), \qquad r_{\perp} = r'_{\perp}$$

$$ct = \gamma\left(ct' + \frac{v}{c}r'_{\parallel} \right)$$

Generalize the derivation of (2.7) to show that

$$\frac{dr_{\parallel}}{dt} = \frac{\dfrac{dr'_{\parallel}}{dt'} + v}{1 + \dfrac{v}{c^2}\dfrac{dr'_{\parallel}}{dt'}}$$

$$\frac{dr_{\perp}}{dt} = \frac{\dfrac{dr'_{\perp}}{dt'}\sqrt{1 - v^2/c^2}}{1 + \dfrac{v}{c^2}\dfrac{dr'_{\parallel}}{dt'}}$$

Note that although the transverse component of **r** is invariant under a Lorentz transformation, the transverse component of $d\mathbf{r}/dt$ is not.

2.6. The diagram shows a coordinate system in which both the sun and the star Polaris are stationary. Photons from Polaris come down making an angle θ with the x-direction. Now suppose that Polaris is viewed from the earth, which moves with speed v in the x-direction. Find the apparent angle to Polaris, as seen by

an observer on earth. *Hint:* do Problem 2.5 first.

Answer: $\theta' = \arccos\left[\dfrac{\cos\theta + \dfrac{v}{c}}{1 + \dfrac{v}{c}\cos\theta} \right].$

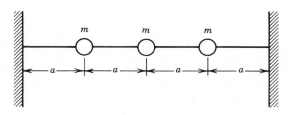

2.7. Two particles are initially at rest and separated by a distance $\Delta\chi$. Due to an attractive force between them they move towards each other and collide in time Δt. If instead of being initially at rest they were moving parallel to each other with speed v, in a direction perpendicular to the line between them, how long would it take them to collide? *Hint:* do Problem 2.5 first.

Answer: $\dfrac{\Delta t}{\sqrt{1 - v^2/c^2}}.$

2.8. Suppose a quasar moves away from us with speed $0.6c$. Its lifetime, measured in its own rest frame is 10^6 years. Over what total span of earth-time would radiation from it be received at the earth?

Answer: 10^6 years $\times \sqrt{\dfrac{c + .6c}{c - .6c}} = 2 \times 10^6$ years.

2.9. The energy of the first excited state of the ^{57}Fe nucleus is 14.4 keV. A nucleus at rest, in this excited state, makes a transition to the ground state by emitting a γ-ray. What is the laboratory energy of this γ-ray?

Answer: $(14.4 - 2 \times 10^{-6})$ keV.

2.10. Consider γ-ray absorption by an ^{57}Fe nucleus at rest in its ground state, leading to the 14.4 keV excited state referred to in Problem 2.9. What must be the laboratory energy of the incident γ-rays? What does the difference between this answer and the answer to Problem 2.9 have to do with the Mossbauer effect?

Answer: $(14.4 + 2 \times 10^{-6})$ keV.

2.11. An astronomer compares the light she receives from opposite ends of a diameter of the sun, the diameter perpendicular to the axis of the sun's rotation. She observes a difference of .078 Å in the wavelength of a 5873 Å line in sodium when she compares spectra taken at the two ends of this diameter. The radius of the sun is 1.4×10^{11} cm. Use this information to calculate the period of the sun's rotation about its axis. *Answer:* 4.4×10^6 s = 51 days.

2.12. A source at the origin of a Lorentz frame sends out spherical electromagnetic wave fronts with frequency f, so that the nth wave front has the equation

$$r \equiv \sqrt{x^2 + y^2 + z^2} = c\left(t - \frac{n}{f}\right)$$

A moving observer describes these wave fronts with coordinates x', y', z', t' such that

$$x = \gamma(x' + vt'), \qquad z = z'$$

$$y = y', \qquad t = \gamma\left(t' + \frac{v}{c^2}x'\right)$$

By calculating dn/dt' at fixed $x'y'z'$, show that the frequency of the wave fronts as measured by the moving observer is $f\gamma(1 - (v/c)\cos\theta)$, where θ is the angle between the velocity of the moving observer and his direction from the source, as measured in the Lorentz frame of the source. Verify that this formula reduces to our 1-dimensional Doppler shift formula (2.20) when $\theta = 0$ or π.

CHAPTER 3

ELECTRICITY AND MAGNETISM

In this chapter, we consider a physical system consisting of electrically charged matter and electromagnetic fields. It is important to keep in mind that the fields have real physical existence, and are not merely mathematical devices for expressing the force laws between charges and currents. For example, consider an atom radiating an electromagnetic wave train over a period of about 10^{-8} s. That wave train can travel for billions of years across interstellar space before it is absorbed by another atom, perhaps long after the emitting atom ceased to exist. There can be no doubt about the physical existence of the oscillating electric and magnetic fields that transferred energy and momentum over such vast intervals of time and distance.

The electromagnetic fields are *coupled* to charged matter. This coupling has two consequences: (1) The fields exert forces on the charges that affect their motion; and (2) the charges are the *sources* of the fields, in the sense that they provide the inhomogeneous terms in Maxwell's equations. The entire theory is consistent with the requirements of special relativity.

Several systems of units are in current use for the expression of the laws of electromagnetic theory. The most common are the SI (Système International) and Gaussian systems. We must choose between them, because it would be impractical and confusing to give all the formulas in both systems. We have chosen Gaussian units because we believe that they do a better job of emphasizing the fundamental physical relationships between the different fields. For example, the physical difference between the **B** and **H** fields at any point is associated with the magnetic effects of the matter in the vicinity of that point. Therefore, there is no physical reason why **B** and **H** should differ in vacuum. In the Gaussian system, **B** and **H** are measured in the same units, and are equal in vacuum. In the SI system, they are measured in different units and their physical equivalence in vacuum is much less evident. Similar remarks apply to the **E** and **D** fields. Another convenience of the Gaussian system is that **E** and **B** are measured in the same units; in fact, the **E** and **B** fields associated with a plane wave propagating in empty space have equal magnitude. The SI system has the advantage that some of its units (volts, amps, ohms) are more familiar than their

Gaussian counterparts. Since some problems are posed in terms of SI units, the student should develop facility in going back and forth between the two systems.[1]

3.1 MAXWELL'S EQUATIONS AND THE BOUNDARY CONDITIONS ON E AND B

In Gaussian units, Maxwell's differential equations for the electromagnetic field in free space are

$$\text{div } \mathbf{B} = 0 \tag{3.1a}$$

$$\text{div } \mathbf{E} = 4\pi\rho \tag{3.1b}$$

$$\text{curl } \mathbf{E} + \frac{1}{c}\frac{\partial \mathbf{B}}{\partial t} = 0 \tag{3.1c}$$

$$\text{curl } \mathbf{B} - \frac{1}{c}\frac{\partial \mathbf{E}}{\partial t} = \frac{4\pi}{c}\mathbf{J} \tag{3.1d}$$

Here $\mathbf{E}(\mathbf{r}, t)$ and $\mathbf{B}(\mathbf{r}, t)$ are the electric and magnetic field vectors, $\rho(\mathbf{r}, t)$ is the charge density, $\mathbf{J}(\mathbf{r}, t)$ is the current density, and c is the speed of electromagnetic waves in vacuum (2.998×10^{10} cm/s). If matter with charge density $\rho(\mathbf{r}, t)$ moves with velocity $\mathbf{v}$, the associated current density is given by

$$\mathbf{J}(\mathbf{r}, t) = \rho(\mathbf{r}, t)\mathbf{v} \tag{3.2}$$

The Gaussian unit of charge is the esu (electrostatic unit). There are 2.998×10^9 esu in a coulomb (C). The charge of an electron is

$$e = -4.803 \times 10^{-10} \text{ esu} = -1.603 \times 10^{-19} \text{ C}.$$

In the Gaussian system, ρ has units of esu/cm^3, $\mathbf{J}$ has units of esu/cm^2 s, and $\mathbf{E}$ and $\mathbf{B}$ both have units of esu/cm^2 = statvolt/cm = Gauss.

The equation expressing conservation[2] of charge

$$\frac{\partial \rho}{\partial t} + \text{div }\mathbf{J} = 0 = \frac{\partial \rho}{\partial t} + \text{div}(\rho\mathbf{v}) \tag{3.3}$$

is obtained by adding the time derivative of (3.1b) to c times the divergence of (3.1d). Thus, Maxwell's equations imply charge conservation, and there can be no solution of Maxwell's equations in which charge conservation is violated.

The integral forms of Maxwell's equations are obtained by applying Gauss' theorem to (3.1a and b), and Stokes' theorem to (3.1c and d):

$$\int_{\substack{\text{any closed} \\ \text{volume}}} \mathbf{B} \cdot d\mathbf{a} = 0 \tag{3.4a}$$

$$\int_{\substack{\text{any closed} \\ \text{volume}}} \mathbf{E} \cdot d\mathbf{a} = 4\pi Q \quad \text{(Gauss' law)} \tag{3.4b}$$

---

[1]Good discussions of the relationships between systems of electromagnetic units are given in appendices of the books *Classical Electrodynamics* by J. D. Jackson, John Wiley & Sons, New York, 1975, *Electricity and Magnetism* by E. M. Purcell, McGraw-Hill, New York, 1965.
[2]Cf. the equation expressing conservation of matter, (1.110).

$$\oint_{\substack{\text{any closed} \\ \text{curve}}} \mathbf{E} \cdot d\mathbf{l} = -\frac{1}{c}\frac{\partial}{\partial t}\int \mathbf{B} \cdot d\mathbf{a} \quad \text{(Faraday's law)} \tag{3.4c}$$

$$\oint_{\substack{\text{any closed} \\ \text{curve}}} \mathbf{B} \cdot d\mathbf{l} = \frac{4\pi}{c}\int\left(\mathbf{J} + \frac{1}{4\pi}\frac{\partial \mathbf{E}}{\partial t}\right) \cdot d\mathbf{a} \quad \text{(Ampere's law)} \tag{3.4d}$$

The surface integrals in (3.4c and d) go over any surface spanning the closed curve followed by the line integrals. The direction of $d\mathbf{l}$ around the curve and the direction of the surface normal $d\mathbf{a}$ must be related by the right-hand rule. Equation (3.4a) expresses the fact that there is no magnetic "charge" analogous to the electric charge of (3.4b). Maxwell's displacement current $1/4\pi \times \partial\mathbf{E}/\partial t$ is included in the statement (3.4d) of Ampere's law.

In some situations it is useful to assume that the charge and current distributions are confined to a two-dimensional surface S. In Figure 3.1a, S carries a surface charge density $\sigma(\text{esu/cm}^2)$. Equations (3.4a) and (3.4b) applied to a small "Gaussian pillbox" cut by S yield

$$(\mathbf{B}_1 - \mathbf{B}_2) \cdot \hat{n} = 0 \tag{3.5a}$$

$$(\mathbf{E}_1 - \mathbf{E}_2) \cdot \hat{n} = 4\pi\sigma \tag{3.5b}$$

Here $\hat{n}$ is a unit vector perpendicular to S, pointing into region 1. Thus, the normal component of $\mathbf{B}$ is continuous across any surface, but the normal component of $\mathbf{E}$ suffers a discontinuity determined by the surface-charge density. Similarly, (3.4c) and (3.4d) applied to the rectangular contour of Figure 3.1b yield

$$(\mathbf{E}_1 - \mathbf{E}_2) \cdot \mathbf{l} = 0 \tag{3.5c}$$

$$(\mathbf{B}_1 - \mathbf{B}_2) \cdot \mathbf{l} = \frac{4\pi}{c}\mathbf{K} \cdot (\hat{n} \times \mathbf{l})$$

$$= \frac{4\pi}{c}(\mathbf{K} \times \hat{n}) \cdot \mathbf{l} \tag{3.5d}$$

Here $\mathbf{K}$ is the density (esu/cm s) of the current flowing on S, and $\mathbf{l}$ is any infinitesimal vector tangential to S.

For every specified $\rho(\mathbf{r}, t)$, $\mathbf{J}(\mathbf{r}, t)$, Maxwell's equations (3.1) have infinitely many solutions. The appropriate solution for a given physical situation is determined by the boundary conditions implied by that situation. For example, if $\mathbf{J}(\mathbf{r}, t) = 0$ and $\rho(\mathbf{r}, t)$ corresponds to a point charge fixed at the origin, we know that the equations (3.1) are satisfied by

$$\mathbf{E}(\mathbf{r}, t) = \frac{q}{r^2}\hat{r} \tag{3.6a}$$

$$\mathbf{B}(\mathbf{r}, t) = 0 \tag{3.6b}$$

FIGURE 3.1 Surface S separates regions 1 and 2. The unit vector $\hat{n}$ is perpendicular to S and points into region 1. In (a) a flat Gaussian "pillbox" is cut by S, with the flat surfaces of the box parallel and close to S. In (b) a rectangular contour is cut by S, with $\mathbf{l}$ parallel and close to S.

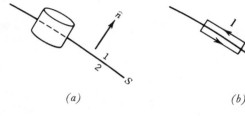

(a)   *(b)*

However, they are also satisfied by

$$E(\mathbf{r}, t) = \frac{q}{r^2}\hat{r} + \hat{x}\alpha \sin\left[\frac{2\pi}{\lambda}(z - ct)\right] \tag{3.7a}$$

$$B(\mathbf{r}, t) = \hat{y}\alpha \sin\left[\frac{2\pi}{\lambda}(z - ct)\right] \tag{3.7b}$$

The difference between (3.6) and (3.7) is a solution of the free-space ($\rho = \mathbf{J} = 0$) set of Maxwell equations. Usually we regard (3.7) as the superposition of two solutions, one associated with q at the origin, since it vanishes if $q = 0$, and the other with a source of electromagnetic radiation far away. From now on, whenever we discuss solutions of Maxwell's equations for some specified ρ, $\mathbf{J}$, we will always imply that we seek the solution that would vanish if $\rho = \mathbf{J} = 0$.

3.2 ELECTROSTATICS

Suppose that we are in a situation in which the fields and charges are constant in time. Then (3.1c) and (3.4c) become

$$\operatorname{curl} \mathbf{E} = 0 \tag{3.8a}$$

$$\oint_{\substack{\text{any closed} \\ \text{curve}}} \mathbf{E} \cdot d\mathbf{r} = 0 \tag{3.8b}$$

The same argument that took us from (1.18a) to (1.20) now allows us to express $\mathbf{E}(\mathbf{r})$ in terms of the gradient of a scalar potential $\phi(\mathbf{r})$

$$\mathbf{E}(\mathbf{r}) = -\nabla\phi(\mathbf{r}) \tag{3.9}$$

If (3.9) is substituted into (3.1b), we get Poisson's equation

$$\operatorname{div} \nabla\phi(\mathbf{r}) \equiv \nabla^2\phi(\mathbf{r}) = -4\pi\rho(\mathbf{r}) \tag{3.10}$$

For the remainder of this section, we assume that all the charge is confined to the interior of a spherical surface of finite (but arbitrarily large) radius R. Thus, for $r > R$, Poisson's equation reduces to Laplace's equation

$$\nabla^2\phi(\mathbf{r}) = 0 \tag{3.11}$$

The general solution of this equation is given by (5.38). We are only interested here in $\mathbf{E}(\mathbf{r})$ fields which vanish as $r \to \infty$ (see the last paragraph of Section 3.1). This means that we should discard the terms in (5.38) that involve positive powers of r, since their gradient will not vanish at $r = \infty$. The remaining expansion has the form

$$\phi(\mathbf{r}) = K + \frac{Q}{r} + \frac{1}{r^2}F(r,\theta,\varphi) \qquad (r > R) \tag{3.12a}$$

Here K and Q are constants, and $F(r,\theta,\phi)$ is bounded as $r \to \infty$.

We now choose to subtract K from the $\phi(\mathbf{r})$ in (3.12a). The new $\phi(\mathbf{r})$

$$\phi(\mathbf{r}) = \frac{Q}{r} + \frac{1}{r^2}F(r,\theta,\varphi) \qquad (r > R) \tag{3.12b}$$

has the same physical content as the old one, since (3.12b) and (3.12a) will yield the same $\mathbf{E}(\mathbf{r})$ when we calculate their gradients (3.9). Thus, we decide to use the arbitrary additive constant implicit in the definition of $\phi(\mathbf{r})$ to arrange for $\phi(\mathbf{r})$ to vanish at $r = \infty$. According to (3.12b), $\phi(\mathbf{r})$ approaches zero for large (but finite) r at least as fast as $1/r$. Correspondingly, $\mathbf{E}(\mathbf{r})$ calculated from (3.9) approaches zero as $r \to \infty$ at least as fast as $1/r^2$. It is also apparent from (3.12b) that as we go farther away from a finite charge distribution, the potential becomes more spherically symmetric. We will soon see that the constant Q in (3.12b) is equal to the total charge in the charge distribution.

The solution of (3.10) whose large-r dependence is given by (3.12b) is

$$\phi(\mathbf{r}) = \int \frac{\rho(\mathbf{r}')}{|\mathbf{r} - \mathbf{r}'|} \, dv' \tag{3.13}$$

The units of potential are statvolts (1 statvolt = 1 erg/esu = 1 esu/cm = 299.8 conventional volts). We now demonstrate three lemmas that will establish that the $\mathbf{E}(\mathbf{r})$ calculated from (3.13) and (3.9) is the only electric field consistent with the given $\rho(\mathbf{r})$ and the requirement that $\mathbf{E}(\mathbf{r})$ vanish at $r = \infty$.

Lemma 1   If $\mathbf{U}(\mathbf{r})$ satisfies

$$\operatorname{curl} \mathbf{U}(\mathbf{r}) = 0 \tag{3.14a}$$

$$\operatorname{div} \mathbf{U}(\mathbf{r}) = 0 \tag{3.14b}$$

$$r^2|\mathbf{U}(\mathbf{r})| \quad \text{is finite as } r \to \infty \tag{3.14c}$$

then

$$\mathbf{U}(\mathbf{r}) = 0 \quad \text{for all } \mathbf{r} \tag{3.14d}$$

Proof   From (3.14a) we know that there exists a $\phi(\mathbf{r})$ such that $\mathbf{U}(\mathbf{r}) = -\nabla\phi(\mathbf{r})$, and (3.14b) then implies that $\phi(\mathbf{r})$ satisfies Laplace's equation $\nabla^2\phi(\mathbf{r}) = 0$. It follows that

$$\operatorname{div}[\phi\nabla\phi] = \nabla\phi \cdot \nabla\phi + \phi\nabla^2\phi = \nabla\phi \cdot \nabla\phi$$

Let us integrate this equation over a volume V bounded by a spherical surface of radius R, and use Gauss' theorem:

$$\int_v \operatorname{div}[\phi\nabla\phi] \, dv = \int_s \phi\nabla\phi \cdot d\mathbf{a}$$

$$= \int_v \nabla\phi \cdot \nabla\phi \, dv \tag{3.15}$$

Now let R become infinitely large. Equation (3.14c) implies that ϕ vanishes at large R at least as fast as $1/R$. Thus, the quantity $\phi\nabla\phi$ in the surface integral of (3.15) vanishes at least as fast as $1/R^3$. This means that as R increases the

FIGURE 3.2 The shaded area represents a localized charge distribution. Equation (3.18*a*) is an approximation to the potential at a distant point **r**. The vector **r'** locates points within the charge distribution.

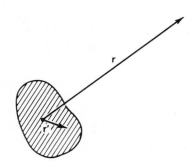

surface integral vanishes at least as fast as $1/R$. But $\nabla\phi \cdot \nabla\phi$ is nonnegative, so if its volume integral over an infinite sphere vanishes, it must be that $\nabla\phi(\mathbf{r})$ vanishes everywhere, which demonstrates the truth of (3.14d).

Lemma 2   If $\mathbf{U}_1(\mathbf{r})$ and $\mathbf{U}_2(\mathbf{r})$ satisfy

$$\operatorname{curl}\mathbf{U}_1(\mathbf{r}) = \operatorname{curl}\mathbf{U}_2(\mathbf{r}) \tag{3.16a}$$

$$\operatorname{div}\mathbf{U}_1(\mathbf{r}) = \operatorname{div}\mathbf{U}_2(\mathbf{r}) \tag{3.16b}$$

$$r^2|\mathbf{U}_1(\mathbf{r})|, \qquad r^2|\mathbf{U}_2(\mathbf{r})| \qquad \text{are finite as } r \to \infty \tag{3.16c}$$

then

$$\mathbf{U}_1(\mathbf{r}) = \mathbf{U}_2(\mathbf{r}) \qquad \text{for all } \mathbf{r} \tag{3.16d}$$

Proof   Define $\mathbf{U}(\mathbf{r}) \equiv \mathbf{U}_1(\mathbf{r}) - \mathbf{U}_2(\mathbf{r})$. This vector field obviously satisfies the requirements of Lemma 1, which implies that it vanishes everywhere.

We see that a vector field that vanishes at infinity at least as fast as $1/r^2$ is determined uniquely by its curl and divergence. An explicit demonstration of this is provided by Lemma 3.

Lemma 3   If $\mathbf{U}(\mathbf{r})$ satisfies

$$\operatorname{curl}\mathbf{U}(\mathbf{r}) = 4\pi\mathbf{a}(\mathbf{r}) \tag{3.17a}$$

$$\operatorname{div}\mathbf{U}(\mathbf{r}) = 4\pi b(\mathbf{r}) \tag{3.17b}$$

$$r^2|\mathbf{U}(\mathbf{r})| \qquad \text{is finite as } r \to \infty \tag{3.17c}$$

then

$$U(\mathbf{r}) = -\nabla\left[\int \frac{b(\mathbf{r}')}{|\mathbf{r} - \mathbf{r}'|}\, dv'\right] + \operatorname{curl}\left[\int \frac{\mathbf{a}(\mathbf{r}')}{|\mathbf{r} - \mathbf{r}'|}\, dv'\right] \tag{3.17d}$$

This is proven by calculating[3] the curl and divergence of (3.17d), and showing that they equal $4\pi\mathbf{a}(\mathbf{r})$ and $4\pi b(\mathbf{r})$, respectively.

If we set $\mathbf{a}(\mathbf{r}) = 0$ and $b(\mathbf{r}) = \rho(\mathbf{r})$ in (3.17), then (3.17d) gives us the solution [(3.9) and 3.12)] for an electrostatic $\mathbf{E}$ field for a specified charge distribution. We also see from (3.17) that a field whose divergence vanishes can be written as a curl of a "vector potential." In Section 3.11 we will see that this is true in general, and not only for fields which vanish at infinity at least as fast as $1/r^2$.

Equation (3.12b) gave the general form of the potential far away from a localized charge distribution. A more explicit expression is[4]

$$\phi(\mathbf{r}) = \frac{Q}{r} + \frac{\mathbf{p} \cdot \mathbf{r}}{r^3} + \cdots \tag{3.18a}$$

$$\begin{aligned}\mathbf{E}(\mathbf{r}) &= -\nabla\phi(\mathbf{r}) \\ &= \frac{Q}{r^2}\hat{r} + \frac{3(\mathbf{p} \cdot \hat{r})\hat{r} - \mathbf{p}}{r^3} + \cdots\end{aligned} \tag{3.18b}$$

---

[3] See, for example, *Classical Electricity and Magnetism* by M. Abraham and R. Becker, Blackie and Son, Limited, London, 1937, p. 37.

[4] Space limitations prevent us from deriving all the formulas we use in this chapter. For more complete presentations, see *Electricity and Magnetism* by E. M. Purcell, McGraw-Hill, New York, 1965, or Volume II of *The Feynman Lectures on Physics*, Addison-Wesley, Reading, Mass., 1963. These two undergraduate-level texts present the formalism of electricity and magnetism, together with interesting discussions of many physical phenomena associated with the behavior of charged matter.

FIGURE 3.3 (a) Dipole in uniform **E** field. A torque acts on the dipole, tending to align it with the field, but there is no net force. (b) Dipole in nonuniform **E** field with **p** parallel to the field. The field at the positive charge is stronger than the field at the negative charge, resulting in a net force that pulls the dipole toward the region of stronger field. (c) Dipole in nonuniform **E** field with **p** antiparallel to the field. Now the stronger force acts on the negative charge, so the dipole is pulled toward the region of weaker field.

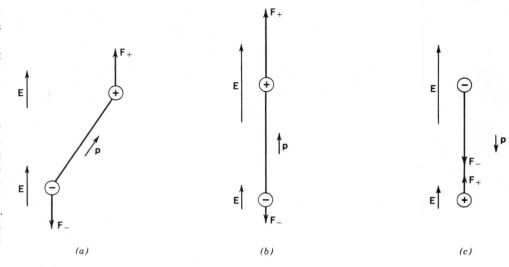

with the monopole moment (total charge) Q and electric dipole moment **p**, defined by

$$Q \equiv \int \rho(\mathbf{r}')\, dv' \qquad \text{(total charge)} \tag{3.18c}$$

$$\mathbf{p} \equiv \int \mathbf{r}'\rho(\mathbf{r}')\, dv' \qquad \text{(electric dipole moment)} \tag{3.18d}$$

The multipole moments of a charge distribution are also useful in the discussion of the forces and torques that a charge distribution experiences when it is placed in an externally generated electric field. If the charge is in the vicinity of the field point labeled by **r**, its energy is approximately

$$U \approx Q\phi(\mathbf{r}) - \mathbf{p} \cdot \mathbf{E}(\mathbf{r}) \tag{3.19a}$$

where **p** is defined relative to the point labeled by **r**. In particular, (3.19a) implies that if we rotate the charge distribution about **r**, we get the lowest energy when **p** lines up parallel to **E**. In fact, one can use (3.19a) to show that the electric field exerts a torque on the charge distribution equal to

$$\boldsymbol{\tau} = \mathbf{p} \times \mathbf{E}(\mathbf{r}) \tag{3.19b}$$

An important special case is when $Q = 0$ but $\mathbf{p} \neq 0$. The simplest realization of this situation is a pair of equal and opposite point charges $\pm q$, separated by a distance d. Application of (3.18d) to this distribution yields a **p** whose magnitude equals qd, and whose direction points from the negative to the positive charge. Figure 3.3 illustrates that such an "electric dipole" has the following properties:

(*a*) In a uniform electric field the dipole experiences a torque, but no net force.

(*b*) In a nonuniform electric field the dipole is pulled into the region of stronger field if **p** and **E** are parallel.

(*c*) In a nonuniform electric field the dipole is pulled into the region of weaker field if **p** and **E** are antiparallel.

These rules can be obtained from (3.19), and hold for any charge distribution with $Q = 0$ and $\mathbf{p} \neq 0$.

PROBLEM 3.2.1 Two point charges q_1 and q_2 are situated 4 cm from one another. Two other charges lie between q_1 and q_2, on the straight line connecting them. These are: a 4-esu charge 2 cm from q_2 and a 2-esu charge 1 cm from q_1. Determine the magnitude and signs of q_1 and q_2 so that $\mathbf{E}(r)$ at large distances r from the charges falls off faster than $1/r^3$ in all directions.

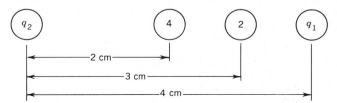

According to (3.18b), both the total charge and total dipole moment of the charge distribution must vanish if $\mathbf{E}$ is to fall off *faster* than $1/r^3$. Thus, we must have

$$q_1 + q_2 + 4 \text{ esu} + 2 \text{ esu} = 0$$
$$2 \times 4 + 3 \times 2 + 4q_1 = 0$$
$$q_1 = -\tfrac{7}{2} \text{ esu}, \qquad q_2 = -\tfrac{5}{2} \text{ esu}$$

Note that we have calculated the dipole moment relative to the location of q_2 as origin. It is easy to verify from (3.18c, 3.18d) that if the total charge Q equals zero, the total dipole moment $\mathbf{p}$ is independent of the origin used for the definition of $\mathbf{r}'$.

PROBLEM 3.2.2 Determine the charge distribution that will give rise to the potential

$$\phi(\mathbf{r}) = W_0 \frac{e^{-\alpha r}}{r} \qquad (W_0, \alpha \text{ constant and } > 0)$$

Verify that your answer satisfies Gauss' law for a sphere of finite radius centered at the origin. Calculate the total charge in the distribution.

Since the potential falls off faster than $1/r$ as $r \to \infty$, it must be that the total charge equals zero. To find $\rho(\mathbf{r})$, we use (3.10),

$$\rho(\mathbf{r}) = -\frac{1}{4\pi} \nabla^2 \phi(\mathbf{r}) = -\frac{1}{4\pi} \nabla^2 \left[\frac{W_0 e^{-\alpha r}}{r} \right]$$

To operate with the Laplacian on a spherically symmetric potential, we use the expression (5.37a) for the Laplacian in spherical polar coordinates, with $l = 0$:

$$\nabla^2 \left[W_0 \frac{e^{-\alpha r}}{r} \right] = W_0 \frac{1}{r^2} \frac{d}{dr} \left[r^2 \frac{d}{dr} \left(\frac{e^{-\alpha r}}{r} \right) \right] = W_0 \alpha^2 \frac{e^{-\alpha r}}{r} \qquad (\text{for } \mathbf{r} \neq 0)$$

$$\rho(\mathbf{r}) = -\frac{W_0}{4\pi} \alpha^2 \frac{e^{-\alpha r}}{r}$$

This would lead to a total charge of

$$\int \rho(\mathbf{r}) \, dv = 4\pi \int_{r=0}^{\infty} \left[-\frac{W_0}{4\pi} \alpha^2 \frac{e^{-\alpha r}}{r} \right] r^2 \, dr$$

$$= -W_0 \alpha^2 \int_{r=0}^{\infty} r e^{-\alpha r} \, dr = -W_0$$

Thus, to achieve a total charge of zero, we add a point charge W_0 at the origin ($r = 0$), leading to a total charge density of [5]

$$\rho(\mathbf{r}) = W_0\left[\delta(\mathbf{r}) - \frac{\alpha^2}{4\pi}\frac{e^{-\alpha r}}{r}\right]$$

To verify Gauss' law, we calculate the flux of $\mathbf{E}$ through a sphere of radius R:

$$\int \mathbf{E} \cdot d\mathbf{a} = -\left.\frac{\partial \phi}{\partial r}\right|_{r=R} \times 4\pi R^2$$

$$= 4\pi W_0[\alpha R + 1]e^{-\alpha R}$$

The total charge contained within this sphere is

$$\int_{r \leq R} \rho(\mathbf{r})\, dv = W_0\int\left[\delta(\mathbf{r}) - \frac{\alpha^2}{4\pi}\frac{e^{-\alpha r}}{r}\right]dv$$

$$= W_0 - W_0\alpha^2\int_0^R e^{-\alpha r}r\, dr$$

$$= W_0(1 + \alpha R)e^{-\alpha R}$$

so that the flux of $\mathbf{E}$ through the sphere is indeed equal to 4π times the charge enclosed.

3.3 SOLVING PROBLEMS USING THE UNIQUENESS THEOREMS OF ELECTROSTATICS

The arguments used in deriving Lemmas 1 and 2 on pages 110–111 can be applied to prove the following uniqueness theorems:

3.3.1 Dirichlet Boundary Conditions

If two functions $\phi_1(\mathbf{r})$ and $\phi_2(\mathbf{r})$ both satisfy Poisson's equation everywhere within a closed surface S,

$$\nabla^2\phi_1(\mathbf{r}) = \nabla^2\phi_2(\mathbf{r}) = -4\pi\rho(\mathbf{r}) \qquad (\mathbf{r} \text{ within } S)$$

and if $\phi_1(\mathbf{r})$ and $\phi_2(\mathbf{r})$ are equal everywhere *on* S, then $\phi_1(\mathbf{r})$ and $\phi_2(\mathbf{r})$ are equal everywhere *within* S. In other words, there is only one function that satisfies Poisson's equation in the interior of a closed surface and has specified values on that surface. Thus, if we find *any* solution of this boundary-value problem, it is *the unique solution*.

3.3.2 Neumann Boundary Conditions

If two functions $\phi_1(\mathbf{r})$ and $\phi_2(\mathbf{r})$ both satisfy Poisson's equation everywhere within a closed surface S, and if the normal derivatives of $\phi_1(\mathbf{r})$ and $\phi_2(\mathbf{r})$ are equal

---

[5]$\delta(\mathbf{r})$ in this equation is a 3-dimensional Dirac δ-function, defined by

$$\delta(\mathbf{r}) = 0 \qquad \text{if } \mathbf{r} \neq 0$$

$$\int_V \delta(\mathbf{r})\, dv = 1 \qquad \text{if the integration volume } V \text{ contains } \mathbf{r} = 0.$$

everywhere on S, then $\phi_1(\mathbf{r})$ and $\phi_2(\mathbf{r})$ differ at most by a constant within S. This implies that there is only one $\mathbf{E}(r)$ that satisfies (3.1b) within a closed surface and has a specified normal component on that surface.

The following are some examples of the use of the uniqueness theorems to solve electrostatics problems.

PROBLEM 3.3.1 A hollow conductor is placed in an externally generated electric field. Show that, at static equilibrium, the electric field within the cavity is zero, irrespective of the shape of the cavity or the complexity of the external field.

At static equilibrium, every point within the conducting material is at the same potential. Otherwise, there would be an electric field in the conducting material and current would flow. Now consider the cavity. Within the cavity $\rho(\mathbf{r}) = 0$, so $\nabla^2\phi(\mathbf{r}) = 0$. On the surface of the cavity, the potential has some constant value, say $\phi = C$. Now define a function $\Phi(\mathbf{r})$ by

$$\Phi(\mathbf{r}) \equiv C \qquad (\text{for } \mathbf{r} \text{ within the cavity})$$

Obviously, $\nabla^2\Phi = 0$ within the cavity, and Φ has the specified value C on the surface of the cavity. It satisfies Poisson's equation and the boundary conditions, and so the first uniqueness theorem says it is the unique solution. Thus, $\phi(\mathbf{r}) = \Phi(\mathbf{r})$ is constant within the cavity, and $\mathbf{E}(\mathbf{r}) = -\nabla\phi(\mathbf{r}) = 0$ within the cavity.

PROBLEM 3.3.2 A copper sphere of radius R contains a spherical cavity of radius a. The center of the cavity is at a distance d from the center of the sphere, and d and a are such that the cavity is entirely within the sphere. There is a total charge of Q on the sphere. Find the electric field (a) within the cavity, and (b) outside the sphere.

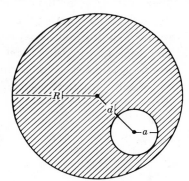

Since $\nabla^2\phi(\mathbf{r}) = 0$ within the cavity and $\phi(\mathbf{r})$ is constant on the surface of the cavity, the previous argument shows that $\mathbf{E}(\mathbf{r}) = 0$ everywhere within the cavity.

Now consider the region outside the sphere. On the surface of the sphere ($r = R$), the potential has some constant value K. The potential vanishes on the sphere at $r = \infty$ (see p. 110). There is no charge when $R < r < \infty$. Thus, $\phi(\mathbf{r})$ must satisfy

$$\phi(R) = K$$

$$\phi(\infty) = 0$$

$$\nabla^2\phi(\mathbf{r}) = 0 \qquad (R \le r \le \infty)$$

The function

$$\phi(\mathbf{r}) = K\frac{R}{r} \qquad (R \le r \le \infty)$$

satisfies these conditions and is thus the unique solution of the problem. To determine the constant K, we apply Gauss' law (3.4b) to a sphere concentric with the copper sphere, but whose radius R' is greater than R. The $\mathbf{E}$ field at $R'\hat{r}$ is

$$\mathbf{E}(R'\hat{r}) = -\nabla\left(K\frac{R}{r}\right)\bigg|_{r=R'} = \frac{KR}{(R')^2}\hat{r}$$

The area of this large sphere is $4\pi(R')^2$, and so the total $\mathbf{E}$-flux through it is $4\pi KR$. Thus, (3.4b) implies that $KR = Q$. We conclude that the electric field outside the copper sphere is independent of the size of the cavity or its location, so long as it doesn't pierce the outer surface of the copper sphere.

PROBLEM 3.3.3   Now suppose that a point charge q is placed at the center of the cavity. What is the effect on the fields within the cavity and outside the sphere?

Since the total charge within a spherical surface of radius $> R$ is now $Q + q$, the above argument shows that the field outside the copper sphere is given by

$$\mathbf{E}(\mathbf{r}) = \frac{Q+q}{r^2}\hat{r} \qquad (r > R)$$

Within the cavity the potential satisfies Poisson's equation with a charge density corresponding to a point charge q at the center of the cavity. The potential is constant on the surface of the cavity. All these conditions are uniquely satisfied by choosing

$$\mathbf{E}(\mathbf{s}) = \frac{q}{s^2}\hat{s} \qquad (s < a)$$

where $\mathbf{s}$ is a vector from the center of the cavity to any point within the cavity.

The method of images is another application of the uniqueness theorem. Suppose that a point charge q is a distance d from an infinite conducting plane. We want to find the potential $\phi(\mathbf{r})$ in the space around q. Thus, we seek a function $\phi(\mathbf{r})$ that has a constant value when $\mathbf{r}$ is on the plane, vanishes as $\mathbf{r}$ moves infinitely far away to the right, and satisfies Poisson's equation to the right of the plane, with ρ appropriate to a point charge $+q$ placed as shown in Figure 3.4a. All these conditions are automatically satisfied by the potential produced to the right of the plane by the pair of charges shown in Figure 3.4b. We can be

FIGURE 3.4 (a) A point charge at a distance d to the right of an infinite conducting plane. (b) The arrow with the solid shank represents the electric field at a point on the midplane due to the combined effects of $\pm q$.

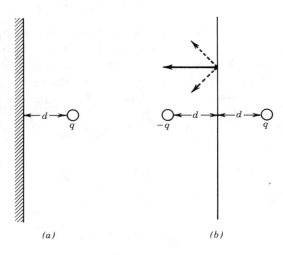

(a)   *(b)*

sure that the plane in Figure 3.4*b* is an equipotential since **E** on the plane is everywhere perpendicular to the plane, so that it can do no work on a charge that moves around in the plane. Since the potential produced by the two charges in Figure 3.4*b* satisfies all the requirements for the physical situation shown in Figure 3.4*a*, we can accept this solution as the unique solution. Thus, in discussing the field outside a conducting plane we can ignore the plane if we add a fictitious "image charge" $-q$, at a distance d behind the plane.

PROBLEM 3.3.4   A charge q is released from rest at a distance d from an infinite conducting plane. How long will it take for the charge to strike the plane?

If the charge is a distance x from the plane, it is at a distance $2x$ from its image. The force on the charge, attracting it to the plane, is thus $q^2/(2x)^2$, and its equation of motion is

$$m\frac{d^2x}{dt^2} = -\frac{q^2}{4x^2}$$

To solve this equation, multiply both sides by dx/dt [cf. (1.17a)]:

$$m\frac{dx}{dt}\frac{d^2x}{dt^2} = -\frac{q^2}{4x^2}\frac{dx}{dt}$$

$$\frac{d}{dt}\left[\frac{1}{2}m\left(\frac{dx}{dt}\right)^2\right] = \frac{d}{dt}\left(\frac{q^2}{4x}\right)$$

$$\frac{1}{2}m\left(\frac{dx}{dt}\right)^2 = \frac{q^2}{4}\left(\frac{1}{x}-\frac{1}{d}\right)$$

We have chosen the integration constant to satisfy the initial condition that $dx/dt = 0$ when $x = d$. Now we must do a second integration,

$$\frac{dx}{dt} = -\sqrt{\frac{q^2}{2m}\left(\frac{1}{x}-\frac{1}{d}\right)}$$

$$\int_0^T dt = -\int_{x=d}^0 \frac{dx}{\sqrt{\frac{q^2}{2m}\left(\frac{1}{x}-\frac{1}{d}\right)}} = \sqrt{\frac{2md}{q^2}}\int_{x=0}^d \frac{\sqrt{x}\,dx}{\sqrt{d-x}} = \frac{\pi}{q}\sqrt{\frac{md^3}{2}}$$

The last integral is conveniently done with the aid of the substitution $x = d\sin^2\theta$.

PROBLEM 3.3.5   An infinitely long charged wire of radius r_0 is parallel to an infinite conducting plane, at a distance h $(h \gg r_0)$ from the surface. The potential difference between the wire and the surface is V. Derive a formula for the magnitude of the electric field just above the surface below the wire.

Suppose that the charge density on the wire is ρ esu/cm. If the conducting plane were not present, the electric field at a distance r from the wire would be calculated from Gauss' law

$$E(r) \cdot 2\pi r = 4\pi\rho$$

$$E(r) = \frac{2\rho}{r} \qquad \text{(due to the wire alone)}$$

If ρ is positive, the direction of the field is radially away from the wire. Now take

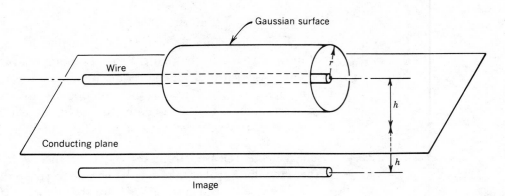

into account the presence of the plane by adding an image wire, with linear charge density $-\rho$ at a distance h below the plane, parallel to the original wire. Then the total electric field at a distance r from the wire, directly below it, is

$$E(r) = \frac{2\rho}{r} + \frac{2\rho}{2h - r}$$

The potential difference between the surface of the wire and the conducting plane is

$$V = \int_{r_0}^{h} E(r)\,dr = 2\rho \int_{r_0}^{h} \left(\frac{1}{r} + \frac{1}{2h - r} \right) dr$$

$$= 2\rho \ln\left(\frac{2h - r_0}{r_0} \right)$$

so the charge density on the wire is

$$\rho = \frac{V}{2\ln\left(\dfrac{2h - r_0}{r_0} \right)} \approx \frac{V}{2\ln\left(\dfrac{2h}{r_0} \right)}$$

Finally, the electric field just above the plane is

$$E(r = h) = 2\rho\left(\frac{1}{h} + \frac{1}{2h - h} \right) = \frac{4\rho}{h}$$

$$\approx \frac{2V}{h \ln\left(\dfrac{2h}{r_0} \right)}$$

3.4 MAXWELL'S EQUATIONS IN THE PRESENCE OF MATERIAL MEDIA

On a macroscopic scale the effect of a material medium on the **E** and **B** fields depends on the

polarization density $\mathbf{P}(\mathbf{r}, t)$ (electric dipole moment/unit volume),
magnetization density $\mathbf{M}(\mathbf{r}, t)$ (magnetic dipole moment/unit volume).

Consider a volume enclosed by a surface S. Within the surface, the effect of the polarization and magnetization densities is to augment the free charge and current densities, $\rho_{\text{free}}(\mathbf{r}, t)$ and $\mathbf{J}_{\text{free}}(\mathbf{r}, t)$, by the effective polarization and magnetization charge and current densities, given by

$$\rho_{\text{pol}}(\mathbf{r}, t) = -\operatorname{div} \mathbf{P}(\mathbf{r}, t) \tag{3.20a}$$

$$\mathbf{J}_{\text{mag}}(\mathbf{r}, t) + \mathbf{J}_{\text{pol}}(\mathbf{r}, t) = c \operatorname{curl} \mathbf{M}(\mathbf{r}, t) + \frac{\partial}{\partial t} \mathbf{P}(\mathbf{r}, t) \tag{3.20b}$$

Thus, the total charge and current densities to be used in the inhomogeneous Maxwell equations are

$$\begin{aligned}
\rho(\mathbf{r}, t) &= \rho_{\text{free}}(\mathbf{r}, t) + \rho_{\text{pol}}(\mathbf{r}, t) \\
&= \rho_{\text{free}}(\mathbf{r}, t) - \operatorname{div} \mathbf{P}(\mathbf{r}, t)
\end{aligned} \tag{3.21a}$$

$$\begin{aligned}
\mathbf{J}(\mathbf{r}, t) &= \mathbf{J}_{\text{free}}(\mathbf{r}, t) + \mathbf{J}_{\text{mag}}(\mathbf{r}, t) + \mathbf{J}_{\text{pol}}(\mathbf{r}, t) \\
&= \mathbf{J}_{\text{free}}(\mathbf{r}, t) + c \operatorname{curl} \mathbf{M}(\mathbf{r}, t) + \frac{\partial \mathbf{P}(\mathbf{r}, t)}{\partial t}
\end{aligned} \tag{3.21b}$$

These equations become

$$\operatorname{div} \mathbf{E} = 4\pi\rho(\mathbf{r}, t) = 4\pi \left[\rho_{\text{free}}(\mathbf{r}, t) - \operatorname{div} \mathbf{P}(\mathbf{r}, t) \right]$$

$$\operatorname{div}(\mathbf{E} + 4\pi\mathbf{P}) = 4\pi\rho_{\text{free}}(\mathbf{r}, t)$$

$$\begin{aligned}
\operatorname{curl} \mathbf{B} - \frac{1}{c} \frac{\partial \mathbf{E}}{\partial t} &= \frac{4\pi}{c} \mathbf{J}(\mathbf{r}, t) \\
&= \frac{4\pi}{c} \left[\mathbf{J}_{\text{free}}(\mathbf{r}, t) + c \operatorname{curl} \mathbf{M}(\mathbf{r}, t) + \frac{\partial \mathbf{P}}{\partial t}(\mathbf{r}, t) \right]
\end{aligned}$$

$$\operatorname{curl}(\mathbf{B} - 4\pi\mathbf{M}) - \frac{1}{c} \frac{\partial}{\partial t}(\mathbf{E} + 4\pi\mathbf{P}) = \frac{4\pi}{c} \mathbf{J}_{\text{free}}(\mathbf{r}, t)$$

Thus, if we define two new fields $\mathbf{D}(\mathbf{r}, t)$ and $\mathbf{H}(\mathbf{r}, t)$ by

$$\mathbf{D}(\mathbf{r}, t) \equiv \mathbf{E}(\mathbf{r}, t) + 4\pi\mathbf{P}(\mathbf{r}, t) \tag{3.22a}$$

$$\mathbf{H}(\mathbf{r}, t) \equiv \mathbf{B}(\mathbf{r}, t) - 4\pi\mathbf{M}(\mathbf{r}, t) \tag{3.22b}$$

the inhomogeneous Maxwell equations can be written as

$$\operatorname{div} \mathbf{D}(\mathbf{r}, t) = 4\pi\rho_{\text{free}}(\mathbf{r}, t) \tag{3.23a}$$

$$\operatorname{curl} \mathbf{H}(\mathbf{r}, t) - \frac{1}{c} \frac{\partial}{\partial t} \mathbf{D}(\mathbf{r}, t) = \frac{4\pi}{c} \mathbf{J}_{\text{free}}(\mathbf{r}, t) \tag{3.23b}$$

The reasoning that led to (3.5b) and (3.5d) now gives

$$(\mathbf{D}_1 - \mathbf{D}_2) \cdot \hat{n} = 4\pi(\sigma_1)_{\text{free}} \tag{3.24a}$$

$$(\mathbf{H}_1 - \mathbf{H}_2) \cdot \mathbf{l} = \frac{4\pi}{c} (\mathbf{K}_{\text{free}} \times \hat{n}) \cdot \mathbf{l} \tag{3.24b}$$

(see Figure 3.1). In particular, if there is no free charge on the surface of discontinuity in Figure 3.1, the perpendicular component of $\mathbf{D}$ is continuous across this surface, and if there is no sheet of free current on the surface of discontinuity, the parallel component of $\mathbf{H}$ is continuous.

Since the homogeneous Maxwell equations are unaffected by the presence of material media, the boundary conditions (3.5a) and (3.5c) on $\mathbf{E}$ and $\mathbf{B}$ continue to apply.

Suppose that $\mathbf{P}(\mathbf{r}, t)$ has a constant value $\mathbf{P}_2$ inside a closed surface, and a different constant value $\mathbf{P}_1$ outside this surface. According to (3.20a), $\rho_{pol}(\mathbf{r}, t)$ is zero everywhere inside and outside the surface. However, this does not mean that the polarization has no effect, since (3.22a) and (3.24a) imply that

$$(\mathbf{E}_1 - \mathbf{E}_2) \cdot \hat{n} = 4\pi \times \left[(\sigma_1)_{free} + (\mathbf{P}_2 - \mathbf{P}_1) \cdot \hat{n} \right]$$

Thus, the discontinuity of $\mathbf{P}$ across the surface can be regarded as producing a polarization surface charge density $(\mathbf{P}_2 - \mathbf{P}_1) \cdot \hat{n}$. Similarly a discontinuity of $\mathbf{M}$ across the surface results in

$$(\mathbf{B}_1 - \mathbf{B}_2) \cdot \mathbf{l} = \frac{4\pi}{c} \left[\mathbf{K}_{free} \times \hat{n} + c(\mathbf{M}_1 - \mathbf{M}_2) \right] \cdot \mathbf{l}$$

$$= \frac{4\pi}{c} \left(\left[\mathbf{K}_{free} + c(\mathbf{M}_2 - \mathbf{M}_1) \times \hat{n} \right] \times \hat{n} \right) \cdot \mathbf{l}$$

We see that the discontinuity in $\mathbf{M}$ has the same effect on the $\mathbf{B}$ fields as a surface current $c(\mathbf{M}_2 - \mathbf{M}_1) \times \hat{n}$. Note that these equivalent polarization and magnetic surface charges and currents are to be understood as alternatives to the use of (3.24a) and (3.24b). If we choose to work with $\mathbf{D}$ and $\mathbf{H}$, the only surface charges and currents that we should use in (3.24) are the free charges and currents.

In vacuum, $\mathbf{P}(\mathbf{r}, t) = \mathbf{M}(\mathbf{r}, t) = 0$, so (3.22) implies that $\mathbf{D}(\mathbf{r}, t) = \mathbf{E}(\mathbf{r}, t)$ and $\mathbf{H}(\mathbf{r}, t) = \mathbf{B}(\mathbf{r}, t)$. In an isotropic medium, for fields that are not too strong, approximate linear relations may exist between the fields:

$$\mathbf{P} = \chi_e \mathbf{E} \qquad (\chi_e \equiv \text{dielectric susceptibility}) \tag{3.25a}$$

$$\mathbf{D} = \varepsilon \mathbf{E} \qquad (\varepsilon = 1 + 4\pi\chi_e \equiv \text{dielectric constant}) \tag{3.25b}$$

$$\mathbf{M} = \chi_m \mathbf{H} \qquad (\chi_m \equiv \text{magnetic susceptibility}) \tag{3.25c}$$

$$\mathbf{B} = \mu \mathbf{H} \qquad (\mu = 1 + 4\pi\chi_m \equiv \text{permeability}) \tag{3.25d}$$

In vacuum, $\chi_e = \chi_m = 0$, and $\varepsilon = \mu = 1$.

PROBLEM 3.4.1 N turns of wire are wrapped around an iron ring in which a small gap has been cut. The circumference of the ring is L and the width of the gap is $W(W \ll L)$. The magnetic permeability of the iron is μ. Find the $\mathbf{B}$ field in the gap when current i flows through the coil.

We assume that the cross section of the ring and the width of the gap are small enough so that the $\mathbf{E}$ and $\mathbf{B}$ fields within them are uniform and tangent to the curve of the ring. The integral form of (3.23b), applied to the dashed contour in Figure 3.5, gives

$$\oint \mathbf{H}(\mathbf{r}) \cdot d\mathbf{r} \approx H(\text{iron}) \times L + H(\text{air}) \times W$$

$$= \frac{4\pi}{c} \int \mathbf{J}_{free} \cdot d\mathbf{a} = \frac{4\pi N i}{c}$$

Here i is the current in the wire, so that Ni is the total current piercing the surface spanned by the dashed contour. Since $\mathbf{B}$ within the gap is approximately perpendicular to the iron faces of the gap, the continuity of $B_\perp$ implies

$$B(\text{iron}) \approx B(\text{air}) \approx H(\text{air})$$

$$= \mu H(\text{iron})$$

FIGURE 3.5 A small gap of width W cut in an iron ring of circumference L.

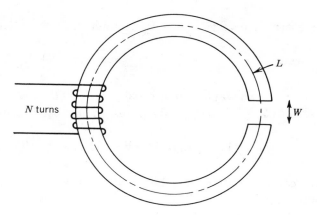

Thus, we have two equations for H(air) and H(iron), from which we obtain

$$H(\text{air}) \approx B(\text{air}) = \frac{\dfrac{4\pi}{c}Ni}{\left(\dfrac{L}{\mu} + W\right)}$$

PROBLEM 3.4.2   An uncharged dielectric sphere of radius R and dielectric constant ε is put into a uniform electric field $\mathbf{E} = E_0\hat{z}$. Find the resultant electric field within and outside the dielectric.

Since there is no free charge, $\operatorname{div}\mathbf{D} = 0$ everywhere. Outside the sphere, $\mathbf{D} = \mathbf{E}$. Inside the sphere $\mathbf{D} = \varepsilon\mathbf{E}$. In both cases, $\operatorname{div}\mathbf{D}$ is proportional to $\operatorname{div}\mathbf{E}$, and the vanishing of $\operatorname{div}\mathbf{D}$ implies the vanishing of $\operatorname{div}\mathbf{E}$,

$$\operatorname{div}\mathbf{E} = 0 \qquad (r \gtrless R)$$

and Laplace's equation is satisfied by the potential $\phi(\mathbf{r})$ both inside and outside the sphere.

Because of the spherical boundary of the dielectric, it is convenient to express the angular dependence of ϕ in terms of spherical harmonics[6]

$$\phi(\mathbf{r}) = \phi(r, \theta, \varphi) = \sum_{l, m} f_{l, m}(r) Y_m^l(\theta, \varphi) \tag{3.26}$$

In Section 5.3.5. it is shown that if (3.26) satisfies Laplace's equation, then the $f_{l, m}(r)$ have the form

$$f_{l, m}(r) = a_{l, m} r^l + b_{l, m} r^{-(l+1)} \tag{3.27}$$

Here $a_{l, m}$ and $b_{l, m}$ are numerical coefficients chosen in order to satisfy the boundary conditions on $\mathbf{E}$ and $\mathbf{D}$.

Since both the shape of the dielectric and the distant $\mathbf{E}$ field are axially symmetric around the $\hat{z}$ axis, the expansion in (3.26) will only require terms with $m = 0$. Equivalently, we can restrict (3.26) to an expansion in Legendre polynomials. We consider the regions $r < R$ and $r > R$ separately. Since we do not

---

[6]A more complete discussion of the properties of spherical harmonics is given in Sections 5.3.4 and 5.3.5.

want $\phi(\mathbf{r})$ to diverge at $r = 0$, we omit the $r^{-(l+1)}$ terms for $r < R$:

$$\phi(r, \theta) = \sum_{l=0}^{\infty} a_l r^l P_l(\cos \theta) \qquad (r < R) \qquad (3.28a)$$

As $r \to \infty$, we want $\mathbf{E}$ to approach $E_0 \hat{z}$, so that $\phi(\mathbf{r})$ should approach $-E_0 z = -E_0 r \cos \theta = -E_0 r P_1(\cos \theta)$. Thus, the external form of the potential must be

$$\phi(r, \theta) = -E_0 r P_1(\cos \theta) + \sum_{l=0}^{\infty} b_l r^{-(l+1)} P_l(\cos \theta) \qquad (r > R) \quad (3.28b)$$

The condition that $D_\perp$ be continuous at $r = R$ implies that $\varepsilon(\partial \phi / \partial r)$ be continuous there:

$$\varepsilon \sum_{l=0}^{\infty} l a_l R^{l-1} P_l(\cos \theta)$$

$$= -E_0 P_1(\cos \theta) - \sum_{l=0}^{\infty} (l + 1) b_l R^{-(l+2)} P_l(\cos \theta) \qquad (3.29)$$

This must be true for all θ. Since the $P_l(\cos \theta)$ are orthogonal for different l over the interval $0 \le \theta < \pi$, (3.29) implies equality of the terms for each separate l:

$$\varepsilon l a_l R^{l-1} = -(l + 1) b_l R^{-(l+2)} \qquad \text{for } l \neq 1$$

$$= -E_0 - 2b_1 R^{-3} \qquad \text{for } l = 1 \qquad (3.30a)$$

Since $E_\parallel = -(1/r) \partial \phi / \partial \theta$, it must be that $\partial \phi / \partial \theta$ is continuous at $r = R$,

$$\sum_{l=0}^{\infty} a_l R^l \frac{d}{d\theta} P_l(\cos \theta) = -E_0 R \frac{d}{d\theta} P_1(\cos \theta) + \sum_{l=0}^{\infty} b_l R^{-(l+1)} \frac{d}{d\theta} P_l(\cos \theta)$$

The derivatives $(d/d\theta) P_l(\cos \theta)$ are also orthogonal for different l, so we can supplement (3.30a) with

$$a_l R^l = b_l R^{-(l+1)}, \qquad \text{for } l \neq 1$$

$$= -E_0 R + b_1 R^{-2}, \qquad \text{for } l = 1 \qquad (3.30b)$$

Equations (3.30a) and (3.30b), imply that

$$a_l = b_l = 0, \qquad \text{if } l \neq 1$$

$$a_1 = -\frac{3}{\varepsilon + 2} E_0$$

$$b_1 = \frac{\varepsilon - 1}{\varepsilon + 2} R^3 E_0$$

Thus, the potential is

$$\phi(r, \theta) = -\frac{3}{\varepsilon + 2} E_0 r \cos \theta = -\frac{3}{\varepsilon + 2} E_0 z \qquad (r < R) \qquad (3.31a)$$

$$= -E_0 z + \frac{\varepsilon - 1}{\varepsilon + 2} E_0 \frac{R^3}{r^2} \cos \theta \qquad (r > R) \qquad (3.31b)$$

$$= -E_0 z + \frac{\varepsilon - 1}{\varepsilon + 2} E_0 \frac{R^3}{r^3} z$$

The **E** field obtained from the gradient of this potential is

$$\mathbf{E} = \frac{3}{\varepsilon + 2} E_0 \hat{z} \qquad (r < R) \tag{3.31c}$$

$$= E_0 \hat{z} + \frac{\varepsilon - 1}{\varepsilon + 2} E_0 R^3 \left(\frac{\hat{z}}{r^3} - \frac{3z}{r^4} \hat{r} \right) \qquad (r > R) \tag{3.31d}$$

Note that if $\varepsilon = 1$ (electrically equivalent to no dielectric sphere present), our solution (3.31) reduces to the constant field $\mathbf{E} = E_0 \hat{z}$ everywhere.

PROBLEM 3.4.3 A spherical "electret" has radius R and uniform polarization density **P** (esu-cm/cm^3). Calculate the potential everywhere.

Choose a coordinate system whose origin is at the center of the sphere and whose z axis points in the direction of **P**. The potential for $r < R$ is still given by (3.28a). However, in this problem there is no external field, so the potential at $r > R$ is given by (3.28b), minus the $-E_0 r P_1(\cos \theta)$ term. The continuity of $\mathbf{E}_{\parallel}$ gives

$$a_l = \frac{b_l}{R^{l+1}} \tag{3.32a}$$

For $r > R$, $\mathbf{D} = \mathbf{E}$, and

$$D_{\perp} = E_{\perp} = -\frac{\partial}{\partial r} \left[\sum_{l=0}^{\infty} b_l r^{-(l+1)} P_l(\cos \theta) \right]$$

For $r < R$, $\mathbf{D} = \mathbf{E} + 4\pi \mathbf{P} = \mathbf{E} + 4\pi P \hat{z}$,

$$D_{\perp} = E_{\perp} + 4\pi P \cos \theta = E_{\perp} + 4\pi P P_1(\cos \theta)$$

$$= -\frac{\partial}{\partial r} \left(\sum_l a_l r^l P_l(\cos \theta) \right) + 4\pi P P_1(\cos \theta)$$

The continuity of $D_{\perp}$ at $r = R$ gives

$$-l a_l R^{l-1} + 4\pi P \delta_{l,1} = (l+1) \frac{b_l}{R^{l+2}} \tag{3.32b}$$

The simultaneous equations (3.32a and b) imply that $a_l = b_l = 0$ unless $l = 1$, and $a_1 = (4\pi/3) P$, $b_1 = (4\pi/3) R^3 P$. Thus, we have

$$\phi(\mathbf{r}) = \frac{4\pi}{3} Pr \cos \theta = \frac{4\pi}{3} Pz \qquad (r < R) \tag{3.32c}$$

$$= \frac{4\pi}{3} P \frac{R^3}{r^2} \cos \theta = \frac{4\pi}{3} PR^3 \frac{\hat{z} \cdot \mathbf{r}}{r^3} \qquad (r > R) \tag{3.32d}$$

The electric field within the sphere is constant and equal to $-(4\pi/3) P \hat{z}$. Comparison with (3.18a) shows that the field outside the sphere is that of a dipole whose dipole moment is the total dipole moment of the sphere, $(4\pi/3) R^3 P \hat{z}$.

3.5 OHM'S LAW AND DC CIRCUITS

In a conductor satisfying Ohm's law, the current density is proportional to the electric field,

$$\mathbf{J}(\mathbf{r}, t) = \sigma \mathbf{E}(\mathbf{r}, t) \qquad (\sigma = \text{conductivity}) \tag{3.33a}$$

$$= \frac{1}{\rho} \mathbf{E}(\mathbf{r}, t) \qquad (\rho = \text{resistivity}) \tag{3.33b}$$

FIGURE 3.6 A segment of uniform wire of length *l*, cross-sectional area *A*, containing constant electric field **E** and current density **J** parallel to the axis of the wire.

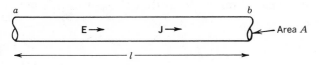

Let us apply (3.33b) to a segment of uniform wire of constant cross section, in which we assume there is a uniform electric field directed parallel to the axis of the wire (cf. Figure 3.6). Let V be the potential difference between the ends of the wire, so that

$$V = \phi_a - \phi_b = El$$

$$= \rho Jl = \frac{\rho l}{A} JA = \frac{\rho l}{A} i \qquad (3.34a)$$

Here $i\ (\equiv AJ)$ is the total current through the wire, and is measured in esu/s in the Gaussian system, and in C/s ($\equiv$ amp) in the conventional system. If we define the *resistance* R of the wire by $R \equiv \rho l/A$, then (3.34a) takes the more familiar form

$$V = iR \qquad (3.34b)$$

The conventional unit of resistance is the ohm ($= 1$ volt/1 amp). The stat-ohm equals

$$\frac{1 \text{ stat-volt}}{1 \text{ esu/s}} = \frac{299.8\,V}{1 \text{ esu/s}} \times \frac{4.803 \times 10^{-10} \text{ esu}}{1.602 \times 10^{-19}C}$$

$$= 8.988 \times 10^{11} \frac{V}{A} \approx 9 \times 10^{11} \text{ ohm}$$

The application of Ohm's law to dc circuit analysis is based on Kirchhoff's rules:

1. Charge is conserved, so that in a steady-state situation the sum of currents flowing into every point of the circuit is zero.

2. The potential is a single-valued function of position in a circuit, so the sum of potential differences around a circuit is zero.

A battery, or cell, in a circuit is symbolized by $\overset{\mathscr{E}}{\dashv\vdash}$. Chemical activity in the battery maintains a potential difference (or EMF) $\mathscr{E}$ between the two terminals. The terminal labeled by the longer line is at higher potential. In an ideal battery, this potential difference will be maintained independent of the magnitude of the current through the battery. In real life, every battery has some internal resistance that decreases the voltage difference across the battery terminals as the current increases.

PROBLEM 3.5.1   In the circuit shown in Figure 3.7, r represents the internal resistance of the battery, whose EMF is $\mathscr{E}$. By changing the variable resistance R, we change the potential difference V read by a voltmeter across the battery, and the current i read by an ammeter in series with the battery. When $V = 4.5$ volts, $i = 0.5$ amps. When $V = 3$ volts, $i = 1$ amp. Find $\mathscr{E}$ and r.

FIGURE 3.7 A circuit designed to measure the EMF $\mathscr{E}$ and internal resistance r of a battery.

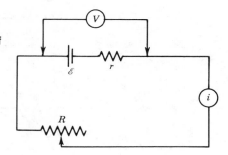

If current i flows through r, there is a potential drop of ir across it. The potential difference read by the voltmeter will be $V = \mathscr{E} - ir$. Thus, we have two equations

$$4.5 = \mathscr{E} - 0.5r$$
$$3 = \mathscr{E} - 1.0r$$

which can be solved simultaneously to yield $\mathscr{E} = 6$ volts, $r = 3$ ohms.

PROBLEM 3.5.2   A current i flows through a battery connected to diagonally opposite corners of a cubic network of twelve r-ohm resistors, as shown in Figure 3.8. What is the voltage V across the battery terminals?

    The symmetry of the network implies that a current of $i/3$ flows along each of the resistors $w–x$, and along each of the resistors $y–z$. The potential drop across each of the resistors is $ir/3$. Furthermore, the current flowing along each $w–x$ resistor divides equally between the two $x–y$ resistors connected to it. Thus, the $x–y$ resistors each carry a current of $i/6$ and support a potential drop of $ir/6$. The total potential drop between w and z is, thus,

$$V = \frac{i}{3}r + \frac{i}{6}r + \frac{i}{3}r = \frac{5}{6}ir$$

and this must be the voltage across the terminals of the battery.

    If the network of Figure 3.8 were put into a black box, with external connections only made to points w and z, measurements outside the box could not distinguish the network from a single resistor whose resistance is $5r/6$. Other examples of electrically equivalent networks are shown in Figure 3.9.

FIGURE 3.8 Each of the 12 resistors has a resistance of r ohms. Corners labeled by the same letter are at the same potential.

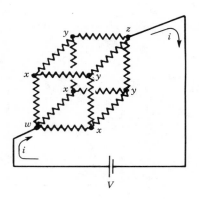

FIGURE 3.9 Some equivalent networks.

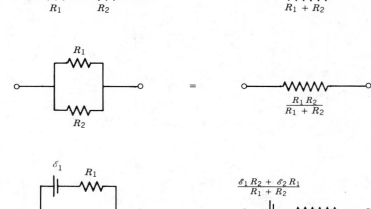

Replacing a component in a circuit by a simpler equivalent component is often a useful step in the analysis of a complicated dc network. This process can sometimes be repeated until the entire circuit is replaced by an equivalent resistance and battery in series. Then (3.34b) can be used to relate the total current through the circuit to the externally applied voltage. Once we know the total current, we can usually find the current through any portion of the circuit. In the following examples we demonstrate the solution of two network problems using Kirchhoff's rules and the method of equivalent circuits.

PROBLEM 3.5.3   Find the current through the resistor R_1 in the circuit shown below.

(a) Method of equivalent circuits:

The current through the battery and R_3 is

$$\frac{\mathscr{E}}{R_3 + \dfrac{R_1 R_2}{R_1 + R_2}}$$

The potential drop across R_1 is

$$\mathscr{E} - \frac{R_3 \mathscr{E}}{R_3 + \dfrac{R_1 R_2}{R_1 + R_2}} = \frac{\mathscr{E} R_1 R_2}{R_1 R_2 + R_2 R_3 + R_1 R_3}$$

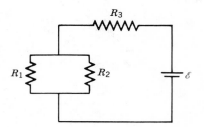

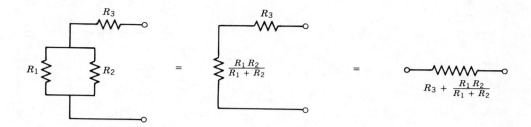

so the current through R_1 is

$$\frac{\mathscr{E}R_2}{R_1R_2 + R_2R_3 + R_1R_3}$$

(b) Method of Kirchhoff's rules:

We choose as our unknowns the currents i_1 and i_2 flowing downwards through resistors R_1 and R_2, respectively. Kirchhoff's first rule (charge conservation) then determines that the current through R_3 is $i_1 + i_2$ as shown. Now we apply Kirchhoff's second rule to two closed loops:

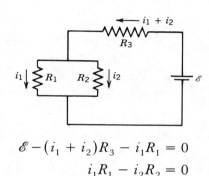

$$\mathscr{E} - (i_1 + i_2)R_3 - i_1R_1 = 0$$
$$i_1R_1 - i_2R_2 = 0$$

Solving these two equations simultaneously for i_1 and i_2 gives

$$i_1 = \frac{\begin{vmatrix} \mathscr{E} & R_3 \\ 0 & -R_2 \end{vmatrix}}{\begin{vmatrix} R_1 + R_3 & R_3 \\ R_1 & -R_2 \end{vmatrix}} = \frac{\mathscr{E}R_2}{R_1R_2 + R_2R_3 + R_1R_3}$$

as before. Note that when we go around a closed loop, the potential falls if we move through a resistor in the same direction as the assumed current, and rises if we move through the resistor in the opposite direction.

PROBLEM 3.5.4 Find the current through the resistor R_3 in the circuit shown on the following page.

(a) Method of equivalent circuits.

If we put $\mathscr{E}_2$ across this circuit, the net EMF across the equivalent resistor is $\mathscr{E}_2 - \mathscr{E}_1R_2/(R_1 + R_2)$. Thus, the current through R_3 is

$$\frac{\mathscr{E}_2 - \dfrac{\mathscr{E}_1R_2}{R_1 + R_2}}{R_3 + \dfrac{R_1R_2}{R_1 + R_2}} = \frac{\mathscr{E}_2(R_1 + R_2) - \mathscr{E}_1R_2}{R_1R_2 + R_2R_3 + R_3R_1}$$

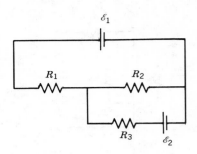

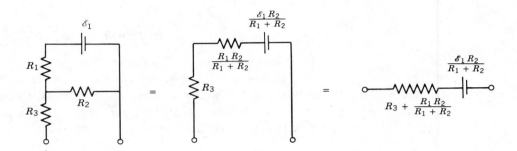

If this is positive the current flows away from the positive terminal of $\mathscr{E}_2$.

    (b) Method of Kirchhoff's rules.

    We take our unknown currents to be i_2 and i_3, as shown. Kirchhoff's second rule gives the two independent equations

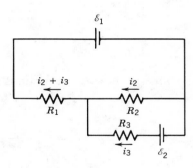

$$\mathscr{E}_2 - i_3 R_3 + i_2 R_2 = 0$$

$$\mathscr{E}_1 + (i_2 + i_3)R_1 + i_2 R_2 = 0$$

whose solution for i_3 is

$$i_3 = \frac{\begin{vmatrix} -R_2 & \mathscr{E}_2 \\ -(R_1 + R_2) & \mathscr{E}_1 \end{vmatrix}}{\begin{vmatrix} -R_2 & R_3 \\ -(R_1 + R_2) & -R_1 \end{vmatrix}} = \frac{\mathscr{E}_2(R_1 + R_2) - \mathscr{E}_1 R_2}{R_1 R_2 + R_2 R_3 + R_3 R_1}$$

We could also have applied Kirchhoff's second rule to a loop around the periphery of the circuit,

$$\mathscr{E}_2 - i_3 R_3 - (i_2 + i_3)R_1 - \mathscr{E}_1 = 0$$

but this equation is not linearly independent of the loop equations we have already used.

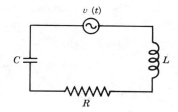

3.6 AC CIRCUITS WITH HARMONIC DRIVING VOLTAGE

Suppose that a resistance R, a capacitance C and an inductance L are connected in a series across an oscillating voltage $v(t) = v_0\cos(\omega t + \phi_0)$. After transient effects have died down, there will be a steady-state current flowing in the circuit, given by $i(t) = i_0\cos(\omega t + \theta_0)$. We wish to determine i_0 and θ_0.

The current $i(t)$, and capacitor charge $q(t)$, satisfy the differential equation

$$L\frac{di}{dt} + Ri + \frac{q}{C} = v(t) \tag{3.35}$$

with $i(t)$ and $q(t)$ related by

$$i(t) = \frac{d}{dt}q(t) \tag{3.36}$$

We can get an equation for $i(t)$ alone by differentiating (3.35),

$$L\frac{d^2i}{dt^2} + R\frac{di}{dt} + \frac{i}{C} = \frac{dv}{dt} = \frac{d}{dt}[v_0\cos(\omega t + \phi_0)] \tag{3.37}$$

The phase difference between $i(t)$ and $v(t)$ makes it more convenient to solve a differential equation with a complex driving voltage

$$L\frac{d^2\tilde{i}(t)}{dt^2} + R\frac{d\tilde{i}}{dt} + \frac{\tilde{i}}{C} = \frac{d}{dt}\tilde{v}(t) \tag{3.38a}$$

in which $\tilde{v}(t)$ is defined by

$$\tilde{v}(t) = v_0 e^{i(\omega t + \phi_0)} \tag{3.38b}$$

The constants L, R, C, v_0, ω, ϕ_0 in (3.38) are all real. Therefore, if $\tilde{i}(t)$ satisfies (3.38), its real part will satisfy (3.37). It is easy to verify by direct substitution that the steady-state solution of (3.38) can be written as

$$\tilde{i}(t) = \frac{1}{Z}\tilde{v}(t) \tag{3.39a}$$

with

$$Z \equiv R + iX \tag{3.39b}$$

$$X \equiv \omega L - \frac{1}{\omega C} \tag{3.39c}$$

The quantity Z is called the impedance of the L-R-C combination, and its imaginary part X is called the reactance. Equation (3.39a) is reminiscent of Ohm's law, with Z the ac generalization of the dc resistance R. In particular, the rules for determining the equivalent impedance of series and parallel combinations,

$$Z = Z_1 + Z_2 \quad \text{(1 and 2 in series)} \tag{3.40a}$$

$$Z = \frac{Z_1 Z_2}{Z_1 + Z_2} \quad \text{(1 and 2 in parallel)} \tag{3.40b}$$

have the same structure as the rules for equivalent dc resistances. The rules (3.40) are easily derived by applying (3.39a) to the circuit combinations.

The physical current $i(t)$ is now obtained as the real part of (3.39a),

$$i(t) = Re \, \tilde{i}(t) = Re\left[\frac{v_0 e^{i(\omega t + \phi_0)}}{R + iX}\right]$$

$$= Re\left[\frac{v_0}{\sqrt{R^2 + X^2}} e^{i(\omega t + \phi_0 - \arctan(X/R))}\right]$$

$$= \frac{v_0}{\sqrt{R^2 + X^2}} \cos(\omega t + \phi_0 - \arctan(X/R)) \qquad (3.41)$$

Thus, the amplitude and phase of the current are

$$i_0 = \frac{v_0}{\sqrt{R^2 + X^2}} \qquad (3.42a)$$

$$\omega t + \theta_0 = \omega t + \phi_0 - \arctan\frac{X}{R} \qquad (3.42b)$$

The current *lags* the voltage by a time interval $(1/\omega) \cdot \arctan(X/R)$. For example, at $t = 0$ the phase of the voltage is ϕ_0, but the current does not achieve this phase until a time t given by

$$\omega t + \phi_0 - \arctan\frac{X}{R} = \phi_0$$

$$t = \frac{1}{\omega}\arctan\frac{X}{R}$$

This current lag will be positive if $X > 0$, which will happen when $\omega L > 1/\omega C$ (see 3.39c). If $\omega L < 1/\omega C$, so that $X < 0$, the lag is negative, and the current is said to *lead* the voltage.

PROBLEM 3.6.1   Find the amplitude and phase of the current through the resistance R in the circuit shown below on the left.

As we did for the dc problems, we begin by replacing the given circuit by a simpler equivalent circuit:

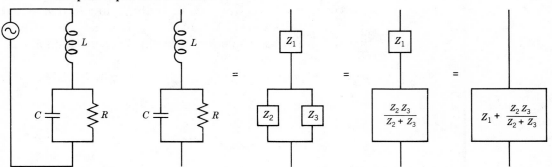

Thus, the total effective impedance seen by the voltage source is

$$Z_{\text{eff}} = Z_1 + \frac{Z_2 Z_3}{Z_2 + Z_3} = \frac{Z_1 Z_2 + Z_2 Z_3 + Z_3 Z_1}{Z_2 + Z_3}$$

The total current through the source is the real part of

$$\tilde{i}(t) = \frac{\tilde{v}(t)}{Z_{\text{eff}}} = \tilde{v}(t)\frac{Z_2 + Z_3}{Z_1 Z_2 + Z_2 Z_3 + Z_3 Z_1}$$

This current is distributed between the impedances Z_2 and Z_3 in inverse proportion to Z_2 and Z_3. Thus, the current through the resistor Z_3 is the real part of

$$\tilde{\imath}(t)\frac{Z_2}{Z_2 + Z_3} = \frac{\tilde{v}(t)Z_2}{Z_1Z_2 + Z_2Z_3 + Z_3Z_1}$$

$$= v_0 e^{i\omega t}\frac{\dfrac{1}{i\omega C}}{\dfrac{i\omega L}{i\omega C} + \dfrac{R}{i\omega C} + i\omega LR}$$

$$= \frac{v_0 e^{i\omega t}}{R(1 - \omega^2 LC) + i\omega L} = \frac{v_0 e^{i(\omega t - \arctan(\omega L/R(1 - \omega^2 LC)))}}{\sqrt{R^2(1 - \omega^2 LC)^2 + \omega^2 L^2}}$$

Hence, the current through R is

$$i(t) = \frac{v_0}{\sqrt{R^2(1 - \omega^2 LC)^2 + \omega^2 L^2}}\cos\left(\omega t - \arctan\frac{\omega L}{R(1 - \omega^2 LC)}\right)$$

An interesting feature of this circuit is that if $\omega^2 = 1/LC$ the current through R becomes

$$i(t) = \frac{v_0}{\omega L}\cos(\omega t - \pi/2) = \frac{v_0}{\omega L}\sin \omega t$$

$$= v_0\sqrt{\frac{C}{L}}\sin\left(\frac{t}{\sqrt{LC}}\right)$$

independent of the value of R.

3.7 POWER CONSUMPTION IN AC CIRCUITS

If the voltage drop across the series L-C-R combination is $v(t)$ while the current is $i(t)$, the rate at which energy is dissipated is

$$P(t) = v(t) \times i(t)$$

$$= \frac{v_0^2}{\sqrt{R^2 + X^2}}\cos(\omega t + \phi_0) \times \cos\left(\omega t + \phi_0 - \arctan\frac{X}{R}\right)$$

$$= \frac{v_0^2}{\sqrt{R^2 + X^2}}\left[\cos^2(\omega t + \phi_0)\cos\left(\arctan\frac{X}{R}\right)\right.$$

$$\left. + \cos(\omega t + \phi_0)\sin(\omega t + \phi_0)\sin\left(\arctan\frac{X}{R}\right)\right]$$

If we average $P(t)$ over an integral number of half-cycles, we get

$$\bar{P} = \frac{v_0^2}{2\sqrt{R^2 + X^2}}\cos\left(\arctan\frac{X}{R}\right) \tag{3.43a}$$

It is convenient to express this in terms of the root-mean-square values of $v(t)$

and $i(t)$:

$$v_{rms} = \sqrt{\overline{\left[v_0\cos(\omega t + \phi_0)\right]^2}} = v_0\sqrt{\overline{\left[\cos(\omega t + \phi_0)\right]^2}} = \frac{v_0}{\sqrt{2}}$$

$$i_{rms} = \sqrt{\overline{\left[i_0\cos\left(\omega t + \phi_0 - \arctan\frac{X}{R}\right)\right]^2}} = \frac{i_0}{\sqrt{2}} = \frac{v_0}{\sqrt{2(R^2 + X^2)}}$$

$$\overline{P} = v_{rms} \times i_{rms} \times \cos\left(\arctan\frac{X}{R}\right)$$

$$= v_{rms} \times i_{rms} \times \frac{R}{\sqrt{R^2 + X^2}} \tag{3.43b}$$

The factor $R/\sqrt{R^2 + X^2}$ in (3.43b) is called the *power factor*. If $X = 0$, so that the impedance is purely resistive, the power factor is equal to 1. However, if $R = 0$ and $X \neq 0$, the power factor vanishes. In this case no work is done on the average as the current passes through the circuit. Energy is shuttled back and forth between the voltage source, the electric field in the capacitor, and the magnetic field in the inductor, but on the average there is no net energy drain from the energy source.[7] But if the impedance of the circuit has a resistive component, some of the energy removed from the voltage source is converted into Joule heat, and this process is irreversible.

3.8 PERMANENT MAGNETS

In steady-state conditions we can set up the following correspondence between $J_{free} = 0$ magnetism and $\rho_{free} = 0$ electrostatics:

$$\operatorname{curl} \mathbf{H} = 0 \leftrightarrow \operatorname{curl} \mathbf{E} = 0, \quad \text{cf. (3.23b) and (3.1c)}$$

$$\operatorname{div} \mathbf{B} = 0 \leftrightarrow \operatorname{div} \mathbf{D} = 0, \quad \text{cf. (3.1a) and (3.23c)}$$

$$\mathbf{B} = \mathbf{H} + 4\pi\mathbf{M} \leftrightarrow \mathbf{D} = \mathbf{E} + 4\pi\mathbf{P}, \quad \text{cf. (3.22b) and (3.22a)}$$

$$\left.\begin{array}{l} \mathbf{B} = \mu\mathbf{H} \\ \mu = 1 + 4\pi\chi_m \end{array}\right\} \begin{array}{l}\text{isotropic}\\\text{materials}\end{array} \left\{\begin{array}{l} \mathbf{D} = \varepsilon\mathbf{E} \\ \varepsilon = 1 + 4\pi\chi_e \end{array}\right. \text{cf. (3.2d) and (3.25c)}$$

Thus, we can make the following correspondence between electric and magnetic quantities:

$$\mathbf{H} \leftrightarrow \mathbf{E} \tag{3.44a}$$

$$\mathbf{B} \leftrightarrow \mathbf{D} \tag{3.44b}$$

$$\mathbf{M} \leftrightarrow \mathbf{P} \tag{3.44c}$$

$$\mu \leftrightarrow \varepsilon \tag{3.44d}$$

$$\chi_m \leftrightarrow \chi_e \tag{3.44e}$$

and use the methods and intuition we developed for electrostatics to solve problems in magnetostatics. In particular, from $\operatorname{curl} \mathbf{H} = 0$ we can conclude that $\mathbf{H}$ can be obtained as the gradient of a single-valued scalar potential, $\mathbf{H} = -\nabla\Phi$, etc.

---

[7]This assumes that the frequency is low enough so that very little energy is radiated away. This requires that $\omega \ll c/l$, where c is the speed of light and l is a length of the order of magnitude of the linear dimensions of the circuit.

PROBLEM 3.8.1 A sphere of radius R, permeability μ, is put into a uniform magnetic field $\mathbf{B} = B_0 \hat{z}$. Find the resultant magnetic field, within and outside the sphere.

Using the correspondence (3.44), we see that this problem is analogous to the problem of a dielectric sphere placed in a uniform electric field. Thus, we can transcribe (3.31) to obtain

$$\mathbf{B} = \frac{3}{\mu + 2} B_0 \hat{z} \qquad (r < R)$$

$$= B_0 \hat{z} + \frac{\mu - 1}{\mu + 1} B_0 R^3 \left(\frac{\hat{z}}{r^3} - \frac{3z}{r^4} \hat{r} \right) \qquad (r > R)$$

PROBLEM 3.8.2 Give the relative directions and magnitudes of the **B** and **H** fields for:

(a) A long rectangular bar magnet with uniform permanent magnetization **M** parallel to its long direction.

According to (3.44c), this is analogous to a long bar with uniform permanent electric polarization **P**. We know that the electrostatic effect of the polarization is reproduced by a surface charge density $\mathbf{P} \cdot d\hat{a}$. This is zero everywhere except at the ends of the bar, where it has the values $\pm P$. Figure 3.10a shows a sketch of the **E** field associated with this effective charge distribution, and (3.44a) allows us to use this same sketch to describe the **H** field of the bar magnet. Note that the convention for labeling the poles of a bar magnet is such that the **H** field points away from the pole labeled "N", and towards the pole labeled "S". To obtain the **B** field, we recall that outside the magnet $\mathbf{M} = 0$, and so $\mathbf{B} = \mathbf{H}$. Moreover, div $\mathbf{B} = 0$ implies that the **B** field lines are everywhere continuous. These considerations lead us to the sketch of the **B** field given in Figure 3.10c.

FIGURE 3.10 (a) The **E** field associated with a uuniformly polarized dielectric bar. The $+$ and $-$ charges on the ends of the rod represent the effective polarization charges associated with the discontinuity in $P_\perp$. (b) The **H** field associated with a uniformly magnetized bar (cf. 3.44a, 3.44c). (c) The **B** field associated with a uniformly magnetized bar. Note that $\mathbf{B} = \mathbf{H}$ outside the bar (where $\mu = 1$), but the **B** field lines are everywhere continuous since div $\mathbf{B} = 0$ everywhere.

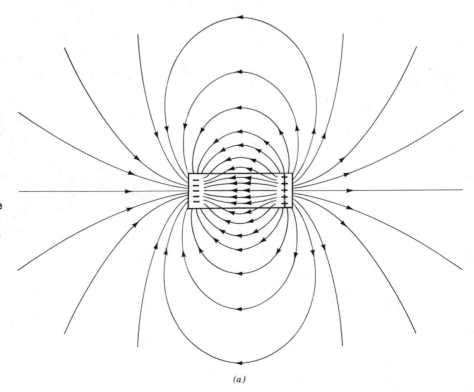

(a)

FIGURE 3.10 (*continued*)

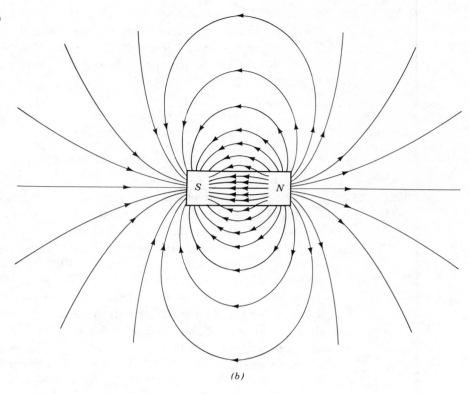

(b)

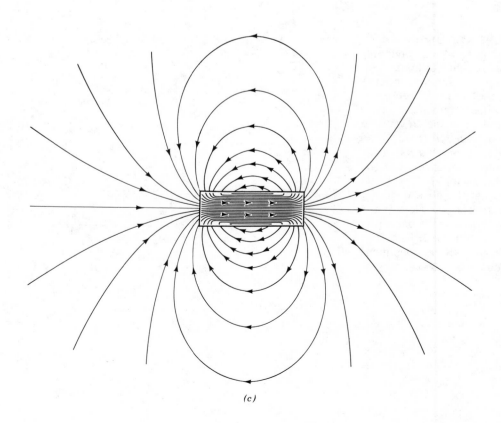

(c)

(b) A continuous iron ring, uniformly and tangentially magnetized.

Since $\mathbf{P} \cdot d\hat{a}$ is everywhere zero in the electrostatic analog, there is no effective polarization charge, and the $\mathbf{E}$ field is zero everywhere. Thus, the $\mathbf{H}$ field of the ring is everywhere zero. Outside the ring, $\mathbf{B} = \mathbf{H} = 0$, and within the iron, $\mathbf{B} = \mathbf{H} + 4\pi\mathbf{M} = 4\pi\mathbf{M}$.

3.9 MAGNETIC FIELDS PRODUCED BY FREE, TIME-INDEPENDENT CURRENTS

3.9.1 The Biot-Savart Law

The steady-state version of Maxwell's equation (3.1d) is

$$\text{curl } \mathbf{B}(\mathbf{r}) = \frac{4\pi}{c}\mathbf{J}(\mathbf{r}) \tag{3.45a}$$

If this equation is coupled with (3.1a) it can be shown that

$$\mathbf{B}(\mathbf{r}) = \frac{1}{c}\int \frac{\mathbf{J}(\mathbf{r}') \times (\mathbf{r} - \mathbf{r}')}{|\mathbf{r} - \mathbf{r}'|^3}\, dv' \tag{3.45b}$$

A simpler form of (3.45b) can be used when the current density is confined to a wire of small cross-sectional area. If we assume that $\mathbf{J}$ is tangent to the wire, and that i, the current, is the total flux of $\mathbf{J}$ through the wire, then (3.45b) can be approximated by

$$\mathbf{B}(\mathbf{r}) = \frac{i}{c}\oint \frac{d\mathbf{r}' \times (\mathbf{r} - \mathbf{r}')}{|\mathbf{r} - \mathbf{r}'|^3} \tag{3.45c}$$

The integration contour is along the closed path of the wire. Note that (3.45c) should only be used for points that are not on the wire. $\mathbf{B}(\mathbf{r})$ of (3.45c) is infinite if $\mathbf{r}$ is on the wire, a consequence of our use of an infinite current density to obtain (3.45c) from (3.45b). If we want the $\mathbf{B}$ field within a real current-carrying wire, we must use (3.45b) and take account of the finite dimensions of the wire and the actual distribution of current density across it.

PROBLEM 3.9.1   Find the $\mathbf{B}$ field produced by current i flowing through a circular wire loop of radius R, at points on a line perpendicular to the plane of the loop passing through its center.

Put the loop in the xy plane, centered at the origin (see diagram on the next page). We need $\mathbf{B}$ at the field point $\mathbf{r} = r\hat{z}$. Points on the loop are labeled by $\mathbf{r}' = R(\hat{x}\cos\theta + \hat{y}\sin\theta)$, so an infinitesimal element of the loop is given by $d\mathbf{r}' = R(-\hat{x}\sin\theta + \hat{y}\cos\theta)d\theta$. Then

$$\mathbf{r} - \mathbf{r}' = r\hat{z} - R(\hat{x}\cos\theta + \hat{y}\sin\theta)$$
$$|\mathbf{r} - \mathbf{r}'|^2 = r^2 + R^2$$

and (3.45c) can be written

$$\mathbf{B}(r\hat{z}) = \frac{i}{c}\int_{\theta=0}^{2\pi} d\theta\, \frac{R(-\hat{x}\sin\theta + \hat{y}\cos\theta) \times [r\hat{z} - R(\hat{x}\cos\theta + \hat{y}\sin\theta)]}{(r^2 + R^2)^{3/2}}$$

$$= \frac{iR}{c(r^2 + R^2)^{3/2}}\int_{\theta=0}^{2\pi} d\theta\,[\hat{x}r\cos\theta + \hat{y}r\sin\theta + R\hat{z}(\sin^2\theta + \cos^2\theta)]$$

$$= \frac{2\pi i R^2}{c(r^2 + R^2)^{3/2}}\hat{z}$$

FIGURE 3.11 A circular loop of wire of radius R in the $x-y$ plane, with its center at the origin. Current i flows counterclockwise.

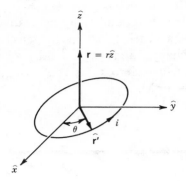

Now suppose that we want **B** at points in the plane of the loop. We still use (3.45c) but set

$$\mathbf{r} = r\hat{x}, \qquad \mathbf{r} - \mathbf{r}' = (r - R\cos\theta)\hat{x} - R\sin\theta\,\hat{y}$$

$$|\mathbf{r} - \mathbf{r}'|^2 = (r - R\cos\theta)^2 + (R\sin\theta)^2$$

$$= r^2 + R^2 - 2rR\cos\theta$$

$$\mathbf{B}(r\hat{x}) = \frac{iR}{c}\int_0^{2\pi} d\theta\, \frac{(-\hat{x}\sin\theta + \hat{y}\cos\theta)\times[(r - R\cos\theta)\hat{x} - R\sin\theta\,\hat{y}]}{(r^2 + R^2 - 2r\cos\theta)^{3/2}}$$

$$= \frac{iR}{c}\int_0^{2\pi} d\theta\, \frac{R\sin^2\theta - \cos\theta(r - R\cos\theta)}{(r^2 + R^2 - 2rR\cos\theta)^{3/2}}\,\hat{z}$$

$$= \frac{iR}{c}\hat{z}\int_0^{2\pi} d\theta\, \frac{R - r\cos\theta}{(r^2 + R^2 - 2rR\cos\theta)^{3/2}}$$

This integral can be expressed in terms of complete elliptic integrals.

3.9.2. Calculating B Fields Using Ampere's Law

Now we turn to the use of Ampere's law to calculate $\mathbf{B}(\mathbf{r})$ when $\mathbf{J}(\mathbf{r})$ has a high degree of symmetry. First of all, we note that Lemma 3 on page 111 guarantees that $\mathbf{J}(\mathbf{r})$ determines $\mathbf{B}(\mathbf{r})$ uniquely, in problems in which $\mathbf{B}(\mathbf{r})$ vanishes at infinity. Thus, if $\mathbf{J}(\mathbf{r})$ exhibits a geometrical symmetry, $\mathbf{B}(\mathbf{r})$ also exhibits that symmetry. For example, suppose that $\mathbf{J}(\mathbf{r})$ is unchanged by any rotation about the $\hat{z}$ axis, so that it has axial symmetry. Now consider two different Cartesian coordinate systems sharing a common origin and $\hat{z}$ axis, with an angle θ between their $\hat{x}$-axes. Since $\mathbf{J}$ is axially symmetric, observers working with these two different coordinate systems would use exactly the same expressions for $\mathbf{J}(\mathbf{r})$. Thus, their versions of (3.1a) and (3.45a) would be identical, and Lemma 3 then implies that they would calculate the same expressions for $\mathbf{B}(\mathbf{r})$. This proves that the **B** field is unchanged by a rotation through θ about the $\hat{z}$ axis, and so it also exhibits axial symmetry. A similar argument can be made if $\mathbf{J}(\mathbf{r})$ is invariant with respect to translations in some direction.

As an example of this method, consider the **B** field of a very long straight wire with a circular cross-section, across which **J** is distributed with axial symmetry. If we are not too close to the ends of the straight section, we make little error if we treat the wire as being infinitely long, with a current distribution which is invariant with respect to translations along its length. Suppose that the wire is centered on the $\hat{z}$ axis. Then the fact that $\mathbf{J}$ is invariant with respect to translations in the $\hat{z}$ direction tells us that

$$\frac{\partial \mathbf{B}}{\partial z} = 0, \qquad \text{(all components of } \mathbf{B}) \tag{3.47}$$

Furthermore, $\mathbf{J}$ points in the $\hat{z}$ direction, so that $J_x = J_y = 0$. Then Ampere's law (3.45a) says that

$$\frac{\partial B_z}{\partial y} - \frac{\partial B_y}{\partial z} = \frac{\partial B_z}{\partial y} = \frac{4\pi}{c} J_x = 0 \tag{3.48a}$$

$$\frac{\partial B_x}{\partial z} - \frac{\partial B_z}{\partial x} = -\frac{\partial B_z}{\partial x} = \frac{4\pi}{c} J_y = 0 \tag{3.48b}$$

Thus, all three Cartesian derivatives of B_z vanish. B_z is constant everywhere, and our assumption that B_z vanishes at infinity then implies that B_z vanishes everywhere.

If we apply (3.4a) to a circular cylinder whose axis is the $\hat{z}$ axis, and use the axial symmetry of $\mathbf{B}$, and our result that $B_z = 0$, we can prove that $\mathbf{B}$ can have no component pointing radially away from the $\hat{z}$ axis. Thus, $\mathbf{B}$ lies in a plane perpendicular to the $\hat{z}$ axis, and is tangent to a circle in this plane centered on the $\hat{z}$ axis. Its magnitude is constant around the circle. Finally we can apply Ampere's law to this circle to get

$$\oint \mathbf{B} \cdot d\mathbf{l} = 2\pi r B(r) = \frac{4\pi}{c} i(r)$$

$$B(r) = \frac{2}{c} \frac{i(r)}{r} \tag{3.49}$$

Here $i(r)$ is the current flowing along the wire within a circle of radius r. $\mathbf{B}$ points in the same direction as the fingers of a right hand grasping the wire with its thumb pointing in the direction of the current flow. Of course, (3.49) could also be obtained by using the Biot-Savart law.

Another example of the use of this kind of reasoning is provided by the $\mathbf{B}$ field produced by the current distribution shown in Figure 3.12 The current is confined to a sheet wrapped around an infinitely long cylinder (not necessarily of circular cross-section) parallel to the $\hat{z}$ axis. This current density is perpendicular to $\hat{z}$ and its magnitude is independent of z. It could be approximated by the current carried in a long, uniform, tightly wound solenoid.

From our hypothesis that $\partial \mathbf{J}/\partial z = 0$, we infer that $\partial \mathbf{B}/\partial z = 0$. Since div $\mathbf{B} = 0$ and $\partial B_z/\partial z = 0$, we have

$$\frac{\partial B_x}{\partial x} + \frac{\partial B_y}{\partial y} = 0. \tag{3.50a}$$

The vanishing of J_z means that

$$\frac{4\pi}{c} J_z = 0 = \frac{\partial B_y}{\partial x} - \frac{\partial B_x}{\partial y} \tag{3.50b}$$

FIGURE 3.12 A current sheet around an infinitely long cylinder whose axis is the *z* axis.

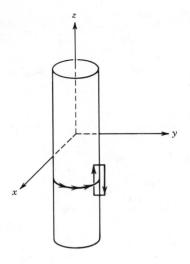

Equations (3.50a) and (3.50b) can be regarded as telling us that the two-dimensional (x and y) divergence and curl of $\hat{x}B_x + \hat{y}B_y$ both vanish everywhere. A two-dimensional version of the Lemma on page 110 then implies that $B_x = B_y = 0$ everywhere. It only remains to find the z-independent value of B_z. Ampere's law for J_x and J_y gives

$$\frac{4\pi}{c}J_x = \frac{\partial B_z}{\partial y} - \frac{\partial B_y}{\partial z} = \frac{\partial B_z}{\partial y} \tag{3.50c}$$

$$\frac{4\pi}{c}J_y = \frac{\partial B_x}{\partial z} - \frac{\partial B_z}{\partial x} = -\frac{\partial B_z}{\partial x} \tag{3.50d}$$

But J_x and J_y vanish, except on the surface of the current sheet. This tells us that B_z is constant everywhere outside the sheet and everywhere inside the sheet. Since B_z vanishes at infinitely large distances from the $\hat{z}$ axis, its constant value outside the current sheet is zero. Its constant value inside the sheet can be gotten from (3.5d), or by applying the integral form of Ampere's law to the rectangular contour shown in Figure 3.12:

$$\oint \mathbf{B} \cdot d\mathbf{l} = \left[B_z(\text{inside}) - B_z(\text{outside}) \right] l = \frac{4\pi}{c}Kl$$

$$B_z(\text{inside}) = \frac{4\pi}{c}K \tag{3.51}$$

with K the current per unit length in the sheet. Thus, we have found that the **B** field of an infinitely long, uniform solenoid is totally confined within the solenoid, is parallel to the axis of the solenoid, and has the constant vlaue (3.51) within the solenoid.

3.10 THE LORENTZ FORCE

The force per unit volume on charged matter, due to externally generated **E** and **B** fields, is

$$\mathbf{F}(\mathbf{r}, t) = \rho(\mathbf{r}, t)\mathbf{E}(\mathbf{r}, t) + \frac{1}{c}\mathbf{J}(\mathbf{r}, t) \times \mathbf{B}(\mathbf{r}, t). \tag{3.52a}$$

Now suppose that the charge distribution is confined to a region of space, centered on **r**, so small that **E** and **B** are nearly constant throughout this region. Let the total charge be q, and its velocity be **v**. If we use (3.2) and integrate (3.52a) over the charge distribution, we get

$$\mathbf{F}_{\text{on } q} = q\left[\mathbf{E}(\mathbf{r}, t) + \frac{\mathbf{v}}{c} \times \mathbf{B}(\mathbf{r}, t)\right] \tag{3.52b}$$

Suppose the patricle speed is small compared to c. Then we can use (3.52b) in Newton's second law of motion:

$$\frac{d}{dt}(m\mathbf{v}) = q\left[\mathbf{E}(\mathbf{r}, t) + \frac{\mathbf{v}}{c} \times \mathbf{B}(\mathbf{r}, t)\right]. \tag{3.53a}$$

In this equation, it is understood that **r** locates the particle at time t. The rate of change of the particle kinetic energy is obtained by taking the scalar product of (3.53a) with **v** [cf. (1.17a)]:

$$\frac{d}{dt}\left(\frac{1}{2}mv^2\right) = q\mathbf{v} \cdot \mathbf{E}(\mathbf{r}, t) \tag{3.53b}$$

If v is comparable to c, we must use the relativistic generalization of Newton's second law. This replaces (3.53) by

$$\frac{d}{dt}(\gamma m\mathbf{v}) = q\left[\mathbf{E}(\mathbf{r}, t) + \frac{\mathbf{v}}{c} \times \mathbf{B}(\mathbf{r}, t)\right] \tag{3.54a}$$

$$\frac{d}{dt}(\gamma mc^2) = \frac{d}{dt}\left[(\gamma - 1)mc^2\right] = q\mathbf{v} \cdot \mathbf{E}(\mathbf{r}, t) \tag{3.54b}$$

In comparing (3.53) and (3.54), it should be recalled that $\gamma m\mathbf{v}$ and $(\gamma - 1)mc^2$ are the relativistic expressions for particle momentum and kinetic energy [see (2.15b) and (2.16)].

Let us consider the important special case in which $\mathbf{E} = 0$ everywhere. Then (3.45b) implies that γ is constant in time, which means that the particle moves at constant speed. Constant γ means that (3.54a) can be written in the form

$$\frac{d\mathbf{v}}{dt} = \frac{q}{\gamma mc}\mathbf{v} \times \mathbf{B} = \boldsymbol{\omega}_B \times \mathbf{v} \tag{3.55a}$$

$$\boldsymbol{\omega}_B = -\frac{q\mathbf{B}}{\gamma mc} \tag{3.55b}$$

If, in addition, **B** is independent of position and time, (3.55) describes a motion in which **v** rotates around $-\mathbf{B}$, with angular speed ω_B. Superposed on this circular motion about $\hat{B}$ is a uniform drift in the $\hat{B}$ direction (since $\hat{B} \cdot d\mathbf{v}/dt = 0$). The orbit of the particle is thus a helix of constant pitch and radius. If this radius is R, the momentum component perpendicular to $\hat{B}$ is

$$p_\perp = \gamma mv_\perp = \gamma m\omega_B R$$

$$= \frac{qBR}{c} \tag{3.55c}$$

Another important application of (3.52a) is to a current carried by a wire of small cross-sectional area. The magnetic force is then

$$\mathbf{F}_{\text{mag}} = \frac{i}{c}\oint d\mathbf{l} \times \mathbf{B} \tag{3.56}$$

Here i is the magnitude of the current, and the line integral goes around the closed wire circuit.

PROBLEM 3.10.1   A square loop with sides of length b carrying current I_1 is attracted to a long straight wire carrying current I_0. The long wire is in the plane of the square and is parallel to two of its sides. The center of the square loop is a distance $D(D > b/2)$ from the wire. Calculate the force of attraction and sketch the relative directions of the currents in the loop and the long wire.

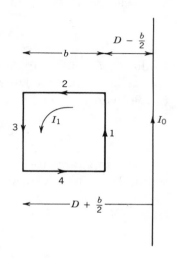

According to (3.49), the **B** field produced by I_0 has magnitude $2I_0/cr$ at a distance r from the wire. **B** is perpendicular to the plane of the paper, pointing out. The force it exerts on segment 1 is $I_1/c \cdot b \cdot 2I_0/c(D - b/2)$ and is directed toward the long wire. The force that the **B** field of I_0 exerts on segment 3 is given by $I_1/c \cdot b \cdot 2I_0/c(D + b/2)$ and is directed away from the long wire. The forces exerted by the **B** field of I_0 on segments 2 and 4 cancel. Thus, the net attractive force between the long wire and the square loop is

$$\frac{2I_1I_0b}{c^2}\left[\frac{1}{D - \dfrac{b}{2}} - \frac{1}{D + \dfrac{b}{2}}\right] = \frac{2I_1I_0b^2}{c^2\left(D^2 - \dfrac{b^2}{4}\right)}.$$

The magnetic dipole moment of a current distribution is defined by

$$\mathbf{m} \equiv \frac{1}{2c}\int \mathbf{r}' \times \mathbf{J}(\mathbf{r}')\, dv' \qquad (3.57a)$$

[cf. (3.18d)]. If current i is confined to a wire loop, this becomes

$$\mathbf{m} = \frac{i}{2c}\oint \mathbf{r}' \times d\mathbf{r}' \qquad (3.57b)$$

If the loop lies in a plane, this can be written as

$$\mathbf{m} = \frac{i\mathbf{a}}{c}$$

Here $|\mathbf{a}|$ is the area of the loop and $\hat{a}$ is perpendicular to the plane of the loop, in a direction related by the right-hand rule to the direction in which current circulates around the loop. The energy of interaction between a magnetic dipole and an external **B** field can be expressed as

$$U = -\mathbf{m} \cdot \mathbf{B}(\mathbf{r}) \tag{3.57c}$$

(cf. 3.19a). The forces and torques that a **B** field exerts on a magnetic dipole are completely analogous to those that an external **E** field exerts on an electric dipole (see p. 112).

PROBLEM 3.10.2 Suppose that the charge density $\rho_e(\mathbf{r}')$ and mass density $\rho_m(\mathbf{r}')$ of an object are such that their ratio is constant:

$$\frac{\rho_e(\mathbf{r}')}{\rho_m(\mathbf{r}')} = \text{constant} = \frac{\int \rho_e(\mathbf{r}') \, dv'}{\int \rho_m(\mathbf{r}') \, dv'} = \frac{e}{M}.$$

Show that the magnetic moment **m** and angular momentum **l** are related by $\mathbf{m} = e\mathbf{l}/2Mc$.

Let $\mathbf{v}(r')$ be the velocity of the material at the point labeled by $\mathbf{r}'$. Then $\mathbf{J}(\mathbf{r}') = \rho_e(\mathbf{r}')\mathbf{v}(\mathbf{r}')$, and (3.57a) yields

$$\mathbf{m} = \frac{1}{2c} \int \mathbf{r}' \times \rho_e(\mathbf{r}')\mathbf{v}(\mathbf{r}') \, dv'$$

$$= \frac{e}{2Mc} \int \mathbf{r}' \times \mathbf{v}(\mathbf{r}')\rho_m(\mathbf{r}') \, dv'$$

$$= \frac{e}{2Mc} \int \mathbf{r}' \times \mathbf{v}(r') \, dM = \frac{e}{2Mc}\mathbf{l} \tag{3.57d}$$

The factor $e/2Mc$ is called the gyromagnetic ratio of the charge distribution.

3.11 POTENTIALS IN TIME-DEPENDENT SITUATIONS

It is often convenient to divide the task of solving Maxwell's equations into two steps:

1). Find the most general **E** and **B** fields that satisfy the homogeneous equations (3.1a) and (3.1c).

2. Pick out the particular **E** and **B** fields that satisfy the inhomogeneous equations (3.1b) and (3.1d), and the boundary conditions.

The first task is accomplished by expressing the **E** and **B** fields in terms of scalar and vector potentials.

Any vector field **B** whose divergence is zero can be written as the curl of another field **A**. For example, if $\mathbf{B}(\mathbf{r}, t)$ satisfies (3.1a) and (x_0, y_0, z_0) locates any fixed point, then the vector field $\mathbf{A}(\mathbf{r}, t)$ defined by

$$A_x \equiv \frac{1}{3}\left\{ \int_{z_0}^{z}\left[B_y(x, y, z', t) + \frac{1}{2}B_y(x, y_0, z', t)\right]dz' \right.$$
$$\left. - \int_{y_0}^{y}\left[B_z(x, y', z, t) + \frac{1}{2}B_z(x, y', z_0, t)\right]dy' \right\}$$

$$A_y \equiv \frac{1}{3}\left\{ \int_{x_0}^{x}\left[B_z(x', y, z, t) + \frac{1}{2}B_z(x', y, z_0, t)\right]dx' \right.$$
$$\left. - \int_{z_0}^{z}\left[B_x(x, y, z', t) + \frac{1}{2}B_x(x_0, y, z', t)\right]dz' \right\}$$

$$A_z \equiv \frac{1}{3}\left\{ \int_{y_0}^{y}\left[B_x(x, y', z, t) + \frac{1}{2}B_x(x_0, y', z, t)\right]dy' \right.$$
$$\left. - \int_{x_0}^{x}\left[B_y(x', y, z, t) + \frac{1}{2}B_y(x', y_0, z, t)\right]dx' \right\}$$

automatically satisfies

$$\mathbf{B}(\mathbf{r}, t) = \text{curl}\,\mathbf{A}(\mathbf{r}, t) \tag{3.58a}$$

This is easily verified by calculating the curl of $\mathbf{A}$, and making use of (3.1a). If (3.58a) is substituted into (3.1c), we get

$$\text{curl}\left[\mathbf{E}(\mathbf{r}, t) + \frac{1}{c}\frac{\partial\mathbf{A}(\mathbf{r}, t)}{\partial t}\right] = 0$$

This implies that the quantity in square brackets is the gradient of a scalar potential. Thus, we write

$$\mathbf{E}(\mathbf{r}, t) = -\nabla\phi(\mathbf{r}, t) - \frac{1}{c}\frac{\partial\mathbf{A}(\mathbf{r}, t)}{\partial t} \tag{3.58b}$$

The $\mathbf{E}$ and $\mathbf{B}$ fields obtained from ϕ and $\mathbf{A}$ by (3.58) automatically satisfy the homogeneous Maxwell equations. If they are substituted into the inhomogeneous Maxwell equations, the result is

$$\text{div}\,\mathbf{E} = \text{div}\left[-\nabla\phi - \frac{1}{c}\frac{\partial\mathbf{A}}{\partial t}\right]$$

$$= -\nabla^2\phi - \frac{1}{c}\frac{\partial}{\partial t}\,\text{div}\,\mathbf{A} = 4\pi\rho \tag{3.59a}$$

$$\text{curl}\,\mathbf{B} - \frac{1}{c}\frac{\partial\mathbf{E}}{\partial t} = \text{curl}\,(\text{curl}\,\mathbf{A}) - \frac{1}{c}\frac{\partial}{\partial t}\left[-\nabla\phi - \frac{1}{c}\frac{\partial\mathbf{A}}{\partial t}\right]$$

$$= -\nabla^2\mathbf{A} + \frac{1}{c^2}\frac{\partial^2\mathbf{A}}{\partial t^2} + \nabla\left[\text{div}\,\mathbf{A} + \frac{1}{c}\frac{\partial\phi}{\partial t}\right] = \frac{4\pi}{c}\mathbf{J} \tag{3.59b}$$

But (3.58a) and (3.58b) do not determine $\mathbf{A}$ and ϕ uniquely. If $\mathbf{A}$ and ϕ are replaced in (3.58) by

$$\mathbf{A}'(\mathbf{r}, t) \equiv \mathbf{A}(\mathbf{r}, t) + \nabla\chi(\mathbf{r}, t) \tag{3.60a}$$

$$\phi'(\mathbf{r}, t) \equiv \phi(\mathbf{r}, t) - \frac{1}{c}\frac{\partial\chi(\mathbf{r}, t)}{\partial t}, \tag{3.60b}$$

then the $\mathbf{E}$ and $\mathbf{B}$ fields calculated there will be the same. The replacements (3.60) are said to define a *gauge transformation*, with $\chi(\mathbf{r}, t)$ the *gauge function*. We can always choose this function in such a way that the transformed potentials satisfy

$$\operatorname{div}\mathbf{A} + \frac{1}{c}\frac{\partial\phi}{\partial t} = 0 \tag{3.61}$$

This is called the *Lorentz condition*. If $\mathbf{A}$ and ϕ satisfy this condition, (3.59a and b) take on the simple forms

$$\nabla^2\phi - \frac{1}{c^2}\frac{\partial^2\phi}{\partial t^2} = -4\pi\rho \tag{3.62a}$$

$$\nabla^2\mathbf{A} - \frac{1}{c^2}\frac{\partial^2\mathbf{A}}{\partial t^2} = -\frac{4\pi}{c}\mathbf{J} \tag{3.62b}$$

A convenient set of particular solutions are the *retarded potentials*

$$\phi(\mathbf{r}, t) = \int \frac{\rho\left(\mathbf{r}', t - \dfrac{|\mathbf{r} - \mathbf{r}'|}{c}\right)}{|\mathbf{r} - \mathbf{r}'|}\, dv' \tag{3.63a}$$

$$\mathbf{A}(\mathbf{r}, t) = \frac{1}{c}\int \frac{\mathbf{J}\left(\mathbf{r}', t - \dfrac{|\mathbf{r} - \mathbf{r}'|}{c}\right)}{|\mathbf{r} - \mathbf{r}'|}\, dv' \tag{3.63b}$$

Note that an event $(\mathbf{r}', t')$ in the source affects the fields at $(\mathbf{r}, t)$ only if $t' = t - |\mathbf{r} - \mathbf{r}'|/c$. The retardation $|\mathbf{r} - \mathbf{r}'|/c$ is the time it takes for an electromagnetic signal to propagate from the source point $\mathbf{r}'$ to the field point $\mathbf{r}$.

If the expression (3.63b) for $\mathbf{A}$ is used in (3.58a), the result is the Biot-Savart law (3.45b).

PROBLEM 3.11.1 Find the magnetic field associated with the vector potential defined by

$$\mathbf{A}(\mathbf{r}, t) = \frac{b}{2}\hat{n} \times \mathbf{r} \qquad (b, \hat{n} \text{ constant})$$

The most straightforward procedure is to write out the components of $\mathbf{A}$,

$$A_x = \frac{b}{2}(n_y z - n_z y)$$

$$A_y = \frac{b}{2}(n_z x - n_x z)$$

$$A_z = \frac{b}{2}(n_x y - n_y x)$$

and then calculate the curl,

$$B_x = \frac{\partial A_z}{\partial y} - \frac{\partial A_y}{\partial z} = \frac{b}{2} n_x + \frac{b}{2} n_x = b n_x, \text{ etc.}$$

The conclusion is that $\mathbf{B} = b\hat{n}$, a uniform field in the $\hat{n}$ direction, of strength b.

PROBLEM 3.11.2 Find the charge and current distributions that would lead to

$$\mathbf{A}(\mathbf{r}, t) = \frac{bt}{r^3} \mathbf{r},$$

$$\phi(\mathbf{r}, t) = 0$$

First we calculate the $\mathbf{E}$ and $\mathbf{B}$ fields:

$$\mathbf{E} = -\nabla\phi - \frac{1}{c}\frac{\partial \mathbf{A}}{\partial t} = -\frac{b}{c}\frac{\mathbf{r}}{r^3}$$

This is the $\mathbf{E}$ field of a point charge $-b/c$ at the origin. The $\mathbf{B}$ field is

$$\mathbf{B} = \text{curl}\,\mathbf{A} = -bt\,\text{curl}\,\nabla\left(\frac{1}{r}\right) = 0$$

Maxwell's equation (3.1d) then implies that $\mathbf{J} = 0$.

In quantum mechanics, the study of the motion of a charged particle in electromagnetic fields requires the construction of the classical Hamiltonian of the system. We now show that

$$H(x, y, z, p_x, p_y, p_z, t) \equiv \frac{1}{2m}\left(\mathbf{p} - \frac{q}{c}\mathbf{A}(x, y, z, t)\right)^2 + q\phi(x, y, z, t) \tag{3.64}$$

can serve as the Hamiltonian of a nonrelativistic particle of charge q and mass m, moving in externally generated fields specified by $\mathbf{A}$ and ϕ. The generalized coordinates are the Cartesian coordinates $r_\alpha(\equiv x, y, z)$ of the particle.

If we use (3.64) as the Hamiltonian, the first set (1.49a) of Hamilton's equations are

$$\frac{\partial H}{\partial p_\alpha} = \frac{1}{m}\left(p_\alpha - \frac{q}{c}A_\alpha\right) = \dot{r}_\alpha = v_\alpha \tag{3.65a}$$

$$p_\alpha = mv_\alpha + \frac{q}{c}A_\alpha \tag{3.65b}$$

The second set (1.48b) are

$$\frac{\partial H}{\partial r_\alpha} = \sum_\beta \frac{\left(p_\beta - \frac{q}{c}A_\beta\right)}{m}\frac{\partial}{\partial r_\alpha}\left(-\frac{q}{c}A_\beta\right) + q\frac{\partial\phi}{\partial r_\alpha}$$

$$= -\frac{d}{dt}p_\alpha = -\frac{d}{dt}(mv_\alpha) - \frac{q}{c}\left(\sum_\beta v_\beta\frac{\partial A_\alpha}{\partial r_\beta} + \frac{\partial A_\alpha}{\partial t}\right) \tag{3.66a}$$

Note that when we calculate the total time derivative of $A_\alpha(\mathbf{r}, t)$, we must take account of the fact that $\mathbf{r}$ locates the particle, so that the motion of the particle contributes to the time dependence of A. Rearranging the terms in (3.66a) leads to

$$\frac{d}{dt}(mv_\alpha) = q\left[-\frac{\partial\phi}{\partial r_\alpha} - \frac{1}{c}\frac{\partial A_\alpha}{\partial t}\right] + \frac{q}{c}\sum_\beta v_\beta\left(\frac{\partial A_\beta}{\partial r_\alpha} - \frac{\partial A_\alpha}{\partial r_\beta}\right) \qquad (3.66b)$$

If $\alpha = x$, the β-sum becomes

$$v_x\left(\frac{\partial A_x}{\partial r_x} - \frac{\partial A_x}{\partial r_x}\right) + v_y\left(\frac{\partial A_y}{\partial r_x} - \frac{\partial A_x}{\partial r_y}\right) + v_z\left(\frac{\partial A_z}{\partial r_x} - \frac{\partial A_x}{\partial r_z}\right)$$

$$= (\mathbf{v} \times \operatorname{curl}\mathbf{A})_x \qquad (3.66c)$$

and corresponding equations can be written for $\alpha = y, z$. Finally, if (3.58a and b) are used to bring in the **E** and **B** fields, we verify that (3.66) is equivalent to the nonrelativistic equation of motion (3.53a).

3.12 LENZ'S LAW

The integral form (3.4c) of Faraday's law implies a connection between the sign of the rate of change of magnetic flux and the sign of the induced EMF. Lenz's law provides a convenient way to remember this connection:

> If current flowed in the direction of the EMF induced by a changing magnetic flux, then the **B** field produced by this current would *oppose* the change in magnetic flux.

PROBLEM 3.12.1 Determine the direction of the current induced in resistor ab when:
(a) The switch S is opened.

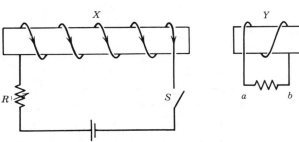

When S is in the closed position, the direction of the current flow in X is as indicated by the arrows. This leads to a **B** field that points to the right through Y. When S opens, this magnetic field collapses. A current flow from a to b would produce a **B** field in Y pointing to the right, which thus opposes the collapse in the B flux through Y.
(b) With S in the closed position, coil Y is brought nearer to coil X.

This move increases the flux to the right through Y. Induced current flowing from b to a would produce a **B** field in Y pointing to the left, which would oppose this increase in B flux through Y.
(c) With S in the closed position, and X and Y fixed, R is decreased.
The effect is the same as in (b) above.

PROBLEM 3.12.2 What are the directions of the force and torque on coil Y when the current i in the solenoid X is increased? (see next page)
The **B** field in Y due to X points to the left. When i is increased, this **B** field increases. To oppose the change in flux, current will be induced in Y in the

direction indicated. This will produce a magnetic moment **m**. The torque exerted on Y by the **B** field of X will tend to align **m** with **B**; thus, it will have the direction indicated below by τ. Since **m** points opposite to the direction of **B**, the force on Y will push it into the region of weaker **B** field, that is, to the right.

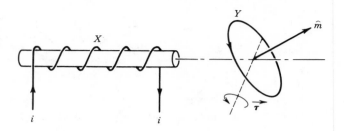

3.13 ENERGY AND MOMENTUM DENSITIES AND FLUX

Equations (3.1c) and (3.1d) can be manipulated to yield

$$-\int_S \frac{c}{4\pi} \mathbf{E} \times \mathbf{B} \cdot d\mathbf{a} = \frac{\partial}{\partial t} \int_V \frac{1}{8\pi}(\mathbf{E} \cdot \mathbf{E} + \mathbf{B} \cdot \mathbf{B})\, dv + \int_V \mathbf{E} \cdot \mathbf{J}\, dv \quad (3.67)$$

The last integral is the rate at which the electromagnetic fields do work on the charge distribution. Thus, if we make the associations

$$\mathbf{S} \equiv \frac{c}{4\pi} \mathbf{E} \times \mathbf{B} \qquad \text{(energy flux density)} \qquad (3.68a)$$

$$u \equiv \frac{1}{8\pi}(\mathbf{E} \cdot \mathbf{E} + \mathbf{B} \cdot \mathbf{B}) \qquad \text{(energy density)} \qquad (3.68b)$$

then (3.67) can be interpreted as saying that the rate at which electromagnetic energy flows into a closed surface S equals the rate at which the electromagnetic energy stored within S increases, plus the rate at which energy is transferred from the electromagnetic field to the matter within S. Similar arguments show that

$$\mathbf{P} \equiv \frac{1}{4\pi c} \mathbf{E} \times \mathbf{B} \qquad (3.69a)$$

is the density of momentum in the electromagnetic field, and

$$T_{\alpha\beta} \equiv \frac{1}{4\pi}\left[\frac{1}{2}\delta_{\alpha\beta}(E^2 + B^2) - E_\alpha E_\beta - B_\alpha B_\beta\right] \qquad (3.69b)$$

is the β-component of the flux of the α-component of this momentum. This means that $\sum_\beta T_{\alpha\beta} ds_\beta$ is the rate at which the α-component of momentum flows across the area element **ds**, in the direction of **ds**. Equivalently, it is the α-component of the force exerted across this area element by the electromagnetic field. Equation (3.69b) defines the Maxwell stress tensor. It provides a convenient and reliable way of calculating forces in electromagnetic field problems.

PROBLEM 3.13.1 The plates of a large parallel plate capacitor are separated by a distance d. The surface charge densities are $\pm\sigma$. What is the force per unit area on each plate?

According to Gauss' law, the **E** field near the plate shown is given by $\mathbf{E} = 4\pi\sigma\hat{x}$. Since $\mathbf{B} = 0$, the stress tensor is given by

$$[T] = \frac{1}{4\pi} \begin{bmatrix} -\frac{1}{2}E^2 & 0 & 0 \\ 0 & \frac{1}{2}E^2 & 0 \\ 0 & 0 & \frac{1}{2}E^2 \end{bmatrix} \tag{3.70a}$$

$$= 2\pi\sigma^2 \begin{bmatrix} -1 & 0 & 0 \\ 0 & 1 & 0 \\ 0 & 0 & 1 \end{bmatrix} \tag{3.70b}$$

The flux of the x component of momentum to the right is thus

$$-2\pi\sigma^2 \frac{\text{momentum}}{(\text{unit time})(\text{unit area})}$$

This means that the x component of momentum flows across the dashed line in Figure 3.13, towards the metal plate, at the rate $2\pi\sigma^2$ per unit area, which means that the force per unit area in the x direction exerted on the plate is $2\pi\sigma^2$. Note that if we tried to calculate this force by multiplying **E** and σ we would get double the correct answer. This is because the **E** field at the plate is due, in part, to the charge on the plate, and a charge does not exert a force on itself. Thus, we must be careful, when we use the Lorentz force, to avoid this kind of self-interaction. However, the stress tensor using the actual fields gives us the true momentum flux and the true forces.

The lower diagonal elements in (3.67b) imply that there is momentum flux across the upper and lower parts of the dashed contour in Figure 3.13, and in the $\pm z$ directions. However, these fluxes cancel each other, and lead to no net forces on the plate.

PROBLEM 3.13.2 What is the pressure on the faces of the gap of the pole pieces in Figure 3.5?

The argument that led to (3.70a) shows that the flux toward each pole face, of the momentum component pointing away from that face, is $B^2/8\pi$. If the current in the coil is i, then the pressure on each face is thus

$$\frac{B^2}{8\pi} = 2\pi \left(\frac{Ni}{c\left[\dfrac{L}{\mu} + W\right]} \right)^2$$

This force tends to pull the faces together.

FIGURE 3.13 To calculate the force on the capacitor plate, we calculate the rate at which it receives momentum from the electromagnetic field.

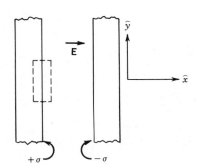

PROBLEM 3.13.3   A parallel plate capacitor, with circular plates of radius a, separation D, is being slowly charged. Calculate the rate of change of the energy stored between the plates, and the rate at which energy flows into this region from the outside.

Suppose that **E** points upward and is increasing. Then a **B** field is induced between the plates given by

$$\text{curl } \mathbf{B} = \frac{1}{c} \frac{\partial \mathbf{E}}{\partial t}$$

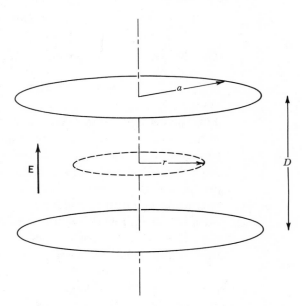

If we integrate this around a circle of radius r, concentric with the plates, we get

$$\int \text{curl } \mathbf{B} \cdot d\mathbf{a} = \oint \mathbf{B} \cdot d\mathbf{l} = \frac{1}{c} \frac{\partial}{\partial t} \int \mathbf{E} \cdot d\mathbf{a}$$

$$2\pi r B = \frac{\pi r^2}{c} \frac{\partial E}{\partial t}$$

$$B(r) = \frac{r}{2c} \frac{\partial E}{\partial t}, \qquad \text{pointing clockwise} \qquad (3.71)$$

We have assumed here that **E** is spatially uniform, and have used the circular symmetry of the problem to infer the tangential direction of B.

Form (3.71) we can infer that the energy stored in the **B** field will be small compared to the energy stored in the **E** field if E changes by a small fraction of itself in time a/c. We suppose this to be so. Then

$$U = \int_V u \, dv = \frac{1}{8\pi} E^2 \pi a^2 D$$

$$\frac{d}{dt} U = \frac{a^2 D}{4} E \frac{\partial E}{\partial t} \qquad (3.72a)$$

The energy flux at the edge of the cylindrical region is

$$\frac{c}{4\pi} E B = \frac{c}{4\pi} E \frac{a}{2c} \frac{\partial E}{\partial t} = \frac{a}{8\pi} E \frac{\partial E}{\partial t} \qquad \text{(pointing inwards)}$$

The rate at which energy enters this region is

$$\int_S \mathbf{S} \cdot d\mathbf{a} = \frac{a}{8\pi} E \frac{\partial E}{\partial t} \cdot 2\pi a D = \frac{a^2 D}{4} E \frac{\partial E}{\partial t} \qquad (3.72b)$$

in agreement with (3.72a).

PROBLEM 3.13.4   Two capacitors (C_1 and C_2) have charges Q_1 and Q_2, respectively. They are then connected, plus side to plus side and minus side to minus side. What is the change in stored electromagnetic energy?

Changing the charge on a capacitor by dQ produces an energy change of

$$V dQ = \frac{Q}{C} dQ = d\left(\frac{Q^2}{2C} \right)$$

Thus, the energy stored in a capacitor can be taken to be $Q^2/2C$. Initially we had total energy

$$\frac{Q_1^2}{2C_1} + \frac{Q_2^2}{2C_2}$$

After connection, the system is essentially a single capacitor, with capacitance $C_1 + C_2$ and charge $Q_1 + Q_2$. Thus, the energy change is

$$\frac{(Q_1 + Q_2)^2}{2(C_1 + C_2)} - \left(\frac{Q_1^2}{2C_1} + \frac{Q_2^2}{2C_2} \right) = -\frac{(Q_1 C_2 - Q_2 C_1)^2}{2 C_1 C_2 (C_1 + C_2)} \le 0$$

The equal sign applies if $Q_1/C_1 = Q_2/C_2$, in which case the plates initially had equal potential differences, and connecting them produces no change.

3.14 INDUCTANCE

Suppose that we have a set of circuits carrying steady currents $i_1, i_2, i_3, \dots$. Each current gives rise to a magnetic field $\mathbf{B}_1, \mathbf{B}_2, \mathbf{B}_3, \dots$. The total magnetic field is

$$\mathbf{B} = \mathbf{B}_1 + \mathbf{B}_2 + \mathbf{B}_3 + \cdots$$

so the total energy stored in the magnetic field is

$$U_B = \int u_B \, dv = \frac{1}{8\pi} \int \mathbf{B} \cdot \mathbf{B} \, dv$$

$$= \frac{1}{8\pi} \int (\mathbf{B}_1 + \mathbf{B}_2 + \mathbf{B}_3 + \cdots) \cdot (\mathbf{B}_1 + \mathbf{B}_2 + \mathbf{B}_3 + \cdots) \, dv \qquad (3.73)$$

The field $\mathbf{B}_j$ due to current i_j in coil j is proportional to i_j. Thus, (3.73) is a homogeneous quadratic form in the i_j, and can be written

$$U_B = \frac{1}{2} \sum_{j, k} M_{jk} i_j i_k \qquad \left(M_{jk} = M_{kj} \right) \qquad (3.74)$$

The coefficients M_{jk} defined by (3.74) depend on the geometrical configurations of the circuits, but are independent of the magnitudes of the currents.

The only way that U_B of (3.73) can be zero is if $\mathbf{B}$ vanishes everywhere. This implies that $\mathbf{B}_1 = \mathbf{B}_2 = \mathbf{B}_3 \cdots = 0$, and $i_1 = i_2 = i_3 \cdots = 0$. Since the quadratic form (3.74) can vanish only if all the i_j vanish, the matrix M is positive definite. One consequence of this is that its determinant is positive.

The diagonal elements of M are called self-inductances, $L_j = M_{jj}$, and the off-diagonal elements are called mutual inductances. The self-inductance L_j depends only on the geometry of circuit j, and not on its relation to other circuits. If we only have circuit j, and i_j changes with time (slowly enough that the current densities remain uniform), (3.74) implies that

$$\frac{d}{dt} U_B = \frac{d}{dt}\left(\frac{1}{2} L_j i_j^2\right) = L_j i_j \frac{di_j}{dt}$$

As the $\mathbf{B}_j$ field changes an $\mathbf{E}$ field is induced, which does work on the moving charges in circuit j at the rate

$$\int_{V_j} \mathbf{E} \cdot \mathbf{J} \, dv = i_j \oint_j \mathbf{E} \cdot d\mathbf{l}$$

$$\equiv i_j \mathscr{E}_j \tag{3.75a}$$

(cf. 3.67, 3.52, 3.52, and 3.45). This equation is true only if $d\mathbf{l}$ points in the direction of the current i_j. The quantity $\mathscr{E}_j$ defined in (3.75a) is the "EMF induced in circuit j." Since positive work done on the moving charges implies a decrease in the magnetic field energy U_B, the conservation of energy requires that

$$i_j \mathscr{E}_j = -\frac{dU_B}{dt} = -L_j i_j \frac{di_j}{dt}$$

$$\mathscr{E}_j = -L_j \frac{di_j}{dt} \tag{3.75b}$$

Similarly, we can infer from (3.74) that a changing current in circuit k causes an EMF

$$\mathscr{E}_j = -M_{jk} \frac{di_k}{dt} \tag{3.75c}$$

to be induced in circuit j.

PROBLEM 3.14.1 Calculate, using the dipole approximation, the mutual inductance of the two circular loops shown in Figure 3.14

Suppose that current i_1 flows in loop 1 as shown. The magnetic moment of the loop is then

$$\frac{i_1 \pi a^2}{c} \hat{z} = \mathbf{m}_1$$

[cf. (3.57c)], and the distant $\mathbf{B}$ field produced by this loop is given by

$$\mathbf{B}(\mathbf{r}) = -\nabla\left(\frac{\mathbf{m}_1 \cdot \mathbf{r}}{r^3}\right) = -i_1 \frac{\pi a^2}{c} \nabla\left(\frac{z}{r^3}\right)$$

$$= -i_1 \frac{\pi a^2}{c}\left(\frac{\hat{z}}{r^3} - \frac{3z\hat{r}}{r^4}\right)$$

In the vicinity of the $\hat{z}$ axis this is approximately equal to

$$\frac{2i_1 \pi a^2}{c} \frac{1}{z^3}$$

FIGURE 3.14 Two circular wire loops, each of radius a, are parallel to the $x - y$ plane, with their centers separated by distance D.

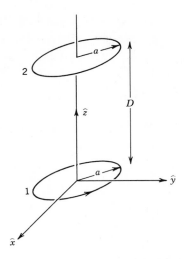

so that the **B** flux through loop 2 due to i_1 is

$$\Phi = -\frac{2i_1}{cD^3}\left(\pi a^2\right)^2$$

If i_1 changes with time, the EMF induced in loop 2 is

$$\mathscr{E}_2 = -\frac{1}{c}\frac{d\Phi}{dt} = -2\frac{\left(\pi a^2\right)^2}{c^2 D^3}\frac{di_1}{dt}$$

Comparison with (3.75b) yields

$$M_{12} = \frac{2\left(\pi a^2\right)^2}{c^2 D^3}$$

PROBLEM 3.14.2 Calculate the self-inductance of a coil consisting of N turns uniformly and tightly wound around a toroid of rectangular cross section, with the dimensions shown in the diagram. The magnetic permeability of the iron is μ.

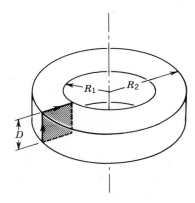

Suppose that current i flows in the coil. Then Ampere's law applied to a circle of radius r ($R_2 \geq r \geq R_1$), concentric with the toroid, gives

$$2\pi r H(r) = \frac{Ni}{c}$$

$$B(r) = \mu H(r) = \frac{\mu Ni}{2\pi cr}$$

The **B** flux through a rectangular cross section of the toroid is

$$\Phi = D \int_{R_1}^{R_2} B(r)\, dr = \frac{D\mu Ni}{2\pi c} \int_{R_1}^{R_2} \frac{dr}{r} = \frac{D\mu Ni}{2\pi c} \ln\left(\frac{R_2}{R_1}\right)$$

The EMF induced in the coil when i changes is

$$\mathcal{E} = -N\frac{1}{c}\frac{d\Phi}{dt} = -\frac{\mu N^2}{2\pi c} D \ln\left(\frac{R_2}{R_1}\right)\frac{di}{dt}$$

so the self-inductance is

$$L = \frac{\mu N^2 D}{2\pi c} \ln\left(\frac{R_2}{R_1}\right)$$

3.15 LORENTZ TRANSFORMATIONS OF THE E AND B FIELDS

In Section 2.1 we discussed an observer, Joe, whose Lorentz frame moved with velocity $v\hat{x}$ relative to the Lorentz frame of Moe. The coordinates x', t' that Joe ascribed to an event were related to the coordinates x, t that Moe ascribed to the *same* event by the Lorentz transformation (2.1). If Moe and Joe observe the same electric and magnetic fields in the space around them, their measurements will be related by [8]

$$E'_x = E_x \qquad\qquad B'_x = B_x \qquad\qquad (3.76a)$$

$$E'_y = \gamma\left(E_y - \frac{v}{c}B_z\right) \qquad B'_y = \gamma\left(B_y + \frac{v}{c}E_z\right) \qquad (3.76b)$$

$$E'_z = \gamma\left(E_z + \frac{v}{c}B_y\right) \qquad B'_z = \gamma\left(B_z - \frac{v}{c}E_y\right) \qquad (3.76c)$$

Here the $\mathbf{E'}, \mathbf{B'}$ fields refer to the same space-time point as do the $\mathbf{E}, \mathbf{B}$ fields. A more general way of writing (3.76) is

$$\mathbf{E'}_\| = \mathbf{E}_\| \qquad \mathbf{B'}_\| = \mathbf{B}_\| \qquad\qquad (3.77a)$$

$$\mathbf{E'}_\perp = \gamma\left(\mathbf{E}_\perp + \frac{1}{c}\mathbf{v}\times\mathbf{B}_\perp\right) \qquad\qquad (3.77b)$$

$$\mathbf{B'}_\perp = \gamma\left(\mathbf{B}_\perp - \frac{1}{c}\mathbf{v}\times\mathbf{E}_\perp\right) \qquad\qquad (3.77c)$$

where $\|$ and $\perp$ refer to the direction of the relative velocity of the two observers.

PROBLEM 3.15.1   Suppose that a large parallel plate capacitor is at rest in the unprimed reference frame, with its plates parallel to the x–y plane. Calculate the electric and magnetic fields measured by an observer moving in the x-direction with speed v.

Let the charge densities on the plates be σ. The unprimed observer measures

$$E_z = 4\pi\sigma \qquad E_x = E_y = 0$$

$$B_x = B_y = B_z = 0$$

---

[8] For a more complete discussion of the behavior of **E** and **B** under Lorentz transformations, see *Classical Electrodynamics*, J. D. Jackson, John Wiley & Sons, 1975, chapter 12.

According to (3.76), a primed observer moving with speed v in the $+\hat{x}$ direction would measure

$$E'_x = E'_y = 0 \qquad E'_z = \gamma E_z = 4\pi\gamma\sigma \tag{3.78a}$$

$$B'_x = B'_z = 0 \qquad B'_y = \gamma\frac{v}{c}E_z = \frac{4\pi\gamma\sigma v}{c} \tag{3.78b}$$

Equation (3.78a) can be interpreted in terms of an increased charge density as seen by the moving observer, due to the Lorentz contraction (Section 2.3, Problem 2.3.1). The primed observer sees two sheets of charge moving in the $-\hat{x}$ direction, with surface current densities $\mp\gamma\sigma v\hat{x}$. Equation (3.78b) is consistent with what he would get by using (3.5d) to calculate the **B** field that this current produces.

PROBLEM 3.15.2   An aluminum disk of radius R, thickness d, conductivity σ and mass density ρ is mounted on a frictionless vertical axis. It passes between the poles of a magnet near its rim, which produces a **B** field perpendicular to the plane of the disk over a small area A of the disk. If the initial angular speed of the disk is Ω_0, how many revolutions will it make before it comes to rest?

An observer on the disk, moving between the pole pieces of the magnet, would feel an **E** field given by (3.76b) or (3.77b):

$$\mathbf{E}' = -\frac{v}{c}B\hat{y} = -\frac{R\omega}{c}B\hat{y}$$

(we assume that the angular speed, ω, is small enough so that $\gamma \approx 1$). This results in a current density **J** given by

$$\mathbf{J} = \sigma\mathbf{E}' = -\frac{R\omega\sigma}{c}B\hat{y}$$

The Lorentz force (3.52a) produced on this current density by the **B** field of the magnet is

$$\mathbf{F} = \frac{1}{c}\mathbf{J} \times \mathbf{B}Ad = -\frac{R\omega\sigma AdB^2}{c^2}\hat{x}$$

which produces a retarding torque about the axis,

$$\tau = -R\hat{y} \times \mathbf{F} = -\frac{R^2\omega\sigma AdB^2}{c^2}\hat{z}$$

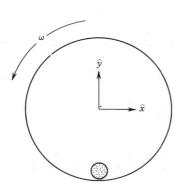

FIGURE 3.15 An aluminum disk rotates about an axis through its center. Near the rim, a magnet produces a **B** field perpendicular to the plane of the disk, over a small area A.

Since the moment of inertia of the disk around its axis is

$$I = \frac{1}{2} MR^2 = \frac{1}{2} \left(\pi R^2 d\rho \right) R^2 = \frac{\pi R^4 \, d\rho}{2}$$

the equation of motion of the disk is

$$\tau_z = - \frac{R^2 \sigma A dB^2}{c^2} \omega \equiv \frac{\pi R^4 \, d\rho}{2} \frac{d\omega}{dt}$$

$$\frac{d\omega}{dt} = - \frac{2\sigma A B^2}{\pi R^2 \rho c^2} \omega$$

whose solution, corresponding to the specified initial conditions, is

$$\omega(t) = \Omega_0 \exp\left(- \frac{2\sigma A B^2}{\pi R^2 \rho c^2} t \right)$$

The number of revolutions of the disk before it comes to rest is thus

$$N = \frac{1}{2\pi} \int_0^\infty \omega(t) \, dt = \frac{\Omega_0}{2\pi} \int_0^\infty \exp\left(- \frac{2\sigma A B^2}{\pi R^2 \rho c^2} t \right) \, dt$$

$$= \frac{\Omega_0 R^2 \rho c^2}{4\sigma A B^2}$$

3.16 PLANE WAVE SOLUTIONS OF MAXWELL'S EQUATIONS IN A HOMOGENEOUS NONCONDUCTING ISOTROPIC MEDIUM

We assume that

$$\left. \begin{matrix} \mathbf{B} = \mu \mathbf{H} \\ \mathbf{D} = \varepsilon \mathbf{E} \end{matrix} \right\} \begin{matrix} \mu, \, \varepsilon \text{ independent} \\ \text{of } \mathbf{r}, \, t \end{matrix}$$

$$\mathbf{J}_{\text{free}} = 0$$

Then (3.1) and (3.23) become

$$\text{div } \mathbf{E} = \text{div } \mathbf{B} = 0 \tag{3.79a}$$

$$\text{curl } \mathbf{E} = - \frac{1}{c} \frac{\partial \mathbf{B}}{\partial t} \tag{3.79b}$$

$$\text{curl } \mathbf{B} = \frac{\varepsilon \mu}{c} \frac{\partial \mathbf{E}}{\partial t} = \frac{n^2}{c} \frac{\partial \mathbf{E}}{\partial t} \tag{3.79c}$$

The parameter $n(= \sqrt{\varepsilon\mu})$ introduced in (3.79c) is called the index of refraction. If we apply the vector identity

$$\text{curl } (\text{curl } \mathbf{v}) = \nabla (\text{div } \mathbf{v}) - \nabla^2 \mathbf{v}$$

to the curls of (3.79b) and (3.79c), we find that $\mathbf{E}$ and $\mathbf{B}$ satisfy the d'Alembert wave equation:

$$\nabla^2 \mathbf{E} - \frac{n^2}{c^2} \frac{\partial^2 \mathbf{E}}{\partial t^2} = 0 = \nabla^2 \mathbf{B} - \frac{n^2}{c^2} \frac{\partial^2 \mathbf{B}}{\partial t^2} \tag{3.80}$$

In Section 1.14e we encountered a scalar version of the d'Alembert equation in connection with pressure fluctuations in fluids. There we investigated plane wave solutions (1.11b). Plane wave solutions of the vector equations (3.80) have the form

$$\mathbf{E}(\mathbf{r}, t) = \mathbf{E}_0 e^{i(\mathbf{k} \cdot \mathbf{r} - \omega t)} \tag{3.81a}$$

$$\mathbf{B}(\mathbf{r}, t) = \mathbf{B}_0 e^{i(\mathbf{k} \cdot \mathbf{r} - \omega t)} \tag{3.81b}$$

where $\mathbf{E}_0$ and $\mathbf{B}_0$ are constant, and $\mathbf{k}$ and ω satisfy

$$k_x^2 + k_y^2 + k_z^2 = k^2 = \frac{n^2 \omega^2}{c^2} \tag{3.82}$$

If (3.81) is to satisfy (3.79a), it must be that $\mathbf{k} \cdot \mathbf{E}_0 = \mathbf{k} \cdot \mathbf{B}_0 = 0$. Thus, the waves are transverse (polarized perpendicular to their direction of propagation). To get the relationship between $\mathbf{E}_0$ and $\mathbf{B}_0$, we compare the curls of (3.81a and b) with (3.79b and c):

$$\text{curl } \mathbf{E} = i\mathbf{k} \times \mathbf{E}_0 e^{i(\mathbf{k} \cdot \mathbf{r} - \omega t)} = \frac{i\omega}{c} \mathbf{B}_0 e^{i(\mathbf{k} \cdot \mathbf{r} - \omega t)}$$

$$\text{curl } \mathbf{B} = i\mathbf{k} \times \mathbf{B}_0 e^{i(\mathbf{k} \cdot \mathbf{r} - \omega t)} = -\frac{in^2 \omega}{c} \mathbf{E}_0 e^{i(\mathbf{k} \cdot \mathbf{r} - \omega t)}$$

Thus, we see that the three vectors $\mathbf{E}_0$, $\mathbf{B}_0$, $\mathbf{k}$ are mutually orthogonal, and that $B_0 = (ck/\omega)E_0 = nE_0$. We summarize by writing

$$\mathbf{E}(\mathbf{r}, t) = E_0 \hat{\varepsilon}_1 e^{i(\mathbf{k} \cdot \mathbf{r} - \omega t)} \tag{3.83a}$$

$$\mathbf{B}(\mathbf{r}, t) = nE_0 \hat{\varepsilon}_2 e^{i(\mathbf{k} \cdot \mathbf{r} - \omega t)} \tag{3.83b}$$

with

$$\hat{\varepsilon}_1 \times \hat{\varepsilon}_2 = \hat{k}, \qquad \hat{k} \times \hat{\varepsilon}_1 = \hat{\varepsilon}_2, \qquad \hat{\varepsilon}_2 \times \hat{k} = \hat{\varepsilon}_1$$

In (3.81) and all the equations that follow it, we imply that we take the real part of any complex expression to find the physical fields. The waves (3.81) and (3.83) are said to be *linearly* polarized since the directions of $\mathbf{E}$ and $\mathbf{B}$ are constant with respect to position and time. The wave propagates in the direction of $\mathbf{k}$ with a phase velocity given by

$$v = \frac{\omega}{k} = \frac{c}{n} \tag{3.84a}$$

The frequency $\nu(= \omega/2\pi)$ and wave length $\lambda(= 2\pi/k)$ are related by

$$\nu\lambda = \frac{\omega}{k} = v \tag{3.84b}$$

Note that the amplitude of $\mathbf{B}$ is n times that of $\mathbf{E}$, so that in vacuum, (where $n = 1$), $\mathbf{E}$ and $\mathbf{B}$ have equal amplitudes. The time-average energy flux associated with (3.83) is

$$\mathbf{S} = \frac{c}{4\pi} (Re\mathbf{E} \times Re\mathbf{B})_{\text{time av}}$$

$$= \frac{c}{8\pi} Re(\mathbf{E} \times \mathbf{B}^*) = \frac{nc}{8\pi} |E_0|^2 \hat{k} \tag{3.85}$$

For *circularly* polarized plane waves propagating in the $\mathbf{k}$ direction, use

$$\mathbf{E}(\mathbf{r}, t) = E_0 \frac{\hat{\varepsilon}_1 \pm i\hat{\varepsilon}_2}{\sqrt{2}} e^{i(\mathbf{k} \cdot \mathbf{r} - \omega t)} \tag{3.86a}$$

$$\mathbf{B}(\mathbf{r}, t) = nE_0 \frac{\hat{\varepsilon}_2 \mp i\hat{\varepsilon}_1}{\sqrt{2}} e^{i(\mathbf{k} \cdot \mathbf{r} - \omega t)} \tag{3.86b}$$

FIGURE 3.16 A linearly polarized electromagnetic wave, with propagation vector **k**.

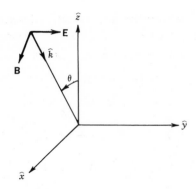

In (3.86), the upper (lower) signs refer to left (right)-hand circularly polarized radiation. An observer toward whom a right-hand circularly polarized wave propagates would see the **E** and **B** vectors rotating in a clockwise direction in the $\hat{\varepsilon}_1 - \hat{\varepsilon}_2$ plane. This can be verified by calculating the real parts of (3.86) in the special case $\hat{\varepsilon}_1 = \hat{x}$, $\hat{\varepsilon}_2 = \hat{y}$, $\hat{k} = \hat{z}$.

PROBLEM 3.16.1 Give the **E** and **B** fields for a wave propagating in a downward direction, which makes an angle θ with the $\hat{z}$ axis, and is linearly polarized with **E** in the $\hat{y}$ direction, and has wave length λ. The index of refraction of the medium is n.

By inspection of Figure 3.16, we see that

$$\hat{k} = -(\hat{z}\cos\theta + \hat{x}\sin\theta)$$

$$\hat{\varepsilon}_1 = \hat{y}$$

so that

$$\hat{\varepsilon}_2 = \hat{k} \times \hat{\varepsilon}_1 = \hat{x}\cos\theta - \hat{z}\sin\theta$$

$$\mathbf{k} \cdot \mathbf{r} = \frac{2\pi}{\lambda}\hat{k} \cdot \mathbf{r} = -\frac{2\pi}{\lambda}(z\cos\theta + x\sin\theta)$$

$$\omega = \frac{ck}{n} = \frac{2\pi}{\lambda}\frac{c}{n}$$

E and **B** are given by (3.83) to be

$$\mathbf{E} = E_0\hat{y}e^{(2\pi i/\lambda)(-[z\cos\theta + x\sin\theta] - ct)}$$

$$\mathbf{B} = nE_0(\hat{x}\cos\theta - \hat{z}\sin\theta)e^{(2\pi i/\lambda)(-[z\cos\theta + x\sin\theta] - ct)}$$

In an important class of problems, n is constant in specified regions of space but changes discontinuously as we cross certain boundary surfaces. To solve these problems we set up appropriate incident, reflected and transmitted waves of the form (3.83) or (3.86) on each side of each boundary surface. Then we apply the required continuity conditions (3.5) or (3.24) across each surface. For waves at optical frequencies ($\sim 10^{15}$ Hz) in dielectrics, the magnetic permeability μ is close to unity and we can set $n = \sqrt{\varepsilon}$, $\mathbf{B} = \mathbf{H}$.

PROBLEM 3.16.2 A beam of monochromatic light of wave length λ in vacuum is incident normally on a nonmagnetic dielectric film of refractive index n. The film thickness is d. Calculate the reflection coefficient (the fraction of the incident energy that is reflected).

We suppose that the film lies between the planes $z = 0$ and $z = d$, and that the wave is incident from below. Thus, for $z < 0$, we have incident and reflected waves

$$\left.\begin{array}{l} \mathbf{E}(\mathbf{r}, t) = \hat{x}\Big(E_i e^{i(kz-\omega t)} - E_r e^{i(-kz-\omega t)} \Big) \\ \mathbf{B}(\mathbf{r}, t) = \hat{y}\Big(E_i e^{i(kz-\omega t)} + E_r e^{i(-kz-\omega t)} \Big) \end{array}\right\} \, z \le 0$$

Within the film we have waves moving in both the $+\hat{z}$ and $-\hat{z}$ directions,

$$\left.\begin{array}{l} \mathbf{E}(\mathbf{r}, t) = \hat{x}\Big(E_1 e^{i(k'z-\omega t)} + E_2 e^{i(-kz'-\omega t)} \Big) \\ \mathbf{B}(\mathbf{r}, t) = n\hat{y}\Big(E_1 e^{i(k'z-\omega t)} - E_2 e^{i(-k'z-\omega t)} \Big) \end{array}\right\} \, 0 \le z \le d$$

and above the film we have only a transmitted wave

$$\left.\begin{array}{l} \mathbf{E}(\mathbf{r}, t) = \hat{x} E_t e^{i(kz-\omega t)} \\ \mathbf{B}(\mathbf{r}, t) = \hat{y} E_t e^{i(kz-\omega t)} \end{array}\right\} \, z \ge d$$

Note that the frequency of the wave has the same value everywhere, but the wavelength is different within the film,

$$\lambda = \frac{c}{\nu} \qquad k = \frac{2\pi}{\lambda} \qquad \text{(vacuum)}$$

$$\lambda' = \frac{\nu}{\nu} = \frac{c}{n\nu} \qquad k' = \frac{2\pi}{\lambda'} = nk \qquad \text{(film)}$$

At $z = 0$ and $z = d$, $E_\parallel$ and $H_\parallel$ ($\approx B_\parallel$) are continuous. This leads to the following four simultaneous equations for E_r, E_1, E_2, and E_t in terms of E_i:

$$E_i - E_r = E_1 + E_2$$

$$E_i + E_r = n(E_1 + E_2)$$

$$E_1 e^{ik'd} + E_2 e^{-ik'd} = E_t e^{ikd}$$

$$n\Big(E_1 e^{ik'd} - E_2 e^{-ik'd} \Big) = E_t e^{ikd}$$

The solution for E_r is

$$E_r = (n^2 - 1)\frac{2i \sin(k'd) \cdot E_i}{(n^2 + 1) \cdot 2i \sin(k'd) - 4n \cos(k'd)}$$

and the ratio of reflected to incident energy flux is

$$\frac{S_r}{S_i} = \frac{|E_r|^2}{|E_i|^2} = \frac{(n^2 - 1)^2 \sin^2(k'd)}{(n^2 + 1)^2 \sin^2(k'd) + 4n^2 \cos^2(k'd)}$$

$$= \frac{(n^2 - 1)^2}{(n^2 + 1)^2 + 4n^2 \cot^2\!\left(\dfrac{2\pi nd}{\lambda}\right)}$$

If the wave moves through a medium of conductivity σ, (3.79c) must be replaced by

$$\text{curl } \mathbf{B} - \frac{\varepsilon\mu}{c}\frac{\partial \mathbf{E}}{\partial t} = \frac{4\pi}{c}\mathbf{J} = \frac{4\pi}{c}\sigma\mathbf{E} \tag{3.87}$$

We can still find plane wave solutions (3.83), but now the propagation vector $\mathbf{k}$ has an imaginary part. This leads to attenuation of the wave as it propagates through the medium. The energy removed from the wave is dissipated in $J^2\sigma$ losses in the conductor.[9]

3.17 SIMPLE MICROSCOPIC MODELS OF ELECTRICAL CONDUCTIVITY AND INDEX OF REFRACTION

Suppose that a plane electromagnetic wave propagates through a medium containing ρ electrons per unit volume, free to move under the influence of the wave. We represent the effect on the electrons of collisions with the atoms of the medium by assuming that an electron with velocity $\mathbf{v}$ experiences a damping force $-mg\mathbf{v}$. We assume that the electron velocity remains small enough so that we can neglect the effect of the $\mathbf{v}/c \times \mathbf{B}$ term in the Lorentz force. Then the equation of motion of an electron is

$$m\frac{d\mathbf{v}}{dt} + mg\mathbf{v} = -e\mathbf{E}$$

$$= -eE_0\hat{\varepsilon}_1 e^{i(\mathbf{k}\cdot\mathbf{r}-\omega t)} \tag{3.88}$$

If we also assume that the wavelength is large compared to the amplitude of oscillation of the electron, we can ignore the time dependence of the $\mathbf{k}\cdot\mathbf{r}$ term in the phase of (3.88). The steady-state solution of (3.88) is

$$\mathbf{v} = -\frac{eE_0}{m(g-i\omega)}e^{i(\mathbf{k}\cdot\mathbf{r}-\omega t)}\hat{\varepsilon}_1$$

This implies a current density $\mathbf{J}$ given by

$$\mathbf{J} = -e\rho\mathbf{v} = \frac{e^2\rho}{m(g-i\omega)}E_0\hat{\varepsilon}_1 e^{i(\mathbf{k}\cdot\mathbf{r}-\omega t)}$$

$$= \frac{e^2\rho}{m(g-i\omega)}\mathbf{E} \tag{3.89a}$$

so that the conductivity σ is

$$\sigma(\omega) = \frac{e^2\rho}{m(g-i\omega)} \tag{3.89b}$$

The damping parameter g is of the order of magnitude of the reciprocal of the time between successive collisions between the electron and the atoms. It is approximately $5 \times 10^{13}\text{s}^{-1}$ for copper. The presence of damping causes the current density given by (3.89a) to lag behind the electric field (3.81a) by a time

---

[9]See, for example, *Classical Electrodynamics*, J. D. Jackson, John Wiley & Sons, 1975, chapter 7.

equal to $(1/\omega) \cdot \arctan(\omega/g)$. This is the physical significance of the imaginary part of the conductivity (3.89b).

Now suppose that each electron is bound harmonically to an atom, with natural frequency ω_0. If we neglect damping, we get the following equation of motion,

$$m\frac{d^2\mathbf{r}}{dt^2} + m\omega_0^2\mathbf{r} = -eE_0 e^{i(\mathbf{k}\cdot\mathbf{r}-\omega t)}\hat{\boldsymbol{\varepsilon}}_1$$

We still assume the wavelength sufficiently long so that $\mathbf{k}\cdot\mathbf{r}$ does not change appreciably over an oscillation. Now the steady-state solution is

$$\mathbf{r}(t) = -\frac{eE_0\hat{\boldsymbol{\varepsilon}}_1}{m(\omega_0^2 - \omega^2)}e^{i(\mathbf{k}\cdot\mathbf{r}-\omega t)}$$

and the current density is

$$\mathbf{J} = -e\rho\mathbf{v}$$

$$= -\frac{i\omega\rho e^2\hat{\boldsymbol{\varepsilon}}_1 E_0}{m(\omega_0^2 - \omega^2)}e^{i(\mathbf{k}\cdot\mathbf{r}-\omega t)} \tag{3.90}$$

To see the effect of the purely imaginary conductivity implied by (3.90), we substitute (3.90) into Maxwell's equation (3.1d) together with the solution (3.81) for $\mathbf{E}$ and $\mathbf{B}$. The result is

$$kB_0 = \left[\frac{4\pi}{c}\frac{\omega e^2\rho}{m(\omega_0^2 - \omega^2)} + \frac{\omega}{c}\right]E_0$$

But (3.1c) still implies that $kE_0 = (\omega/c)B_0$. If we eliminate E_0 and B_0, we find

$$k^2 = \frac{\omega^2}{c^2}\left[1 + \frac{4\pi\rho e^2}{m(\omega_0^2 - \omega^2)}\right]$$

We can then get the effective value of the index of refraction by using (3.84a):

$$n(\omega) = \frac{ck}{\omega} = \sqrt{1 + \frac{4\pi\rho e^2}{m(\omega_0^2 - \omega^2)}}$$

$$\equiv \sqrt{1 + \frac{\omega_p^2}{\omega_0^2 - \omega^2}} \tag{3.91}$$

Because the index of refraction given in (3.91) depends on ω, the medium is said to be *dispersive*. The origin of this term will become clearer in our discussion of refraction in Section 4.4.1.

The plasma frequency

$$\omega_p \equiv \sqrt{\frac{4\pi\rho e^2}{m}}$$

introduced in (3.91) is the frequency of charge-density fluctuations in a system consisting of mobile electrons moving against a background of stationary charge (with total charge density zero).

3.18 ELECTROMAGNETIC WAVES IN CAVITIES AND PIPES WITHIN PERFECT CONDUCTORS

The **E** field is zero within a perfect conductor, since a nonzero **E** field would lead to infinite current density. Furthermore, **B** can have no time-varying part within a perfect conductor, since Faraday's law (3.1c) implies that wherever there is a time-varying **B** field there is also an **E** field. In what follows, we assume that any time-independent **B** field in the conductor has been eliminated. Then **E** = **B** = 0 within a perfect conductor, and the continuity conditions (3.5a,b and 3.24a,b) imply that

$$
\left.
\begin{aligned}
E_{\parallel} &= 0 \\
B_{\perp} &= 0 \\
\varepsilon E_{\perp} &= 4\pi\sigma_{\text{free}} \\
\hat{n} \times \mathbf{H}_{\parallel} &= \frac{4\pi}{c}\mathbf{K}_{\text{free}}
\end{aligned}
\right\}
\begin{aligned}
&\text{immediately} \\
&\text{outside} \\
&\text{a} \\
&\text{perfect} \\
&\text{conductor}
\end{aligned}
$$

$$(3.92\text{a})$$
$$(3.92\text{b})$$
$$(3.92\text{c})$$
$$(3.92\text{d})$$

Here the subscripts $\parallel$ and $\perp$ refer to the surface of the conductor. Our problem is to find fields that are solutions of (3.79) and (3.80) in the space around the conductors, but satisfy the boundary conditions (3.92) at all the surfaces of the conductors.

PROBLEM 3.18.1 Consider an empty rectangular cavity in a perfect conductor. The sides of the cavity have lengths d_x, d_y, d_z. Determine the space and time dependence of the **E** and **B** fields of the lowest frequency mode of oscillation.

Let us try to find a standing wave solution whose **E** field points along the $\hat{z}$ axis. The rectangular shape of the cavity suggests that we can satisfy the boundary conditions with a product function of the form

$$\mathbf{E} = E_0\hat{z}\,\sin(k_x x)\sin(k_y y)\sin(\omega t) \tag{3.93a}$$

FIGURE 3.17 A hollow box whose walls are perfect conductors.

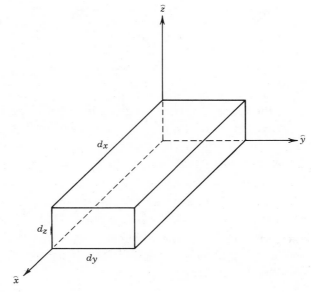

Equation (3.93a) contains no z dependence because we require that (3.79a) be satisfied. If we are also to satisfy (3.92a), **E** must vanish on all the vertical faces ($x = 0$, d_x; $y = 0$, d_y). Thus, we require that

$$k_x = \frac{\pi}{d_x} n_x \qquad n_x = 1, 2, 3, \ldots \tag{3.94a}$$

$$k_y = \frac{\pi}{d_y} n_y \qquad n_y = 1, 2, 3, \ldots \tag{3.94b}$$

To find the **B** field associated with (3.93a) we use (3.79b),

$$-\frac{1}{c}\frac{\partial \mathbf{B}}{\partial t} = \operatorname{curl} \mathbf{E}$$

$$= E_0 \big\{ \hat{x} k_y \sin(k_x x)\cos(k_y y)$$

$$- \hat{y} k_x \cos(k_x x)\sin(k_y y)\big\}\sin(\omega t)$$

which is satisfied by

$$\mathbf{B} = E_0 \frac{c}{\omega}\big\{ \hat{x} k_y \sin(k_x x)\cos(k_y y)$$

$$- \hat{y} k_x \cos(k_x x)\sin(k_y y)\big\}\cos(\omega t) \tag{3.93b}$$

It is easily verified that this **B** satisfies (3.79a), that $B_x = 0$ at $x = 0$ and d_x, and that $B_y = 0$ at $y = 0$ and d_y. Thus, (3.93a and b) satisfy the field equations and the boundary conditions, provided that k_x and k_y belong to the discrete sets of values listed in (3.94).

Since (3.93) is a solution of the wave equation (3.78), ω is given by

$$\omega^2 = \frac{c^2}{n^2}\big(k_x^2 + k_y^2 \big)$$

$$= \frac{c^2 \pi^2}{n^2}\left[\left(\frac{n_x}{d_x} \right)^2 + \left(\frac{n_y}{d_y} \right)^2 \right]$$

To get the mode of lowest frequency, set $n_x = n_y = 1$, and choose d_x and d_y to be the longest dimensions of the box.

The procedure just described is similar to the procedures used in the discussion of standing waves on a string (Section 1.14d) and sound waves in a box (Section 1.14e). In all cases we have to solve the d'Alembert equation with space-dependent, time-independent boundary conditions. The method of separation of variables leads to a Helmholtz spatial equation. The boundary conditions select a discrete set of solutions to this spatial problem, and then the d'Alembert equation associates a definite oscillation frequency with each allowed spatial solution. One of Erwin Schrödinger's great contributions was the recognition that a similar procedure could be used to explain the discrete set of energies observed for bound quantum-mechanical systems [see (5.7), (5.8) and Section 5.4].

Next we consider wave propagation inside an infinitely long pipe with rectangular cross-section. The walls of the pipe are parallel to the $\hat{x}$ axis. We replace (3.93) by a traveling wave

$$\mathbf{E} = E_0 \hat{z} \sin(k_y y) e^{i(k_x x - \omega t)} \tag{3.95a}$$

$$\mathbf{B} = E_0 \frac{c}{\omega} \left[-i\hat{x} k_y \cos(k_y y) \right. $$

$$\left. - \hat{y} k_x \sin(k_y y) \right] e^{i(k_x x - \omega t)} \tag{3.95b}$$

It is easily verified that these fields satisfy the boundary conditions (3.92a and b) on the walls of the pipe if k_y is given by (3.94b). They satisfy the field equations (3.79) and (3.80) if

$$k_x^2 + k_y^2 = \frac{n^2 \omega^2}{c^2}$$

$$\omega^2 = \frac{c^2}{n^2} \left[k_x^2 + \frac{\pi^2 n_y^2}{d_y^2} \right] \tag{3.96}$$

If the wave is to propagate along the pipe without changing its amplitude, k_x in (3.95) must be real. Then (3.96) implies that waves will only propagate in the n_y mode if their frequency exceeds a *cut-off* frequency $\omega_c(n_y)$:

$$\omega > \omega_c(n_y) \equiv \frac{\pi c}{n} \frac{n_y}{d_y} \tag{3.97}$$

The lowest value of ω_c corresponds to $n_y = 1$. Alternatively, we can use (3.96) to express the wavelength in terms of ω,

$$\lambda = \frac{2\pi}{k_x} = \frac{2\pi}{\sqrt{\dfrac{n^2 \omega^2}{c^2} - \dfrac{\pi^2 n_y^2}{d_y^2}}}$$

If (3.97) is not satisfied, λ is imaginary, corresponding to an exponentially decaying or growing wave.

The wave (3.95) has its **E** vector aligned perpendicular to the direction of propagation, but the **B** vector has a component in the direction of propagation. This is called a *transverse electric* wave. One can also construct waves in which the **B** vector is transverse (but not the **E** vector). It is not possible to have both **E** and **B** transverse if the cross section of the pipe is a simply-connected region.[10]

Once we have determined the **E** and **B** fields at the surface of the conductor, as in (3.93) or (3.95), we can determine the surface charge and current densities by using (3.92c and d). For example, the charge density on the $z = d_z$ surface of the cavity shown in Figure 3.17 is given by (3.92c) and (3.93a):

$$\sigma_{\text{free}}(x, y, z = d_z) = \frac{\varepsilon}{4\pi} E_0 \sin(k_x x) \sin(k_y y) \sin(\omega t)$$

---

[10] See *Classical Electrodynamics*, J. D. Jackson, John Wiley & Sons, 1975, Section 8.2.

and the surface current density is given by (3.92d) and (3.93b):

$$
\begin{aligned}
\mathbf{K}_{\text{free}}(x, y, z = d_z) &= \frac{c}{4\pi} E_0 \frac{c}{\omega}(-\hat{z}) \\
&\quad \times \left\{ \hat{x} k_y \sin(k_x x)\cos(k_y y) - \hat{y} k_x \cos(k_x x)\sin(k_y y) \right\} \cos(\omega t) \\
&= \frac{c^2}{4\pi} \frac{E_0}{\omega} \left\{ -\hat{y} k_y \sin(k_x x)\cos(k_y y) \right. \\
&\quad \left. - \hat{x} k_x \cos(k_x x)\sin(k_y y) \right\} \cos(\omega t)
\end{aligned}
$$

3.19 RADIATION BY NONRELATIVISTIC ACCELERATING CHARGES

Equations (3.58) and (3.63) enable us to calculate the **E** and **B** fields produced by any specified charge and current distributions. In particular, these equations can be used[11] to find the fields produced by moving point charges.

We first consider the motion of a single nonrelativistic point charge q. Let $\mathbf{s}(t)$ specify the location of q at time t. Since electromagnetic effects propagate in free space with finite speed c, the fields at point $\mathbf{r}$ at time t depend on the location and motion of the charge at a "retarded" time t', given by

$$
t' = t - \frac{|\mathbf{r} - \mathbf{s}(t')|}{c} \tag{3.98}
$$

Figure 3.18 illustrates how (3.98) can be solved for the t' that must be used to determine the fields at $\mathbf{r}$ and t. Once this is done, the fields can be calculated from

$$
\mathbf{B}(\mathbf{r}, t) = \frac{q}{c^2 R(t')} \ddot{\mathbf{s}}(t') \times \hat{R}(t') \tag{3.99a}
$$

$$
\mathbf{R}(t') \equiv \mathbf{r} - \mathbf{s}(t') \tag{3.99b}
$$

$$
\mathbf{E}(\mathbf{r}, t) = \mathbf{B}(\mathbf{r}, t) \times \hat{R}(t') \tag{3.99c}
$$

These expressions are valid far away ($R \gg c^2/|\ddot{\mathbf{s}}|$) from a nonrelativistic charge ($|\dot{\mathbf{s}}| \ll c$).

Figure 3.19 shows the geometric relations between $\mathbf{E}$, $\mathbf{B}$, $\ddot{\mathbf{s}}$ and $\hat{R}$. It is seen that $\mathbf{E}$ lies in the plane determined by $\ddot{\mathbf{s}}$ and $\hat{R}$, whereas $\mathbf{B}$ is perpendicular to this plane. The energy flux density (3.68a) far away from the accelerating charge is

$$
\begin{aligned}
\mathbf{S}(\mathbf{r}, t) &= \frac{c}{4\pi} \mathbf{E}(\mathbf{r}, t) \times \mathbf{B}(\mathbf{r}, t) \\
&= \frac{c}{4\pi} (\mathbf{B} \cdot \mathbf{B}) \hat{R} \\
&= \frac{c}{4\pi} \left[\frac{q|\ddot{\mathbf{s}}(t')|}{c^2 R(t')} \right]^2 \sin^2\theta \, \hat{R} \\
&= \frac{q^2 |\ddot{\mathbf{s}}(t')|^2}{4\pi c^3 R^2} \sin^2\theta \, \hat{R}
\end{aligned} \tag{3.100}
$$

---

[11] See *Classical Electrodynamics*, J. D. Jackson, John Wiley & Sons, 1975, chapter 14.

FIGURE 3.18 Finding the retarded time. The curve $s_x(t')$ represents the world-line of the charge. The lines $x - x' = \pm c(t - t')$ are on the "retarded light-cone" of (x, t). The retarded time t' that satisfies (3.98) is the t' coordinate of P, the intersection of the world-line of the charge with the retarded light-cone.

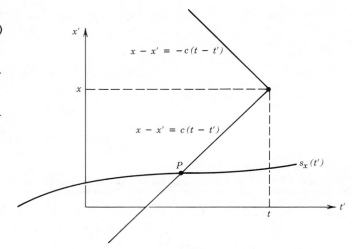

Here θ is the angle between $\ddot{\mathbf{s}}(t')$ and $\hat{R}(t')$. We do not distinguish between $R(t)$ and $R(t')$ in the denominator of (3.100), since they differ by a small factor, of the order of $|\dot{\mathbf{s}}|/c$. The fact that the power/unit area is proportional (at large R) to $1/R^2$ implies that the power per unit solid angle is independent of R:

$$\frac{dP}{d\Omega} = \frac{q^2 |\ddot{\mathbf{s}}(t')|^2}{4\pi c^3} \sin^2\theta \tag{3.101}$$

The total radiated power is obtained by integrating (3.101) over all directions

$$P = \int_{\substack{\theta = 0 \to \pi \\ \phi = 0 \to 2\pi}} \frac{dP}{d\Omega} \sin\theta \, d\theta \, d\phi = \frac{2\pi q^2 |\ddot{\mathbf{s}}|^2}{4\pi c^3} \int_0^\pi \sin^3\theta \, d\theta$$

$$= \frac{2}{3} \frac{q^2 |\ddot{\mathbf{s}}|^2}{c^3} \tag{3.102}$$

PROBLEM 3.19.1   A particle of mass m, charge e, is suspended from a light spring of spring constant k. If it is set into oscillation, how long will it take the oscillator to radiate away half its energy?

FIGURE 3.19 The polarization of the **E** and **B** radiation fields produced by a non-relativistic charge with acceleration $\ddot{\mathbf{s}}$.

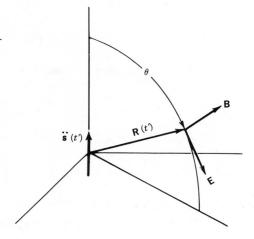

We assume that the radiation has a very small effect on the motion of the charge (we will check this later). Thus, we use the formulas for a free oscillator to determine the acceleration and the radiation rate. Suppose that the oscillator has amplitude A, so that its displacement is given by

$$x(t) = A \sin \omega_0 t, \qquad \omega_0 = \sqrt{k/m}$$

Then the acceleration is

$$\ddot{x}(t) = \frac{d^2 x}{dt^2} = -A\omega_0^2 \sin \omega_0 t$$

and the mean value of $\ddot{x}^2$ (averaged over a cycle) is $A^2 \omega_0^4 / 2$. Thus, power is radiated away at the rate

$$P = \frac{2}{3} \frac{e^2}{c^3} \frac{1}{2} A^2 \omega_0^4 = \frac{1}{3} \frac{e^2 \omega_0^4}{c^3} A^2 \tag{3.103a}$$

The mechanical energy of an oscillator is

$$E = \frac{1}{2} mv^2 + \frac{1}{2} kx^2 = \frac{1}{2} kA^2 = \frac{1}{2} m\omega_0^2 A^2 \tag{3.103b}$$

Thus, (3.103a) and (3.103b) imply that

$$P = -\frac{dE}{dt} = \frac{1}{3} \frac{e^2 \omega_0^4}{c^3} \frac{2E}{m\omega_0^2} = \frac{2}{3} \frac{e^2 \omega_0^2}{mc^3} E$$

and we see that the energy of the oscillator decays with time according to the formula

$$E(t) = E(0) \times e^{-\frac{2}{3} \frac{e^2 \omega_0^2}{mc^3} t}$$

Thus, it takes a time $t_{1/2}$ given by

$$\frac{2}{3} \frac{e^2 \omega_0^2}{mc^3} t_{1/2} = \ln 2$$

$$t_{1/2} = \frac{3}{2} \frac{mc^3}{e^2 \omega_0^2} \ln 2 = \frac{3 \ln 2}{4\pi} \frac{mc^3}{e^2 \omega_0} \text{ periods}$$

for the oscillator to radiate away half its initial energy. In particular, we find $t_{1/2} \approx 2 \times 10^{27}$ periods for a 1 cm diameter copper ball charged to 1000 V and oscillating with a frequency of 1000 Hz. This long half-life confirms our assumption that the radiation has a very small effect on the motion of the charge.

PROBLEM 3.19.2   A proton is uniformly accelerated to an energy of 20 MeV over a distance of 20 m. How much energy is radiated away?

Since 20 MeV $\ll$ 938 MeV, we can use nonrelativistic mechanics. If the final kinetic energy is E and the acceleration length is d, the accelerating force is E/d and the magnitude of the acceleration is E/md. The power radiated away is

$$P = \frac{2}{3} \frac{e^2}{c^3} \ddot{x}^2 = \frac{2}{3} \frac{e^2}{c^3} \left(\frac{E}{md} \right)^2$$

The travel time t is given by $d = 1/2 at^2$, $t = \sqrt{2d/a} = d\sqrt{2m/E}$. Thus, the total energy radiated is $Pt = 2\sqrt{2} e^2 \cdot (E/mc^2)^{3/2}/(3d)$. For $E = 20$ MeV, d = 20 m, this is

$$\frac{2\sqrt{2}}{3} \frac{(4.8x10^{-10})^2 \, \text{esu}^2}{2000 \, \text{cm}} \times \left(\frac{20 \, \text{MeV}}{938 \, \text{MeV}} \right)^{3/2}$$

$$= 3.4 \times 10^{-25} \, \text{erg} = 2 \times 10^{-13} \, \text{eV}$$

PROBLEM 3.19.3   An electron moves in a circular orbit of radius R about a nucleus of charge Ze. At what rate does it radiate?

The electron undergoes a centripetal acceleration a_c given by

$$a_c = \frac{Ze^2/R^2}{m}$$

Thus, the radiated power is

$$P = \frac{2}{3} \frac{e^2}{c^3} \left(\frac{Ze^2}{mR^2} \right)^2$$

For an electron in the smallest Bohr orbit of a hydrogen atom, $R = .529 \times 10^{-8}$cm, and P is approximately 0.5 erg/s. This result is inconsistent with the observed stability of the hydrogen atom. The need to resolve this contradiction provided much of the motivation for the invention of quantum mechanics.

Now suppose that there are several accelerating charges $q_1, q_2, \ldots,$ with respective world-lines $s_1(t), s_2(t), \ldots$. The total **E** and **B** fields observed at **r** and t will now be the vector sum of terms such as (3.99a and c), one for each contributing charge. Note, however, that the retarded times associated with the different charges will generally be different. This is illustrated in Figure 3.20 for the particular case of two charges. The necessity of evaluating each acceleration at a different time complicates the analysis. But the problem simplifies if the charges are close together. If they are spread over a distance of the order of L, the difference between their retarded times is of the order of L/c. We now make the additional assumption that the accelerations change by small fractions of themselves during this time interval,

$$|\dddot{\mathbf{s}}| \times \frac{L}{c} \ll |\ddot{\mathbf{s}}| \tag{3.104}$$

In this case we can make the approximation of using the *same* retarded time for

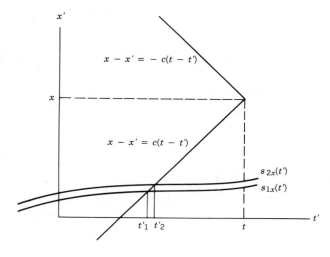

FIGURE 3.20 With a source consisting of two point charges, there are two world-lines intersecting the retarded light-cone of (x, t), and thus two retarded times t_1' and t_2'.

all the charges. This is called the *dipole approximation*. We can then generalize (3.101) and (3.102) to

$$\frac{dP}{d\Omega} = \frac{1}{4\pi c^3}\left|\sum_j q_j\ddot{\mathbf{s}}_j(t')\right|^2 \sin^2\theta \tag{3.105}$$

$$P = \frac{2}{3c^3}\left|\sum_j q_j\ddot{\mathbf{s}}_j(t')\right|^2 \tag{3.106}$$

We will apply (3.105) and (3.106) to a set of charges which oscillate with a common frequency about fixed equilibrium positions:

$$\mathbf{s}_j(t) = \mathbf{a}_j + \mathbf{b}_j e^{-i\omega t} \tag{3.107}$$

As usual, the actual position of charge j is obtained by calculating the real part of (3.107). By allowing the $\mathbf{b}_j$ to be complex, we can vary the oscillation phase from one charge to another. To obtain the time-averaged radiated power we calculate

$$\ddot{\mathbf{s}}_j(t) = -\omega^2\mathbf{b}_j e^{-i\omega t}$$

$$\left[Re\left(\sum_j q_j\ddot{\mathbf{s}}_j(t)\right)\right]^2_{\text{time avg}} = \frac{\omega^4}{2}\left|\sum_j q_j\mathbf{b}_j\right|^2$$

[cf (3.85)]. Note that if we write the electric dipole moment associated with the set (3.107) as the sum of time-independent and time-varying parts,

$$\mathbf{p} = \sum_j q_j\mathbf{s}_j(t) = \mathbf{p}_0 + \mathbf{p}_1 e^{-i\omega t} \tag{3.108a}$$

$$\mathbf{p}_1 \equiv \sum_j q_j\mathbf{b}_j \tag{3.108b}$$

then (3.105) and (3.106) can be expressed as

$$\left[\frac{dP}{d\Omega}\right]_{\text{time avg}} = \frac{\omega^4}{8\pi c^3}|\mathbf{p}_1|^2\sin^2\theta \tag{3.109}$$

$$[P]_{\text{time avg}} = \frac{1}{3}\frac{\omega^4}{c^3}|\mathbf{p}_1|^2 \tag{3.110}$$

If we require that (3.107) should satisfy the criterion (3.104), we find we must have

$$\frac{\omega L}{c} = \frac{2\pi\nu L}{c} = 2\pi\frac{L}{\lambda} \ll 1 \tag{3.111}$$

Thus, in the case of an oscillating source, we can say that the dipole approximation is valid when the wavelength of the emitted radiation is much larger than the dimensions of the source.

If conditions are such that the dipole approximation is not valid, or if the dipole amplitude (3.108b) vanishes identically, then a discussion of the radiation must include consideration of the variation of the retarded time across the source. One way of treating this problem involves an expansion of the radiation fields in powers of the parameter L/λ. The dipole approximation described above is the first term in the expansion. A complete discussion of this subject is given by Jackson.[12]

---

[12]*Classical Electrodynamics*, J. D. Jackson, John Wiley & Sons, 1975, chapters 9 and 16.

FIGURE 3.21 $\hat{\varepsilon}_1$ and $\hat{\varepsilon}_2$ are two orthogonal unit vectors in the plane of **E** and **B**.

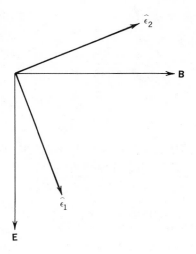

Finally, we consider the energy flux measured by a detector sensitive only to a certain type of polarization. We can regard such a device as a consisting of a filter which passes only the specified polarization, followed by a detector that measures total energy flux. Figure 3.21 shows the **E** and **B** vectors of Figure 3.19, as seen by a viewer looking towards the source. Two orthogonal unit vectors, $\hat{\varepsilon}_1$ and $\hat{\varepsilon}_2$, are in the plane of **E** and **B**, perpendicular to $\hat{R}$. We can write

$$\mathbf{E} = (\mathbf{E} \cdot \hat{\varepsilon}_1)\hat{\varepsilon}_1 + (\mathbf{E} \cdot \hat{\varepsilon}_2)\hat{\varepsilon}_2$$

$$\mathbf{B} = (\mathbf{B} \cdot \hat{\varepsilon}_1)\hat{\varepsilon}_1 + (\mathbf{B} \cdot \hat{\varepsilon}_2)\hat{\varepsilon}_2$$

Now suppose we have a device which will absorb the $\hat{\varepsilon}_2$ component of **E** and the $\hat{\varepsilon}_1$ component of **B**, but has no effect on the other components. We will call such a device a *linear polarization filter*, *with orientation* $\hat{\varepsilon}_1$. The **E** and **B** fields transmitted by this filter are

$$\mathbf{B} = \frac{(\ddot{\mathbf{p}} \times \hat{R}) \cdot \hat{\varepsilon}_2}{c^2 R}\hat{\varepsilon}_2 = \frac{(\hat{R} \times \hat{\varepsilon}_2) \cdot \ddot{\mathbf{p}}}{c^2 R}\hat{\varepsilon}_2 = -\frac{\hat{\varepsilon}_1 \cdot \ddot{\mathbf{p}}}{c^2 R}\hat{\varepsilon}_2$$

$$\mathbf{E} = \mathbf{B} \times \hat{R} = -\frac{\hat{\varepsilon}_1 \cdot \ddot{\mathbf{p}}}{c^2 R}\hat{\varepsilon}_1$$

and the transmitted energy flux per unit solid angle is

$$\left[\frac{dP}{d\Omega}\right]_{\text{time avg},\hat{\varepsilon}_1} = \frac{\omega^4}{8\pi c^3}|\mathbf{p}_1 \cdot \hat{\varepsilon}_1|^2 \qquad \text{(linear polarization only)} \tag{3.112}$$

[cf (3.99), (3.109)].

A similar analysis can be performed for circular polarization. Two complex unit vectors

$$\hat{\varepsilon}_l \equiv \frac{1}{\sqrt{2}}\left(\hat{\varepsilon}_1 + i\hat{\varepsilon}_2\right) \tag{3.113a}$$

$$\hat{\varepsilon}_r \equiv \frac{1}{\sqrt{2}}\left(\hat{\varepsilon}_1 - i\hat{\varepsilon}_2\right) \tag{3.113b}$$

were used in (3.86) to describe left and right-hand circularly polarized radiation. Since

$$\hat{\varepsilon}_l^* \cdot \hat{\varepsilon}_l = \hat{\varepsilon}_r^* \cdot \hat{\varepsilon}_r = 1$$

$$\hat{\varepsilon}_l^* \cdot \hat{\varepsilon}_r = \hat{\varepsilon}_r^* \cdot \varepsilon_l = 0$$

we can write **E** and **B** in the form

$$\mathbf{E} = (\mathbf{E} \cdot \hat{\varepsilon}_l^*)\hat{\varepsilon}_l + (\mathbf{E} \cdot \hat{\varepsilon}_r^*)\hat{\varepsilon}_r \qquad (3.114a)$$

$$\mathbf{B} = (\mathbf{B} \cdot \hat{\varepsilon}_l^*)\hat{\varepsilon}_l + (\mathbf{B} \cdot \hat{\varepsilon}_r^*)\hat{\varepsilon}_r \qquad (3.114b)$$

A device that transmits only left-hand circular polarized radiation will absorb the $\hat{\varepsilon}_r$ components of (3.114). If it does not affect the $\hat{\varepsilon}_l$ components, the transmitted fields are

$$\mathbf{B} = \frac{(\ddot{\mathbf{p}} \times \hat{R}) \cdot \hat{\varepsilon}_l^*}{c^2 R}\hat{\varepsilon}_l = \frac{(\hat{R} \times \hat{\varepsilon}_l^*) \cdot \ddot{\mathbf{p}}}{c^2 R}\hat{\varepsilon}_l = i\frac{\hat{\varepsilon}_l^* \cdot \ddot{\mathbf{p}}}{c^2 R}\hat{\varepsilon}_l$$

$$\mathbf{E} = \mathbf{B} \times \hat{R} = -\frac{\hat{\varepsilon}_l^* \cdot \ddot{\mathbf{p}}}{c^2 R}\hat{\varepsilon}_l$$

In this case, the time-averaged energy flux is determined by $\mathbf{E} \times \mathbf{B}^*$ (cf. 3.85). The energy flux per unit solid angle transmitted by this filter is then

$$\left(\frac{dP}{d\Omega}\right)_{\text{time avg},\hat{\varepsilon}} = \frac{\omega^4}{8\pi c^3}|\hat{\mathbf{p}}_1 \cdot \hat{\varepsilon}^*|^2 \qquad (3.115)$$

with $\hat{\varepsilon} = \hat{\varepsilon}_l$ of (3.113a). It is clear that we can also use (3.115) if the filter transmits only right-hand circularly polarized radiation, if $\hat{\varepsilon}$ is set equal to $\hat{\varepsilon}_r$ of (3.113b). Moreover, (3.115) can also be used for linear polarization, if $\hat{\varepsilon}$ is set equal to the real unit vector that points in the direction of the orientation of the filter [cf. (3.112)].

PROBLEM 3.19.4 Suppose the dipole moment is given by

$$\mathbf{p}(t) = p(\hat{x}\cos\omega t + \hat{y}\sin\omega t)$$

Find the angular distribution of emitted radiation as measured by (a) a detector sensitive only to right-hand circularly polarized radiation, and (b) a detector sensitive only to the polarization $\hat{\varepsilon}_1$ illustrated in the drawing.

In this case, $\mathbf{p}_1$ of (3.108a) is $p \cdot (\hat{x} + i\hat{y})$. By inspection of the drawing, we see that

$$\hat{\varepsilon}_1 = \cos\theta(\cos\varphi\hat{x} + \sin\varphi\hat{y}) - \sin\theta\hat{z}$$

$$\hat{\varepsilon}_2 = -\sin\varphi\hat{x} + \cos\varphi\hat{y}.$$

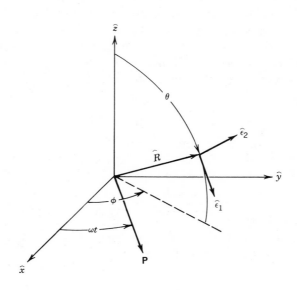

For the right-hand circular polarization, we need

$$\hat{\varepsilon}_r^* \cdot \mathbf{p}_1 = p\frac{\hat{\varepsilon}_1 + i\hat{\varepsilon}_2}{\sqrt{2}} \cdot (\hat{x} + i\hat{y})$$

$$= \frac{p}{\sqrt{2}}\left[\cos\theta\cos\varphi - i\sin\varphi + i\cos\theta\sin\varphi - \cos\varphi\right]$$

$$= \frac{p}{\sqrt{2}}(\cos\theta - 1)(\cos\varphi + i\sin\varphi)$$

$$\left(\frac{dP}{d\Omega}\right)_{\text{time avg},\hat{\varepsilon}_r} = \frac{\omega^4 p^2}{16\pi c^3}|(\cos\theta - 1)(\cos\varphi + i\sin\varphi)|^2$$

$$= \frac{\omega^4 p^2}{16\pi c^3}(\cos\theta - 1)^2 \tag{3.116a}$$

Note that the radiation pattern has a maximum in the $-\hat{z}$ direction (where $\theta = \pi$), but that no radiation is emitted in the $+\hat{z}$ direction (where $\theta = 0$). In Section 5.12 we will relate this asymmetry to the angular momentum of the circularly polarized radiation. On the other hand, for linearly polarized radiation $\hat{\varepsilon}_1$, we need

$$\hat{\varepsilon}_1^* \cdot \mathbf{p} = p\hat{\varepsilon}_1 \cdot (\hat{x} + i\hat{y}) = p\cos\theta(\cos\varphi + i\sin\varphi)$$

$$\left(\frac{dP}{d\Omega}\right)_{\text{time avg},\hat{\varepsilon}_1} = \frac{\omega^4 p^2}{8\pi c^3}(\cos\theta)^2 \tag{3.116b}$$

which is symmetric with respect to reflection across the $\hat{x}$–$\hat{y}$ plane. This symmetry is related to the impossibility of giving a prescription, using electrodynamics alone, that would tell someone how to construct a coordinate system of definite handedness.

REVIEW PROBLEMS

3.1. An uncharged hollow conducting spherical shell, with inner and outer radii equal to R_1 and R_2 respectively, is placed into an otherwise uniform electric field of strength $\mathbf{E}_0$. Calculate the charge density induced on the inner and outer surfaces of the shell, and the electric field everywhere.

Answer: $\sigma = \dfrac{3}{4\pi}E_0\cos\theta,$ on outer surface
$= 0,$ on inner surface

$\mathbf{E} = E_0\left[\hat{E}_0 + \dfrac{3\hat{r}\cos\theta - \hat{E}_0}{r^3}R_2^3\right],$ for $r > R_2$
$= 0,$ for $r < R_2$.

3.2. A conductor of arbitrary shape is surrounded by an electric field $\mathbf{E}(\mathbf{r})$. Show that the net force exerted by the field on the conductor is

given by

$$\mathbf{F} = \frac{1}{2}\int_{\text{surface}} \sigma\mathbf{E}\,ds$$

where σ is the surface charge density and $\mathbf{E}$ is the electric field immediately outside the surface. *Hint*: Use Maxwell's stress tensor.

3.3. Two uniform dielectrics are separated by the x–y plane. The dielectric constant is ε_1 for $z > 0$ and ε_2 for $z < 0$. A point charge q is embedded in the upper dielectric, on the z-axis, a distance d above the interface. Find the electric field at all points. Hint: For $z > 0$ the field can be obtained by *supplementing* the real charge q by an image charge q' on the z-axis at $z = -d$. For $z < 0$ the field can be obtained by *replacing* the real charge q by a fictitious charge q'' on the z-axis at $z = d$. Find q' and q''.

Answer:

$$E = \frac{q}{\varepsilon_1} \left[\frac{\rho\hat{\rho} + (z - d)\hat{z}}{\left[\rho^2 + (z-d)^2\right]^{3/2}} \right.$$

$$\left. + \frac{\varepsilon_1 - \varepsilon_2}{\varepsilon_1 + \varepsilon_2} \frac{\rho\hat{\rho} + (z + d)\hat{z}}{\left[\rho^2 + (z+d)^2\right]^{3/2}} \right] \quad \text{for } z > 0$$

$$= \frac{2q}{\varepsilon_1 + \varepsilon_2} \frac{\rho\hat{\rho} + (z - d)\hat{z}}{\left[\rho^2 + (z-d)^2\right]^{3/2}} \quad \text{for } z < 0.$$

3.4. Find the force on the charge q in the previous problem.

$$\text{Answer: } F = \frac{\varepsilon_1 - \varepsilon_2}{\varepsilon_1(\varepsilon_1 + \varepsilon_2)} \left(\frac{q}{2d}\right)^2 \hat{z}.$$

3.5. The conducting plates of a capacitor are long co-axial cylinders of radii R_1 and R_2 ($R_2 > R_1$). The space between the plates is filled with a material which is a perfect insulator with dielectric constant ε if the electric field is less than a critical value E_c, but is a perfect conductor if the electric field exceeds E_c.

(a) What is the largest potential difference that can be maintained between the cylindrical plates if the field between them is to be less than E_c at all points?

$$\text{Answer: } V_{max} = E_c R_1 \ln\left(\frac{R_2}{R_1}\right).$$

(b) What is the largest potential difference that can be maintained between the cylindrical plates if no current is to flow between them?

$$\text{Answer: } V_{max} = E_c \frac{R_2}{2.718\ldots},$$
$$\text{if } R_1 < \frac{R_2}{2.718\ldots}$$
$$= E_c R_1 \ln\left(\frac{R_2}{R_1}\right), \quad \text{if } R_1 > \frac{R_2}{2.718\ldots}.$$

3.6. A large piece of dielectric has uniform polarization P, except for a spherical hole of radius R. Find the electric field within the hole and in the surrounding dielectric. Hint: Try to make use of the solution given in Problem 3.4.3 for a uniformly polarized dielectric sphere in empty space.

$$\text{Answer: } E = \frac{4\pi}{3} P, \quad \text{within the hole}$$
$$= \frac{4\pi}{3} \frac{R^3}{r^3} [P - 3(P \cdot \hat{r})\hat{r}], \quad \text{outside the hole.}$$

3.7. Let us represent the electrons in an atom by a uniform spherical cloud of total charge $-Ze$ and radius R, with a point nucleus of charge $+Ze$ at its center. A uniform external field E_0 is applied. What is the induced dipole moment? Assume that the electronic charge distribution remains spherical.

$$\text{Answer: } P = R^3 E_0.$$

3.8. Let P_1 and P_2 be the static electric dipole moments of two small objects of zero net charge. The vector R goes from the center of object 1 to the center of object 2.

(a) Find the force and torque acting on object 2.

Answer:

$$F = 3\left[\left(\frac{P_1 \cdot P_2}{R^4} - 5\frac{(P_1 \cdot R)(P_2 \cdot R)}{R^6}\right)\hat{R} \right.$$
$$\left. + \frac{P_1(P_2 \cdot R) + P_2(P_1 \cdot R)}{R^5}\right)\right]$$
$$\tau = \frac{3(P_1 \cdot \hat{R})(P_2 \times \hat{R}) - P_2 \times P_1}{R^3}.$$

(b) Apply your formulae to the special case $P_1 = P\hat{z}$, $P_2 = P\hat{x}$, $R = R\hat{y}$.

$$\text{Answer: } F = 0$$
$$\tau = \frac{P^2 \hat{y}}{R^3}.$$

3.9. Twelve identical capacitors are connected in such a way that each, together with its leads, forms one edge of a cube. Each capacitor is rated at 0.1 μf, with a 600 V maximum voltage.

(a) Compute the effective capacitance between diagonally opposite corners of the cube. *Answer:* 0.12 μf.

(b) Compute the maximum voltage that can be applied to diagonally opposite corners of the cube such that the voltage across each capacitance does not exceed the rated maximum. *Answer:* 1500 V.

3.10. Suppose you have a galvanometer with an internal resistance of 8 ohms, which shows full-scale deflection when it carries a current of 4 milliamps.

(a) How could you use this galvanometer to make a voltmeter whose full-scale deflection will be 10 volts?
Answer: Put a 2492 Ω resistor in series with the galvanometer.

(b) How could you use this galvanometer to make an ammeter whose full-scale deflection will be 10 amps?
Answer: Put a .032 Ω resistor in parallel with the galvanometer.

3.11. C_1 and C_2 represent two closed contours.

(a) Calculate the magnetic flux through C_1 due to a steady current i flowing through C_2. Express your answer in terms of a double line integral around C_1 and C_2.
Answer: $\Phi = \dfrac{i}{c} \oint_{C_1} \oint_{C_2} \dfrac{d\mathbf{r}_1 \cdot d\mathbf{r}_2}{|\mathbf{r}_1 - \mathbf{r}_2|}$.

(b) What is the mutual inductance of C_1 and C_2?
Answer: $M_{12} = \dfrac{1}{c^2} \oint_{C_1} \oint_{C_2} \dfrac{d\mathbf{r}_1 \cdot d\mathbf{r}_2}{|\mathbf{r}_1 - \mathbf{r}_2|} = M_{21}$.

3.12. Two small circular wire loops, of radii R_1 and R_2 respectively, lie in the same plane separated by distance R. Assume that R is so much larger than R_1 and R_2 that the dipole approximation is valid, and calculate the mutual inductance of the two loops.
Answer: $\dfrac{1}{c^2} \dfrac{\pi^2 R_1^2 R_2^2}{R^3}$.

3.13. A proton moves with constant speed v (almost equal to c) along a straight-line orbit a distance b from a stationary electron. Calculate the impulse felt by the electron. Compare your result with the impulse felt by the electron when the proton is nonrelativistic (Problem 3.3.2).
Answer: $\dfrac{2e^2}{vb}$, transverse. The same as for a nonrelativistic proton.

3.14. A toroidal iron ring with permeability $\mu = 1000$ is wound with a uniform coil of 10 turns per centimeter, carrying a current of 5 amperes.

(a) Compute the magnitude of the B and H fields inside the iron.
Answer: $H = 5$ Gauss, $B = 5000$ Gauss.

(b) The current is turned off, but B remains unchanged due to permanent magnetization of the iron. What is H?
Answer: $H = 0$.

(c) The ring is cut and opened into a straight bar. Describe the resulting $\mathbf{B}$ and $\mathbf{H}$ fields qualitatively.   *Answer*: Figure 3.10.

3.15. A conducting rod of length l and resistance R moves in contact with two conducting rails which are connected to a ballistic galvanometer. A uniform magnetic field $\mathbf{B}$ is perpendicular to the rod and to its direction of motion.

(a) How much charge will flow through the galvanometer when the rod moves a distance d?
Answer: $\dfrac{Bl}{c} \cdot \dfrac{d}{R}$.

(b) What force is required to make the rod move with constant speed v?
Answer: $\dfrac{B^2}{c^2} \cdot \dfrac{vl^2}{R}$.

3.16. The plates of a large capacitor cover the planes $z = 0$ and $z = d$. Between the plates are constant fields $\mathbf{E} = -E_0 \hat{z}$ and $\mathbf{B} = B_0 \hat{x}$. An electron (charge $-e$) is released from rest at the origin.

(a) Find the orbit of the electron.
Answer: $x(t) = 0$
$$y(t) = c\frac{E_0}{B_0}\left[\frac{mc}{eB_0}\sin\left(\frac{eB_0}{mc}t\right) - t\right]$$
$$z(t) = c\frac{E_0}{B_0}\frac{mc}{eB_0}\left[1 - \cos\left(\frac{eB_0}{mc}t\right)\right].$$

(b) For a given E, what condition must be satisfied by B_0 if the electron is to reach the $z = d$ plate?
Answer: $B_0 < \sqrt{\dfrac{mc^2 E_0}{2ed}}$.

3.17. A circular loop of radius R carries current i in a uniform magnetic field B perpendicular to

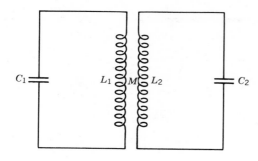

the plane of the loop. An observer looking along the field lines says the current flow is clockwise.

(a) Are the magnetic forces directed radially inwards or outwards? *Answer*: outwards.

(b) What is the tension (or compression) in the wire? *Answer*: $\dfrac{iBR}{c}$.

3.18. A plane electromagnetic wave is scattered by a free electron. Calculate the total scattering cross-section

$$\sigma_T = \frac{\text{total scattered power}}{\text{incident energy flux}}$$

You may assume the wave length of the radiation is long compared to the amplitude of the oscillation of the electron, and that the electron speed remains small enough so that the effect of the incident **B** field is negligible.

$$Answer: \sigma_T = \frac{8\pi}{3}\left(\frac{e^2}{mc^2}\right)^2.$$

3.19. An electromagnetic wave with circular frequency ω propagates in a medium of dielectric constant ε, magnetic permeability μ, and conductivity σ. Show that there is a plane wave solution in which the amplitude of the **E** and **B** fields decreases exponentially along the direction of propagation, and find the characteristic decay length. You can assume that σ is great enough so that $4\pi\sigma/(\varepsilon\omega) \gg 1$.

$$Answer: \frac{c}{\sqrt{2\pi\mu\omega\sigma}}$$

3.20. The drawing shows two L–C circuits which are coupled via their mutual inductance, M. Find the natural oscillation frequencies of this system. Neglect resistance. *Answer*:

$$\omega_\pm = \left[\frac{L_1C_1 + L_2C_2 \pm \left[(L_1C_1 - L_2C_2)^2 + 4C_1C_2M^2\right]}{2C_1C_2(L_1L_2 - M^2)}\right.$$

CHAPTER 4

OPTICS

Optics is the study of the propagation of light, and light is a form of electromagnetic radiation. Thus, the answer to every optics problem can be found, in principle, by solving Maxwell's equations with boundary conditions appropriate to the physical system under consideration. Unfortunately, this program is too difficult to carry out for any real optical system. The subject of optics has evolved instead by developing approximate procedures, whose physical predictions are very nearly equal to those obtained from a full solution of Maxwell's equations. Fraunhofer's diffraction theory is one of the most useful of these procedures. An essential part of this theory is that the radiant energy flux at any observation point is obtained by calculating a coherent sum of complex amplitudes associated with different parts of the system, and then squaring the magnitude of this sum. A similar procedure is used in quantum mechanics to calculate a probability from the square of the magnitude of a sum of complex "probability amplitudes." The application of the Fraunhofer prescription is rather straightforward. Its derivation from Maxwell's equations is more difficult, and is beyond the scope of this book.[1]

A further simplification occurs when the wavelength is much smaller than the distance over which there are appreciable changes in the optical properties of the media carrying the radiation. In this limit, wave optics reduces to ray optics, and problems can be solved by the application of well-defined geometrical constructions. This situation also has a quantum analogue. We will see in Section 5.1 that classical mechanics is obtained from quantum mechanics in the limit in which the "deBroglie wavelength" is small compared to the distances over which the potentials change appreciably.

4.1 FRAUNHOFER DIFFRACTION THEORY

Figure 4.1 shows a plane monochromatic light wave propagating in the z direction, incident from below on an opaque screen lying in the $x-y$ plane. This

---

[1] For a thorough treatment of this subject, see *Principles of Optics*, 6th ed., M. Born and E. Wolf, Pergamon Press, Elmsford, New York, 1980, chapters 8 and 11.

174

FIGURE 4.1 A plane mono-chromatic wave incident from below on a screen perforated by one or more holes.

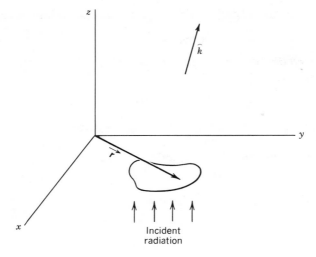

incident wave has energy flux density I_0 (energy/area-time) and wavelength λ. The screen is perforated by one or more holes near the origin. Let $J(\hat{k})$ be the energy flux per unit solid angle scattered into the direction $\hat{k}$. Then, according to the Fraunhofer approximation,

$$J(\hat{k}) = I_0 \times \left| \frac{1}{\lambda} \int_{\text{holes}} e^{-i\mathbf{k}\cdot\mathbf{r}} d^2r \right|^2 \tag{4.1}$$

The integral in (4.1) is a two-dimensional integral over the holes in the opaque screen, and $\mathbf{r}$ locates points in the holes relative to the origin. The vector $\mathbf{k}$ is in the direction of observation $\hat{k}$, and has magnitude $2\pi/\lambda$. The ratio $J(\hat{k})/I_0$, which has the dimensions of area, is called the *differential cross section*, and is written

$$\frac{d\sigma}{d\Omega}(\hat{k}) \equiv \frac{J(\hat{k})}{I_0} = \left| \frac{1}{\lambda} \int_{\text{holes}} e^{-i\mathbf{k}\cdot\mathbf{r}} d^2r \right|^2 \tag{4.2}$$

Finally, the *scattering amplitude* $f(\hat{k})$ is defined by

$$f(\hat{k}) = \frac{1}{\lambda} \int_{\text{holes}} e^{-i\mathbf{k}\cdot\mathbf{r}} d^2r \tag{4.3a}$$

in terms of which (4.2) can be written

$$\frac{d\sigma}{d\Omega}(\hat{k}) = |f(\hat{k})|^2 \tag{4.3b}$$

The scattering amplitude has the dimensions of length; it is the two-dimensional Fourier transform of the pattern of holes in the screen. Since $\mathbf{r}$ in (4.3a) is confined to the x–y plane, $\hat{z} \cdot \mathbf{r} = 0$, and the forward scattering amplitude $f(\hat{z})$ is given by (4.3a) to be

$$f(\hat{z}) = \frac{1}{\lambda} \int_{\text{holes}} d^2r = \frac{\text{area of holes}}{\lambda} \tag{4.4}$$

Let R be the distance between the region of the holes in the screen and the point of observation, and let a be the order of magnitude of the linear dimensions of the holes. Then the Fraunhofer approximation is valid when

$$\sqrt{R\lambda} \gg a$$

If this condition is not satisfied, but

$$\sqrt{R\lambda} \gg a \cdot \frac{a}{R}$$

then Fresnel scattering theory may be used.[2]

PROBLEM 4.1.1   Find the Fraunhofer diffraction from a rectangular slit.

Suppose that there is one hole in the screen, covering the region

$$-\frac{a}{2} \le x \le \frac{a}{2}$$

$$-\frac{b}{2} \le y \le \frac{b}{2}$$

Then the scattering amplitude (4.3a) equals

$$f(\hat{k}) = \frac{1}{\lambda} \int\limits_{x=-a/2}^{a/2} \int\limits_{y=-b/2}^{b/2} e^{-i(k_x x + k_y y)} \, dx \, dy$$

$$= \frac{1}{\lambda} \int\limits_{-a/2}^{a/2} e^{-ik_x x} \, dx \int\limits_{-b/2}^{b/2} e^{-ik_y y} \, dy = \frac{4}{\lambda} \left[\frac{\sin(k_x a/2)}{k_x} \right]\left[\frac{\sin(k_y b/2)}{k_y} \right] \qquad (4.5)$$

This expression drops from ab/λ in the forward direction to zero as k_x and/or k_y increases to $2\pi/a$ or $2\pi/b$, respectively.

If $b \gg a$ (long narrow slit), we get a more rapid falloff of intensity with increasing k_y than with increasing k_x. In other words, the diffraction pattern is more spread out in the x direction than in the y direction. To focus our attention on the x spreading, we rewrite (4.5) in terms of the polar coordinates of $\hat{k}$,

$$k_x = \frac{2\pi}{\lambda} \sin \alpha \cos \beta \qquad k_y = \frac{2\pi}{\lambda} \sin \alpha \sin \beta$$

$$d\Omega = \sin \alpha \, d\alpha \, d\beta$$

and calculate, near $\beta = 0$, the diffracted flux per unit angle α, per unit observation distance in the y direction (see Figure 4.2). The result is

$$\frac{\text{diffracted flux}}{d\alpha \cdot dy} = I_0 \cdot \frac{4b^2}{\pi^2 R} \left[\frac{\sin\left(\dfrac{\pi a}{\lambda} \sin \alpha \right)}{\sin \alpha} \right]^2 \qquad (4.6)$$

Here R is the distance to the observing screen, and enters (4.6) through the relation

$$dy = R \sin \alpha \, d\beta$$

valid near $\beta = 0$. According to (4.6) the diffracted flux falls from a minimum at $\sin \alpha = 0$ to zero at $\sin \alpha = \lambda/a$.

---

[2] For a discussion of the distinction between Fraunhofer and Fresnel diffraction, see *Introduction to Modern Optics*, G. R. Fowles, Holt, Rinehart and Winston, New York, 1968, chapter 4.

FIGURE 4.2 Circular hole in an opaque screen. r and ϕ are plane polar coordinates of points in the hole, while α and β are spherical polar coordinates of **k**.

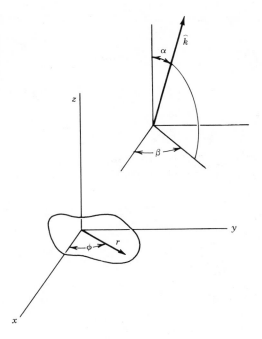

PROBLEM 4.1.2 Find the Fraunhofer diffraction by a circular hole of radius a.

In this case the scattering amplitude (4.3a) is given by

$$f(\hat{k}) = \frac{1}{\lambda} \int_0^a r\,dr \int_0^{2\pi} e^{-i\mathbf{k}\cdot\mathbf{r}}\,d\phi \tag{4.7}$$

(cf. Figure 4.2). The polar coordinates of **r**, which remains in the x–y plane, are $(\pi/2, \phi)$. If we continue to use (α, β) for the polar coordinates of $\hat{k}$, we have

$$\mathbf{r} = r[\hat{x}\cos\phi + \hat{y}\sin\phi]$$

$$\mathbf{k} = k[\hat{x}\sin\alpha\cos\beta + \hat{y}\sin\alpha\sin\beta + \hat{z}\cos\alpha]$$

$$\mathbf{k}\cdot\mathbf{r} = kr[\sin\alpha(\cos\phi\cos\beta + \sin\phi\sin\beta)]$$

$$= kr\sin\alpha\cdot\cos(\phi - \beta)$$

and (4.7) becomes

$$f(\alpha,\beta) = \frac{1}{\lambda}\int_0^a r\,dr \int_0^{2\pi} e^{-ikr\sin\alpha\cdot\cos(\phi-\beta)}\,d\phi = f(\alpha,0) \tag{4.8}$$

Thus, the scattering is axially symmetric, as expected from the symmetry of the circular hole.

The integral in (4.8) can be expressed in terms of Bessel functions. A convenient starting point is the expression[3]

$$e^{ix\cos\phi} = \sum_{m=0}^{\infty} \frac{2}{1 + \delta_{m0}}\,i^m J_m(x)\cos m\phi \tag{4.9a}$$

---

[3]A useful compilation of the properties of Bessel functions, and other special functions of interest in physics, is contained in *Handbook of Mathematical Functions*, M. Abramowitz and I.A. Stegun, National Bureau of Standards Applied Mathematics Series, 1964.

from which we obtain

$$J_m(x) = \frac{i^{-m}}{2\pi} \int_0^{2\pi} e^{ix \cos \phi} \cos m\phi \, d\phi \tag{4.9b}$$

If we set $m = 0$, we get an integral similar to the ϕ integral in (4.8). Thus,

$$f(\alpha, \beta) = \frac{2\pi}{\lambda} \int_0^a J_0(-kr \sin \alpha) r \, dr \tag{4.10}$$

By differentiating (4.9b) with respect to x, and by integrating it partially with respect to ϕ, it can be shown that

$$x^m J_{m-1}(x) = \frac{d}{dx} \left[x^m J_m(x) \right]$$

In particular,

$$x J_0(x) = \frac{d}{dx} \left[x J_1(x) \right]$$

enables us to do the r integration in (4.10), to obtain

$$f(\alpha, \beta) = a \frac{J_1(ka \sin \alpha)}{\sin \alpha} \tag{4.11}$$

Using the limit $J_1(x) \underset{x \to 0}{\to} x/2$, we can check that (4.11) gives the correct answer (4.4) for the forward scattering amplitude.

$J_1(x)$ vanishes at $x = 0, 3.812, 7.016, \ldots$, etc. Thus, when α satisfies

$$\sin \alpha = \frac{3.812}{ka} = .61 \frac{\lambda}{a} = 1.22 \frac{\lambda}{2a}$$

the scattering amplitude (4.11) vanishes. The angular width of the central maximum of the diffraction pattern of a circular hole of diameter D is thus equal to $\arcsin(1.22 \, \lambda/D)$.

If two beams of light making a small angle θ with each other are diffracted by a hole of diameter D, the central maxima of their diffraction patterns will overlap if $\theta < \arcsin(1.22 \, \lambda/D)$. But, if $\theta > \arcsin(1.22 \, \lambda/D)$, the central maxima of the diffractions will not overlap, and the illuminated spots on the viewing screen will be clearly resolved from each other. The resolution criterion

$$\theta > \arcsin(1.22\lambda/D) \approx 1.22\lambda/D \qquad (\lambda/D \ll 1) \tag{4.12}$$

is called the *Rayleigh criterion*.

PROBLEM 4.1.3   A diffraction-limited telescope, designed for use in space, has an aperture of 1 m and a focal length of 20 m. At the focus is a TV detector whose "resolution element" size is 10^{-5} m. At what wavelength is the telescope optimal in the sense that 10^{-5} m corresponds to the diffraction limit?

Let $\Delta \alpha$ be the angular width of the central maximum of the diffraction pattern produced by the 1-m aperture. This produces a spot whose size is (20 m) $\times \Delta \alpha$ in the focal plane. If this equals the 10^{-5} m resolution element, then

$$\Delta \alpha = 1.22\lambda/D = 10^{-5}/20$$

$$\lambda = \frac{10^{-5}}{20} \cdot \frac{D}{1.22} = \frac{10^{-5}}{20} \times \frac{1 \text{ m}}{1.22} = 4 \times 10^{-7} \text{ m}$$
$$= 4,000 \text{ Å}$$

PROBLEM 4.1.4 A plane light wave is diffracted through an irregularly shaped hole in an opaque flat screen placed perpendicular to the direction of propagation. Behind the screen we measure the angular distribution of diffracted light intensity. The Fraunhofer pattern shows point symmetry—any two points, diametrically opposite from the center of the pattern, have equal intensity. Derive this result.

Let us resolve the propagation vector $\mathbf{k}$ into components $\mathbf{k}_\parallel$ and $\mathbf{k}_\perp$, parallel and perpendicular to the screen, respectively:

$$\mathbf{k} = \mathbf{k}_\parallel + \mathbf{k}_\perp$$

Since $\mathbf{r}$ in (4.3a) refers to points in the plane of the screen,

$$\mathbf{k} \cdot \mathbf{r} = \mathbf{k}_\parallel \cdot \mathbf{r}$$

$$f(\hat{k}) = \frac{1}{\lambda} \int\limits_{\text{holes}} e^{-i\mathbf{k}_\parallel \cdot \mathbf{r}} \, d^2r$$

The propagation vector for the point on the opposite side of the center of the pattern is

$$\mathbf{k}' = -\mathbf{k}_\parallel + \mathbf{k}_\perp$$

$$f(\hat{k}') = \frac{1}{\lambda} \int\limits_{\text{holes}} e^{+i\mathbf{k}_\parallel \cdot \mathbf{r}} = [f(\hat{k})]^*$$

Thus,

$$\frac{d\sigma}{d\Omega}(\hat{k}) = |f(\hat{k})|^2 = |f(\hat{k}')|^2 = \frac{d\sigma}{d\Omega}(\hat{k}')$$

4.2 DIFFRACTION BY SEVERAL IDENTICAL HOLES

Suppose that the opaque screen has N identical holes, located at positions $\mathbf{r}_1, \mathbf{r}_2, \ldots, \mathbf{r}_N$. Points within the jth hole are located by $\mathbf{r} = \mathbf{r}_j + \mathbf{t}$ (see Figure 4.3). Since the holes are identical, $\mathbf{t}$ takes on the same range of values for each

FIGURE 4.3 Several identical holes in an opaque screen. The vectors $\mathbf{r}_j$ ($j = 1, 2, \ldots$) go from the origin to the corresponding points in the various holes. The vector $\mathbf{t}$ relates points in the jth hole to $\mathbf{r}_j$.

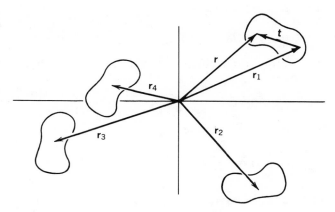

hole. The integral in the scattering amplitude (4.3a) is then a sum of integrals, one for each hole:

$$
\begin{aligned}
f(\hat{k}) &= \frac{1}{\lambda} \sum_{j=1}^{N} \int e^{-i\mathbf{k}\cdot(\mathbf{r}_j + \mathbf{t})} \, d^2t \\
&= \left(\sum_{j=1}^{N} e^{-i\mathbf{k}\cdot\mathbf{r}_j} \right) \cdot \left(\frac{1}{\lambda} \int_{\text{hole}} e^{-i\mathbf{k}\cdot\mathbf{t}} \, d^2t \right) \\
&= g(\hat{k}) \cdot f_{\text{one-hole}}(\hat{k})
\end{aligned}
\tag{4.13a}
$$

Thus, the presence of several identical holes means that the one-hole scattering amplitude is multiplied by an *interference* factor

$$
g(\hat{k}) \equiv \sum_{j=1}^{N} e^{-i\mathbf{k}\cdot\mathbf{r}_j}
\tag{4.13b}
$$

which depends on the locations of the holes, but not on their individual shape. On the other hand, the one-hole amplitude in (4.13a) depends only on the shape of the hole, and is independent of the number or locations of the holes. Corresponding to (4.13a), the angular distribution of the diffracted flux can be written as a product of interference and one-hole diffraction patterns:

$$
\frac{d\sigma}{d\Omega}(\hat{k}) = |g(\hat{k})|^2 \left[\frac{d\sigma(\hat{k})}{d\Omega} \right]_{\text{one-hole}}
\tag{4.14}
$$

In the forward direction,

$$
\mathbf{k}\cdot\mathbf{r}_j = k\hat{z}\cdot\mathbf{r}_j = 0 \qquad (\text{all } j)
$$

Then each term in the interference factor (4.13b) equals unity, and we get

$$
g(\hat{z}) = N
\tag{4.15a}
$$

$$
\frac{d\sigma(\hat{z})}{d\Omega} = N^2 \left[\frac{d\sigma(\hat{z})}{d\Omega} \right]_{\text{one-hole}}
\tag{4.15b}
$$

This is an example of *constructive interference* between the wave amplitudes from the individual holes.

Constructive interference can occur even if $\hat{k} \neq \hat{z}$ if all the holes are arranged in a periodic array. For example, suppose we have a *diffraction grating* with N long slits, each of width a and with distance d between corresponding points of adjacent slits. Then we can write

$$
\left.
\begin{aligned}
\mathbf{r}_j &= (j-1)\cdot d \cdot \hat{x} \\
\mathbf{k}\cdot\mathbf{r}_j &= (j-1)\cdot d \cdot k_x = (j-1)\cdot d \cdot k \sin \alpha
\end{aligned}
\right\} j = 1, 2, \ldots, N
\tag{4.16}
$$

If the observation angle α has a value α_m such that

$$
\sin \alpha_m = \frac{2m\pi}{kd} = m\frac{\lambda}{d} \qquad (m = 0, 1, 2, \ldots,)
\tag{4.17}
$$

the sum (4.13b) becomes

$$
g(\alpha_m) = \sum_{j=1}^{N} e^{-i\cdot(j-1)\cdot 2m\pi} = \sum_{j=1}^{N} 1 = N
$$

and we again have a situation of complete constructive interference. Actually, for the uniform spacing (4.16) we can evaluate $g(\alpha)$ for any value of α:

$$g(\alpha) = \sum_{j=1}^{N} e^{-i(j-1)kd\sin\alpha} = \sum_{j=0}^{N-1} e^{-ijkd\sin\alpha}$$

$$= \sum_{j=0}^{N-1} \left(e^{-ikd\sin\alpha}\right)^{j} = \frac{1 - e^{-iNkd\sin\alpha}}{1 - e^{-ikd\sin\alpha}}$$

$$= e^{-i(N-1)kd\sin\alpha/2} \frac{\sin\left(\dfrac{Nkd}{2}\sin\alpha\right)}{\sin\left(\dfrac{kd}{2}\sin\alpha\right)}$$

$$= e^{-i(N-1)\pi d\sin\alpha/\lambda} \frac{\sin\left(\dfrac{N\pi d}{\lambda}\sin\alpha\right)}{\sin\left(\dfrac{\pi d}{\lambda}\sin\alpha\right)} \tag{4.18}$$

FIGURE 4.4 Diffracted intensity (4.19) for a grating with 5 slits, each of width equal to λ and a center-to-center spacing of 3λ. The abscissa is the diffraction angle α. The numbers written at the bottom of the graph are values of $\sin\alpha$. Note the principal maxima which occur when $\sin\alpha = m\lambda/d$ ($m = 0, 1, 2, \ldots$), and the $N-2$ subsidiary maxima which occur between them. The overall decrease in amplitude with increasing α is due to the single-slit diffraction pattern (4.6).

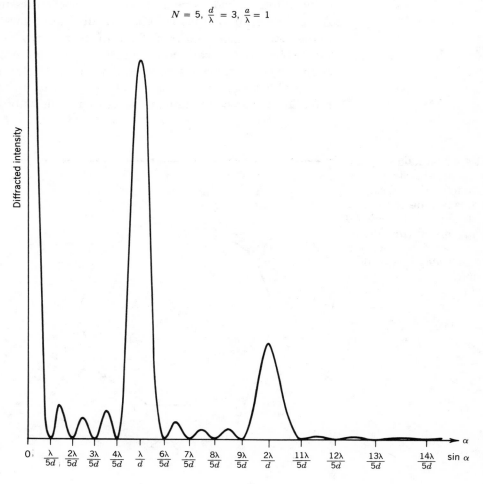

$N = 5, \ \frac{d}{\lambda} = 3, \ \frac{a}{\lambda} = 1$

It is easy to verify that (4.18) approaches N as α approaches each of the α_m defined in (4.17). If we combine (4.18) with the single-slit diffraction pattern (4.6), we get an N-slit pattern given by

$$\frac{\text{diffracted flux}}{d\alpha \, dy} = I_0 \frac{4b^2}{\pi^2 R} \left[\frac{\sin\left(\frac{N\pi d}{\lambda} \sin \alpha \right)}{\sin\left(\frac{\pi d}{\lambda} \sin \alpha \right)} \right]^2 \left[\frac{\sin\left(\frac{\pi \alpha}{\lambda} \sin \alpha \right)}{\sin \alpha} \right]^2 \qquad (4.19)$$

Figure 4.4 is a plot of (4.19) for $N = 5$, $d/\lambda = 3$, and $a/\lambda = 1$.

The strong peaks at the α-values given by the α_m of (4.17) are called *principal maxima*. To see how sharp these peaks are, we determine how far α must deviate from α_m before (4.18) and (4.19) vanish. By comparing

$$\frac{Nkd}{2} \sin \alpha_m = Nm\pi$$

with

$$\frac{Nkd}{2} \sin(\alpha_m + \delta) = (Nm + 1)\pi$$

we see that the half-width δ of the mth principal maximum is approximately

$$\delta \simeq \frac{2\pi}{Nkd \cos \alpha_m} = \frac{\lambda}{Nd \cos \alpha_m} \qquad (\text{for } \delta \ll 1) \qquad (4.20)$$

Thus, by using a grating whose length, Nd, is much greater than λ we can achieve very sharp principal maxima. This result can be understood geometrically if the individual terms in (4.13b) are represented by unit vectors in an Argand diagram (see Figure 4.5). The diagram shows that $|g(\alpha)|^2$ goes from N^2 to zero as $kd \sin \alpha$ goes from $2m\pi$ to $2m\pi + 2\pi/N$. The situation depicted in Figure 4.5b is an example of complete *destructive interference*, as opposed to the complete constructive interference depicted in Figure 4.5a

A diffraction grating is sometimes used to ascertain whether the incident radiation is monochromatic, or whether it consists of two components separated

FIGURE 4.5 Argand diagram representations of the sum (4.13b) for $g(\hat{k})$. In each case 0 labels the origin, and $N = 10$. In (a) $\sin \alpha = m\lambda / d$ (integral m). In (b), $\sin \alpha = m\lambda / d + \lambda / Nd$. In (c), the 10 phases vary in a random manner.

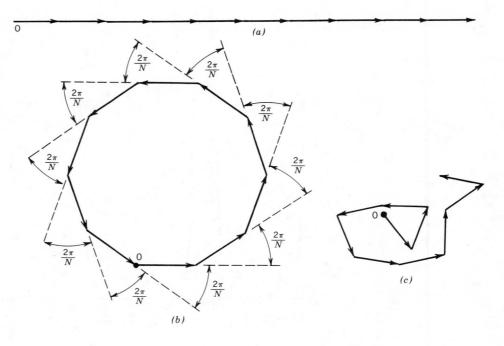

in wavelength by $\Delta\lambda$. According to (4.17), two components separated by $\Delta\lambda$ will produce mth-order principal maxima separated by

$$\Delta\alpha_m = \frac{m\,\Delta\lambda}{d\cdot\cos\alpha_m}$$

It will be possible to resolve these two peaks if $\Delta\alpha_m$ is comparable to, or greater than their individual widths. This requires that

$$\frac{m\,\Delta\lambda}{d\cdot\cos\alpha_m} \gtrsim \delta = \frac{\lambda}{N\cdot d\cdot\cos\alpha_m}$$

$$\frac{\Delta\lambda}{\lambda} \gtrsim \frac{1}{m\cdot N} \equiv \frac{1}{R_m} \tag{4.21}$$

The quantity R_m defined by (4.21) is called the *resolving power* of the mth-order principal maximum. We saw in (4.20) that if we want a grating with sharp principal maxima we should get a long one. We see in (4.21) that if the grating is to have high spectral resolving power, it should have many lines.

Next suppose that the slits are located at random points x_j in an interval of length L, so that $0 \le x_j \le L$ for $j = 1, 2, \ldots, N$. If α is small enough so that $\sin\alpha \ll \lambda/L$, then each term in (4.13b) will be close to unity and we will still get the constructive interference given by (4.15) and illustrated in Figure 4.5a. However, if $\sin\alpha > \lambda/L$, the angles ($kx_j\sin\alpha$) will be distributed at random over the entire unit circle. Then the Argand diagram representation of (4.13b) is a sum of N unit vectors pointing in random directions (Figure 4.5c). The different slits still contribute coherently to the total scattering amplitude, but now the interference is neither completely constructive nor completely destructive.[4]

PROBLEM 4.2.1 Suppose we have a large ensemble of screens, each with N holes distributed at random and the same average number of holes per unit area. All the holes have the same shape, size and orientation. Find an expression for $d\sigma/d\Omega(\hat{k})$ at a given $\hat{k}$, but averaged over the ensemble. Assume that $\hat{k}$ is sufficiently far from the forward direction that all phases are equally probable.

Let a bar over a quantity represent an ensemble average. We need to calculate

$$
\overline{|g(\hat{k})|^2} = \overline{\left|\sum_{j=1}^{N} e^{-i\mathbf{k}\cdot\mathbf{r}_j}\right|^2} = \overline{\left|\sum_{j=1}^{N}\cos(\mathbf{k}\cdot\mathbf{r}_j) - i\sum_{j=1}^{N}\sin(\mathbf{k}\cdot\mathbf{r}_j)\right|^2}
$$

$$
= \overline{\left[\sum_{j=1}^{N}\cos(\mathbf{k}\cdot\mathbf{r}_j)\right]^2} + \overline{\left[\sum_{j=1}^{N}\sin(\mathbf{k}\cdot\mathbf{r}_j)\right]^2}
$$

$$
= \sum_{j=1}^{N}\overline{\left(\left[\cos(\mathbf{k}\cdot\mathbf{r}_j)\right]^2 + \left[\sin(\mathbf{k}\cdot\mathbf{r}_j)\right]^2\right)}
$$

$$
+ \sum_{j\ne l}\overline{\left[\cos(\mathbf{k}\cdot\mathbf{r}_j)\cos(\mathbf{k}\cdot\mathbf{r}_l) + \sin(\mathbf{k}\cdot\mathbf{r}_j)\sin(\mathbf{k}\cdot\mathbf{r}_l)\right]}
$$

$$
= \sum_{j=1}^{N} 1 + \sum_{j\ne l}\overline{\cos\left[\mathbf{k}\cdot(\mathbf{r}_j - \mathbf{r}_l)\right]} = N
$$

---

[4] For an interesting analysis of the statistical distribution of $|g(\hat{k})|^2$ when the phases of the individual terms in (4.13b) are random, see the article by E. Merzbacher, J. M. Feagin and T.-H. Wu, *Am. J. Phys.* **45** (1977) 964.

The ensemble average of $\cos[\mathbf{k} \cdot (\mathbf{r}_j - \mathbf{r}_l)]$ vanishes since the phase $\mathbf{k} \cdot (\mathbf{r}_j - \mathbf{r}_l)$ varies at random from one member of the ensemble to another. Finally the ensemble-averaged diffracted intensity at $\hat{k}$ is

$$\overline{\frac{d\sigma(\hat{k})}{d\Omega}} = N \left[\frac{d\sigma(\hat{k})}{d\Omega} \right]_{\text{one-hole}}$$

which is the formula we would get if we made the (incorrect) assumption that each of the N holes contributes incoherently to the total diffraction pattern.

PROBLEM 4.2.2 Compare the angular resolutions of the following two optical systems:

(a) The 200-inch diameter telescope on Mt. Palomar for 5000 Å light (assuming that this telescope is diffraction limited).

(b) Two radio telescopes, operated as an interferometer, 140 feet and 85 feet in diameter, separated by 2000 miles, for 18 cm radio waves.

According to (4.12) the 200-inch diameter telescope will be able to resolve two sources with an angular separation $\Delta\alpha$ of

$$\Delta\alpha = 1.22 \times \frac{5000 \times 10^{-8} \text{ cm}}{200 \text{ in} \times 2.54 \text{ cm/in}} = 1.2 \times 10^{-7} \text{ radians}$$

According to (4.20), $\Delta\alpha$ for the two radio telescopes is approximately

$$\Delta\alpha \approx 2\delta = 2 \times \frac{\lambda}{2d} = \frac{18 \text{ cm}}{2000 \text{ miles} \times 1.6 \times 10^5 \text{ cm/mile}}$$

$$= 6 \times 10^{-8} \text{ radians} \qquad (\text{for } m = 0)$$

Thus the radio telescopes, operated as an interferometer, have better angular resolution than the optical telescope.

PROBLEM 4.2.3 Monochromatic electromagnetic radiation is incident on a plane opaque sheet with two long parallel slits. Each slit has a width equal to two wavelengths of the radiation, and the centers of the slits are separated by five wavelengths. What pattern of radiation is expected at large distances beyond the plane?

We use the general formula (4.19) with $N = 2$, $a = 2\lambda$, $d = 5\lambda$:

$$I(\alpha) \propto \left[\frac{\sin(10\pi \sin\alpha)}{\sin(5\pi \sin\alpha)} \cdot \frac{\sin(2\pi \sin\alpha)}{\sin\alpha} \right]^2 = \left[2\cos(5\pi \sin\alpha) \cdot \frac{\sin(2\pi \sin\alpha)}{\sin\alpha} \right]^2$$

$$\underbrace{}_{\substack{\text{interference} \\ \text{between slits}}} \underbrace{\phantom{\frac{\sin(2\pi \sin\alpha)}{\sin\alpha}}}_{\substack{\text{diffraction} \\ \text{by a single slit}}}$$

4.3 LIGHT FROM REAL SOURCES

The plane wave solution (3.83) of Maxwell's equations provides a convenient way of relating many optical phenomena to electromagnetic theory. For example, the laws of reflection and refraction at the interface between media having different refractive indices can be obtained by using plane waves to represent incident, reflected, and transmitted radiation, and then applying the appropriate boundary conditions (3.5) to the $\mathbf{E}$ and $\mathbf{B}$ fields at the interface. However, it is clear that the $\mathbf{E}$ and $\mathbf{B}$ fields produced by a real light source, such as an incandescent filament or a gaseous discharge, have much more complex space- and time-dependences than those given by a plane wave. The fields observed at a given point and time are superpositions of radiation emitted independently by many different atoms. Thus, statistical methods are required to describe the results of optical intensity measurements made by macroscopic devices over finite time intervals.

A somewhat simplified picture that exhibits the interference between the fields of different atoms is the following:

1. Each of the N radiating atoms contributes fields

$$\left.\begin{array}{l} \mathbf{E}_m = \hat{x}E_0 e^{i(kz - \omega t + \alpha_m)} \\ \mathbf{B}_m = \hat{y}B_0 e^{i(kz - \omega t + \alpha_m)} \end{array}\right\} m = 1, 2, 3, \ldots, N$$

2. It takes an atom about 10^{-8} s to emit a wave train. Thus, the phases α_m remain constant for a time of the order of 10^{-8} s. At optical frequencies ($\sim 10^{15}$ Hz) this is long enough for about 10^7 oscillations of the field.

3. The atoms radiate independently of each other. Therefore, at any time the phases α_m (for different m) are uncorrelated with each other, and α_m at one time is uncorrelated with α_m a time interval t later, if $t > 10^{-8}$ s.[5]

The total fields at $\mathbf{r}, t$ are thus

$$\mathbf{E}(\mathbf{r}, t) = \sum_{m=1}^{N} \mathbf{E}_m(\mathbf{r}, t) = \hat{x}E_0 \sum_m e^{i(kz - \omega t + \alpha_m)}$$

$$= \hat{x}E_0 e^{i(kz - \omega t)} \sum_m e^{i\alpha_m} \tag{4.22a}$$

$$\mathbf{B}(\mathbf{r}, t) = \sum_{m=1}^{N} \mathbf{B}_m(\mathbf{r}, t) = \hat{y}B_0 e^{i(kz - \omega t)} \sum_{m=1}^{N} e^{i\alpha_m} \tag{4.22b}$$

and the energy flux at $\mathbf{r}, t$ is

$$\mathbf{S}(\mathbf{r}, t) = \frac{c}{4\pi} Re\,\mathbf{E}(\mathbf{r}, t) \times Re\,\mathbf{B}(\mathbf{r}, t) \tag{4.23}$$

Now we average $\mathbf{S}(\mathbf{r}, t)$ over a time interval of about 10^{-8} s, during which the α_m remain constant, but ωt goes through many cycles. As we saw in (3.85), this time average of (4.23) can be written as

$$\overline{\mathbf{S}(\mathbf{r})} = \frac{c}{8\pi} Re(\mathbf{E} \times \mathbf{B}^*)$$

$$= \frac{c}{8\pi} E_0 B_0 Re \sum_{m,n=1}^{N} e^{i\alpha_m} e^{-i\alpha_n} \hat{z} = \frac{c}{8\pi} E_0 B_0 Re \sum_{m,n=1}^{N} e^{i(\alpha_m - \alpha_n)} \hat{z}$$

$$= \frac{c}{8\pi} E_0 B_0 Re \left[\sum_{m=1}^{N} 1 + \sum_{m \neq n} e^{i(\alpha_m - \alpha_n)} \right] \hat{z}$$

$$= \frac{c}{8\pi} E_0 B_0 \left[N + \sum_{m \neq n} \cos(\alpha_m - \alpha_n) \right] \hat{z} \tag{4.24}$$

This gives the result of a measurement that averages the light intensity over a 10^{-8}-second interval. However, most laboratory measurements of light inten-

---

[5] If the light source is a laser, the α_m remain constant for much longer periods than 10^{-8} s. This is because light emission in a laser cavity involves the collective action of many atoms, which are coupled together by the electromagnetic field in the cavity.

sity average it over many 10^{-8}-second intervals. If we average $\cos(\alpha_m - \alpha_n)$ in (4.24) over these many intervals, we get zero, since we have assumed that there is no correlation between α_m and α_n when $m \neq n$. Thus, a long-interval time average of (4.24) gives

$$\bar{\bar{\mathbf{S}}}(\mathbf{r}) = \hat{z} \cdot \frac{c}{8\pi} E_0 B_0 N \qquad (4.25)$$

which is an incoherent sum of the contributions from each atom.

PROBLEM 4.3.1   Show that the magnitude of the flux average $\overline{\mathbf{S}(\mathbf{r})}$, given by (4.24), is equal to the square of the distance gone by a drunkard taking N steps, each of magnitude $\sqrt{(c/8\pi)E_0B_0}$, in directions $\alpha_1, \ldots, \alpha_N$.

Let the mth step be represented by the vector $\mathbf{r}_m = \sqrt{(c/8\pi)E_0B_0}\,\hat{r}_m$. The total displacement of the drunkard is

$$\mathbf{R} = \sum_{m=1}^{N} \mathbf{r}_m = \sqrt{\frac{c}{8\pi}E_0B_0}\sum_{m=1}^{N}\hat{r}_m$$

and the square of the magnitude of $\mathbf{R}$ is

$$\mathbf{R} \cdot \mathbf{R} = \frac{c}{8\pi}E_0B_0\sum_{m,n=1}^{N}\hat{r}_m \cdot \hat{r}_n = \frac{c}{8\pi}E_0B_0\sum_{m,n}\cos(\alpha_m - \alpha_n)$$

$$= \frac{c}{8\pi}E_0B_0\left[N + \sum_{m \neq n}\cos(\alpha_m - \alpha_n)\right]$$

in agreement with (4.24).

PROBLEM 4.3.2   Let $\left|\overline{\mathbf{S}(\mathbf{r})}\right|$ signify the magnitude of $\overline{\mathbf{S}(\mathbf{r})}$. According to (4.25), the mean of repeated measurements of this quantity is $(c/8\pi)E_0B_0N$. Calculate the root-mean-square (rms) dispersion of $\left|\overline{\mathbf{S}(\mathbf{r})}\right|$ about this mean.

The rms dispersion of $\left|\overline{\mathbf{S}(\mathbf{r})}\right|$ about its mean is

$$\sigma = \sqrt{\overline{\left(\left|\overline{\mathbf{S}(\mathbf{r})}\right| - \frac{c}{8\pi}E_0B_0N\right)^2}}$$

$$= \frac{c}{8\pi}E_0B_0\sqrt{\overline{\left(\sum_{m \neq n}\cos(\alpha_m - \alpha_n)\right)^2}}$$

The mean-square sum is

$$\overline{\left(\sum_{m \neq n}\cos(\alpha_m - \alpha_n)\right)^2} = \sum_{\substack{m_1, m_2, n_1, n_2 \\ m_1 \neq n_1, m_2 \neq n_2}}\overline{\cos(\alpha_{m_1} - \alpha_{n_1})\cos(\alpha_{m_2} - \alpha_{n_2})}$$

$$= \frac{1}{2}\sum_{\substack{m_1, m_2, n_1, n_2 \\ m_1 \neq n_1, m_2 \neq n_2}}\left[\overline{\cos(\alpha_{m_1} - \alpha_{n_1} + \alpha_{m_2} - \alpha_{n_2})}\right.$$

$$\left. + \overline{\cos(\alpha_{m_1} - \alpha_{n_1} - \alpha_{m_2} + \alpha_{n_2})}\right]$$

The mean of $\cos(\alpha_{m_1} - \alpha_{n_1} + \alpha_{m_2} - \alpha_{n_2})$ equals unity if $m_1 = n_2$ and $n_1 = m_2$; otherwise it vanishes. Of all the m_1, m_2, n_1, n_2 sets included in the sum in (4.22), there are $N^2 - N$ ways in which $m_1 = n_2$ and $n_1 = m_2$ (without allowing

$m_1 = n_1, m_2 = n_2$). A similar analysis applies to $\cos(\alpha_{m_1} - \alpha_{n_1} - \alpha_{m_2} + \alpha_{n_2})$. Thus, we conclude that

$$\overline{\left(\sum_{m \neq n} \cos(\alpha_m - \alpha_n) \right)^2} = N^2 - N$$

and the root-mean-square dispersion of $|\bar{\mathbf{S}}|$ about its mean is

$$\sigma = \frac{c}{8\pi} E_0 B_0 \sqrt{N^2 - N} \simeq \frac{c}{8\pi} E_0 B_0 N \qquad \text{(for large } N\text{)} \qquad (4.26)$$

We see from (4.25) and (4.26) that if we measured light intensity repeatedly for 10^{-8}-second intervals (and this is possible with modern detecting systems), then the measurements would fluctuate about the mean (4.25) with an rms dispersion, σ, comparable to this mean. If the detecting system averaged over $\nu \cdot 10^{-8}$-second intervals, the mean would still be given by (4.25), but now the rms dispersion between successive measurements would be $\sigma/\sqrt{\nu}$. For example, repeated measurements of one-second average intensities would have an rms dispersion that is approximately $\sqrt{10^8} = 10^4$ times smaller than the long-term average intensity.

Now consider the interference fringes observed in a Michelson interferometer. This interference is produced by the superposition of two beams which have a common origin but different path lengths. A path length difference of Δl introduces a phase difference between the beams of $2\pi \Delta l / \lambda \equiv \phi$. But if Δl is of the order of a meter or less, then the travel times of the two beams will be associated with the same set of atomic phases α_m. Thus, the observed intensity, averaged over $\sim 10^{-8}$ s, is

$$\left| \sum_m e^{i\alpha_m} + \sum_m e^{i(\alpha_m + \phi)} \right|^2 = \left| \sum_m e^{i\alpha_m}(1 + e^{i\phi}) \right|^2$$

$$= \left| \sum_m e^{i\alpha_m} \right|^2 \left| e^{i\phi/2}(e^{i\phi/2} + e^{-i\phi/2}) \right|^2$$

$$= 4 \left| \sum_m e^{i\alpha_m} \right|^2 \cos^2 \frac{\phi}{2} \qquad (4.27)$$

We see that the interference between the beams is constructive or destructive, depending on the value of $\cos^2 \phi/2$.

Suppose that the two beams in the interferometer were from different light sources. Then (4.27) would be replaced by

$$\left| \sum_m e^{i\alpha_m} + \sum_m e^{i(\bar{\alpha}_m + \phi)} \right|^2$$

$$= \left| \sum_m e^{i\alpha_m} \right|^2 + \left| \sum_m e^{i\bar{\alpha}_m} \right|^2 + \sum_{m_1, m_2} \left[e^{i(\alpha_{m_1} - (\bar{\alpha}_{m_2} + \phi))} + e^{-i(\alpha_{m_1} - (\bar{\alpha}_{m_2} + \phi))} \right]$$

$$= \left| \sum_m e^{i\alpha_m} \right|^2 + \left| \sum_m e^{i\bar{\alpha}_m} \right|^2 + 2 \sum_{m_1, m_2} \cos(\alpha_{m_1} - \bar{\alpha}_{m_2} - \phi) \qquad (4.28)$$

Here we have used α_m and $\bar{\alpha}_m$ to label the different atomic phases of the mth components in the two beams. We assume that the light sources are independent, so α_m and $\bar{\alpha}_m$ are uncorrelated. Then, if our observation averages (4.28) over many 10^{-8}-second intervals, each cosine term will average to zero. What remains is

$$\left| \sum_m e^{i\alpha_m} \right|^2 + \left| \sum_m e^{i\bar{\alpha}_m} \right|^2$$

an incoherent sum of contributions from each beam. This is independent of the path-difference-related phase ϕ, and there are no interference fringes if the beams are from different light sources. However, if (4.28) is measured over a time interval comparable to 10^{-8} s (for incandescent sources), then the third term does not vanish and interference effects are observed. If the light sources are two lasers, one can detect interference effects in measurements averaged over longer time intervals.

If the beams are from the same incandescent light source, but Δl is greater than a few meters, then the travel times of the two beams will differ by more than 10^{-8} s, and the fields at the viewing screen from the two beams will, at any time, be associated with different sets of atomic phases. The situation is similar to that produced by independent sources for the two beams, and no interference fringes will be produced. Of course, in a real measurement the fringes will gradually disappear as we go from $\Delta l < 1$ m to $\Delta l/\lambda >$ several meters.

We can use (4.22) to describe unpolarized light if we allow each component to have its own linear polarization

$$\hat{\varepsilon}_m = \hat{x} \cos \theta_m + \hat{y} \sin \theta_m$$

with the θ_m distributed uniformly and randomly between 0 and 2π. Then (4.22) and (4.24) are replaced by

$$\mathbf{E}(\mathbf{r}, t) = E_0 e^{i(kz - \omega t)} \sum_{m=1}^{N} \hat{\varepsilon}_m e^{i\alpha_m} \tag{4.29a}$$

$$\mathbf{B}(\mathbf{r}, t) = B_0 e^{i(kz - \omega t)} \sum_{m=1}^{N} \hat{z} \times \hat{\varepsilon}_m e^{i\alpha_m} \tag{4.29b}$$

$$\overline{\mathbf{S}(\mathbf{r}, t)} = \frac{c}{8\pi} E_0 B_0 Re \sum_{m_1, m_2} \hat{\varepsilon}_{m_1} \times \left(\hat{z} \times \hat{\varepsilon}_{m_2} \right) e^{i(\alpha_{m_1} - \alpha_{m_2})}$$

$$= \hat{z} \frac{c}{8\pi} E_0 B_0 Re \sum_{m_1, m_2} \hat{\varepsilon}_{m_1} \cdot \hat{\varepsilon}_{m_2} e^{i(\alpha_{m_1} - \alpha_{m_2})}$$

$$= \hat{z} \frac{c}{8\pi} E_0 B_0 \left[N + \sum_{m_1 \neq m_2} \hat{\varepsilon}_{m_1} \cdot \hat{\varepsilon}_{m_2} \cos(\alpha_{m_1} - \alpha_{m_2}) \right] \tag{4.30}$$

Since the polarization vectors $\hat{\varepsilon}_m$ are independent of the α_m, each term $\hat{\varepsilon}_{m_1} \cdot \hat{\varepsilon}_{m_2} \cos(\alpha_{m_1} - \alpha_{m_2})$ will average to zero if we average (4.30) over many 10^{-8}-second intervals. This leaves a net energy flux of magnitude $(c/8\pi)E_0 B_0 N$, as we obtained for linearly polarized light.

Now suppose that the unpolarized light of (4.29) passes through a polarizing filter that only transmits the component of the $\mathbf{E}$ field parallel to some direction $\hat{p}$ in the x–y plane. On the far side of the filter the fields will be

$$\mathbf{E}(\mathbf{r}, t) = \hat{p} E_0 e^{i(kz - \omega t)} \sum_{m=1}^{N} \hat{p} \cdot \hat{\varepsilon}_m e^{i\alpha_m} \tag{4.31a}$$

$$\mathbf{B}(\mathbf{r}, t) = \hat{z} \times \hat{p} B_0 e^{(kz - \omega t)} \sum_{m=1}^{N} \hat{p} \cdot \hat{\varepsilon}_m e^{i\alpha_m} \tag{4.31b}$$

and the energy flux, averaged over 10^{-8} s, is

$$\overline{\mathbf{S}(\mathbf{r})} = \frac{c}{8\pi} \hat{p} \times (\hat{z} \times \hat{p}) E_0 B_0 \left| \sum_{m=1}^{N} \hat{p} \cdot \hat{\varepsilon}_m e^{i\alpha_m} \right|^2$$

$$= \hat{z} \frac{c}{8\pi} E_0 B_0 \sum_{m_1, m_2 = 1}^{N} (\hat{p} \cdot \hat{\varepsilon}_{m_1})(\hat{p} \cdot \hat{\varepsilon}_{m_2}) e^{i(\alpha_{m_1} - \alpha_{m_2})}$$

$$= \hat{z} \frac{c}{8\pi} E_0 B_0 \left[\sum_{m=1}^{N} (\hat{p} \cdot \hat{\varepsilon}_m)^2 + \sum_{m_1 \neq m_2} (\hat{p} \cdot \hat{\varepsilon}_{m_1})(\hat{p} \cdot \hat{\varepsilon}_{m_2}) \cos(\alpha_{m_1} - \alpha_{m_2}) \right]$$

As before, the $m_1 \neq m_2$ terms do not survive averaging over many 10^{-8}-second intervals. The average value of each $(\hat{p} \cdot \hat{\varepsilon}_m)^2$ term is $1/2$, since the $\hat{\varepsilon}_m$ are distributed at random about any direction $\hat{p}$. Thus, the long-time-averaged flux that emerges from the polarizing filter has magnitude $(c/16\pi)E_0 B_0$, which is one-half the magnitude of the incident flux.

PROBLEM 4.3.3 Light of intensity $(c/16\pi)E_0 B_0$ emerges from a polaroid sheet, linearly polarized in the $\hat{p}$ direction. What experiment could you do to determine whether the light incident on the other side of the sheet is

(a) unpolarized, of intensity $(c/8\pi)E_0 B_0$, or

(b) linearly polarized parallel to $\hat{p}$, of intensity $(c/16\pi)E_0 B_0$, or

(c) linearly polarized at 45° to $\hat{p}$, of intensity $(c/8\pi)E_0 B_0$?

Rotate the polaroid sheet. If (a) is true, the transmitted intensity will be unaffected. If (b) is true, the transmitted intensity will decrease no matter which way the sheet is rotated. If (c) is true, the transmitted intensity will decrease if the sheet is rotated in one direction, and increase if it is rotated in the opposite direction.

4.4 GEOMETRICAL OPTICS

4.4.1 Fermat's Principle of Least Time

The path taken by a light ray between points A and B is such that the time of travel along this path is stationary with respect to small deviations from this path.[6] Usually, the time along the path is an absolute minimum.

Figure 4.6a shows a light beam traveling from A to B after reflection from a plane surface. The travel time for the indicated path is

$$T(x) = \frac{1}{v}\sqrt{h_1^2 + x^2} + \frac{1}{v}\sqrt{h_2^2 + (D - x)^2}$$

Here v is the speed of propagation of light in the medium above the plane. $T(x)$ will be a minimum when

$$0 = \frac{dT}{dx} = \frac{1}{v}\left[\frac{x}{\sqrt{h_1^2 + x^2}} - \frac{D - x}{\sqrt{h_2^2 + (D - x)^2}} \right]$$

---

[6] For the relation between Fermat's principle and the wave and photon interpretations of optics, see *The Feynman Lectures on Physics*, R. P. Feynman, R. B. Leighton, and M. Sands, Addison-Wesley, Reading, Mass., 1964, vol. I, chapter 26.

FIGURE 4.6 (a) A light path from *A* to *B* via reflection from a plane surface, (b) a light path from *A* to *B* via refraction at a plane surface.

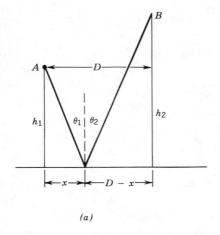

(a)

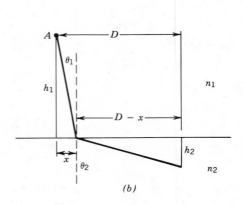

(b)

Thus,

$$\sin \theta_1 = \sin \theta_2, \qquad \theta_1 = \theta_2 \qquad (4.33)$$

and the angle of incidence equals the angle of reflection.

In Figure 4.6*b* the points *A* and *B* are on opposite sides of a plane surface separating media with refractive indices n_1 and n_2. Now the travel time is

$$T(x) = \frac{n_1}{c}\sqrt{h_1^2 + x^2} + \frac{n_2}{c}\sqrt{h_2^2 + (D - x)^2}$$

with *c* the speed of light in vacuum. To minimize $T(x)$, we set

$$0 = \frac{dT}{dx} = \frac{1}{c}\left[\frac{n_1 x}{\sqrt{h_1^2 + x^2}} - \frac{n_2(D - x)}{\sqrt{h_2^2 + (D - x)^2}} \right]$$

$$n_1 \sin \theta_1 = n_2 \sin \theta_2 \qquad (4.34)$$

This is Snell's law of refraction. Equations (4.33) and (4.34) can also be obtained by applying Maxwell's equations to waves propagating above and below the plane, and using the appropriate boundary conditions to match the **E** and **B** vectors in the incident, reflected, and refracted waves (cf. Section 3.16).

In general, the index of refraction of a medium depends on the frequency of the radiation (cf. Section 3.17). Thus, if the incident light in Figure 4.6*b* has several frequency components, they will emerge at several different angles of refraction θ_2, even though they all had the same angle of incidence θ_1. The dependence of the ratio n_2/n_1 on frequency has produced a *dispersion* of the incident light. We have already encountered the phenomenon of dispersion in connection with diffraction gratings (cf. 4.17).

PROBLEM 4.4.1   An idealized planet has a transparent atmosphere whose index of refraction *n* at a certain wavelength varies with height according to

$$n(h) = 1 + Ne^{-h/H}$$

where

> *h* = distance above the planet's spherical surface.
>
> *N* = a constant, which is much smaller than unity.
>
> *H* = a constant which is much smaller than the planet's radius *R*.

Find the approximate height at which a light ray can travel in a circular path around the planet (i.e., along $h = $ constant). What condition on N, H, R is required for there to be a solution with $h > 0$? Use Fermat's principle.

The time it takes for a light beam to circle the planet at radius r is

$$T = \oint \frac{n(r)}{c} \sqrt{(r\,d\theta)^2 + (dr)^2}$$

$$= \int_0^{2\pi} \frac{n(r(\theta))}{c} \sqrt{r^2 + \left(\frac{dr}{d\theta}\right)^2} \, d\theta$$

The Euler-Lagrange condition for this to be stationary, subject to the restrictions $r(0) = r(2\pi) = r_0$, is

$$\frac{d}{d\theta} \frac{\partial L}{\partial \left(\dfrac{dr}{d\theta}\right)} = \frac{\partial L}{\partial r}$$

with

$$L\left(r, \frac{dr}{d\theta}\right) = n(r) \sqrt{r^2 + \left(\frac{dr}{d\theta}\right)^2}$$

If we carry out the differentiations, we see that a solution with r constant is possible only if

$$n(r) + r \frac{dn(r)}{dr} = 0$$

If we apply this condition to

$$n(r) = 1 + N e^{-(r-R)/H}$$

we find that

$$e^{(r-R)/H} = N\left[\frac{r}{H} - 1\right]$$

Since r and R are comparable, and much greater than H, this equation is

FIGURE 4.7 A bundle of rays
emerging from A and
focused at B.

approximately the same as

$$e^{(r-R)/H} = \frac{NR}{H}$$

$$r - R = h = H \ln\left(\frac{NR}{H}\right)$$

If h is to be positive, it must be that $NR > H$.

4.4.2 The Idea of a Focus

If the rays in a bundle emanating from A are focused at B, it must be that the travel times along all the paths of the rays in the bundle are equal (Figure 4.7). If the travel time along one path were less than the travel time along its neighbors, Fermat's principle would require that the light follows that path alone. B is said to be the real image or focus of A.

PROBLEM 4.4.2 The index of refraction of glass can be increased by diffusing in impurities. It is then possible to make a lens of constant thickness. Given a disk of radius a and thickness d, find the radial variation of the index of refraction $n(r)$ that will bring rays emitted from A in the diagram below to a focus at B. Assume that the lens is thin ($d \ll a, b$).

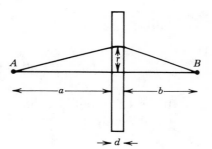

Compare the two paths shown between A and B. The requirement of equal time along the two paths gives

$$\sqrt{a^2 + r^2} + d \cdot n(r) + \sqrt{b^2 + r^2} = a + d \cdot n(0) + b$$

Thus

$$n(r) = n(0) + \frac{a + b - \sqrt{a^2 + r^2} - \sqrt{b^2 + r^2}}{d}$$

4.4.3 Refraction at a Spherical Surface

In Figure 4.8 a homogeneous region of refractive index n_1 is to the left of a spherical surface of radius R, and a homogeneous region of refractive index n_2 is to the right of this surface. Suppose that point O' is the focus of the light emitted from point O. Then the travel time along paths OPO' and OQO' are equal. Thus,

$$n_1 s_1 + n_2 s_2 = n_1 \cdot OV + n_2 \cdot O'V \tag{4.35}$$

If

$$h \ll s_1, s_2, R \qquad \text{(paraxial rays)} \tag{4.36}$$

then

$$s_1 = \sqrt{OQ^2 + h^2} = OQ\sqrt{1 + \left(\frac{h}{OQ}\right)^2} \simeq OQ\left[1 + \frac{1}{2}\frac{h^2}{OQ^2}\right]$$

$$s_1 \simeq OQ + \frac{1}{2}\frac{h^2}{OQ} \tag{4.37a}$$

$$s_2 \simeq O'Q + \frac{1}{2}\frac{h^2}{O'Q} \tag{4.37b}$$

$$R \simeq CQ + \frac{1}{2}\frac{h^2}{CQ} \tag{4.37c}$$

If these approximations are used in (4.35), we get

$$\frac{n_1}{OQ} + \frac{n_2}{O'Q} = \frac{n_2 - n_1}{CQ}$$

If (4.36) is true, this last equation is also approximately equivalent to

$$\frac{n_1}{OV} + \frac{n_2}{O'V} = \frac{n_2 - n_1}{R} \tag{4.38}$$

Equation (4.38) is appropriate to the situation shown in Figure 4.8, in which the center of curvature of the spherical surface is in the medium of larger n, and in which there is a real object to the left of the spherical surface. We can continue to use (4.38) in other situations if we obey the following conventions:

1. R is negative if the center of the spherical surface is in the region of smaller n.

2. A negative object distance OV refers to a virtual object at a distance $-OV$ to the right of V. This means that the light traveling in medium 1 appears to be converging to a point $-OV$ to the right of V.

FIGURE 4.8 A spherical surface, centered at C and of radius R, separates homogeneous media with refractive indices n_1 and n_2. This diagram refers to a situation in which $n_2 > n_1$.

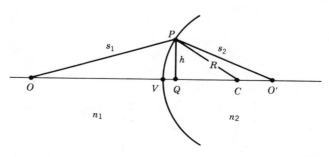

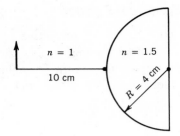

3. A negative image distance $O'V$ refers to a virtual image at a distance $-O'V$ to the left of V. This means that the light traveling in medium 2 appears to be diverging from a point $-O'V$ to the left of V.

PROBLEM 4.4.3 A hemispherical lens has a 4-cm radius of curvature, and is made of glass of refractive index 1.5. An object is placed on the axis of the lens, 10 cm from the curved face. Where is the final image? (See Figure 4.9).

We first apply (4.38) to the refraction at the spherical surface:

$$\frac{1}{10 \text{ cm}} + \frac{1.5}{O'V} = \frac{1.5 - 1}{4 \text{ cm}}$$

The solution of this equation gives $O'V = 60$ cm, and the image of this first refraction is 60 cm to the right of the point where the axis pierces the curved surface. This means that this image, which is the object of the refraction at the plane surface, is 56 cm to the right of the plane surface. Thus, for this second refraction, we have

$$\frac{1.5}{-56 \text{ cm}} + \frac{1}{O'V} = \frac{1 - 1.5}{\infty}$$

$$O'V = \frac{56 \text{ cm}}{1.5} = +37.33 \text{ cm}$$

The final image is 37.33 cm to the right of the plane surface of the lens.

4.4.4 Refraction by a Thin Lens with Two Spherical Surfaces

If we apply (4.38) to the two surfaces of the lens shown in Figure 4.10, and neglect the thickness of the lens, we obtain the lensmaker's formula

$$\frac{1}{o} + \frac{1}{i} = (n - 1)\left(\frac{1}{R_1} - \frac{1}{R_2}\right) \equiv \frac{1}{f} \tag{4.39}$$

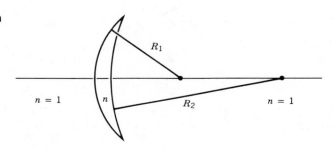

Here o and i are the object and image distances, respectively, n is the refractive index of the glass, R_1 and R_2 are the radii of curvature of the surfaces, and f is the *focal length* of the lens. If either surface curves in the opposite direction to that shown in Figure 4.10, the sign of the corresponding R should be negative. Negative values of o and/or i imply virtual object and/or image. The lateral magnification of the lens is given by

$$m = -\frac{i}{o} \tag{4.40}$$

The minus sign signifies that a real image of a real object is inverted, whereas a virtual image of a real object is erect.

PROBLEM 4.4.4   A lens must be selected to photograph a cubic spark chamber 10 inches on a side. The image size is to be 1 cm (so as to fit three images on 35 mm film). The distance between the spark chamber and film is 3 m. Assume, where necessary, that the sparks and light used for illumination will be in the visible red portion of the spectrum.

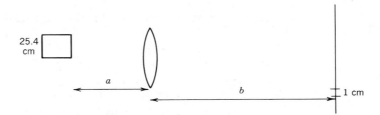

(a) What focal length lens should be chosen?

If the 25.4 cm box is to have a 1 cm image, then $a/b = 25.4/1$; also $a + b = 300$ cm. Thus, a $= 25.4/26.4 \times 300$ cm, $b = 1/26.4 \times 300$ cm, and the focal length f is given by

$$\frac{1}{f} = \frac{1}{a} + \frac{1}{b}, \qquad f = \frac{ab}{a + b} = 10.9 \text{ cm}$$

(b) If the F-stop of the lens is 5.6 (the F number is the ratio f/D, where f is the focal length and D the diameter of the lens), compare the resolution limit imposed by diffraction with that imposed by the 3 μm grain size.

Equation (4.12) expresses the diffraction limit of a lens in terms of its diameter D. We can calculate D from

$$D = \frac{f}{(f/D)} = \frac{10.9 \text{ cm}}{5.6} = 1.95 \text{ cm}$$

Then (4.12) gives

$$\Delta\alpha = 1.22\frac{\lambda}{D} \simeq 1.22 \times \frac{6500 \text{ Å}}{1.95 \times 10^8 \text{ Å}} = 4.1 \times 10^{-5} \text{ radians}$$

This corresponds to an uncertainty in position on the screen of about

$$4.1 \times 10^{-5}b = 5 \times 10^{-4} \text{ cm} = 5 \text{ } \mu\text{m}$$

Thus, the diffraction produced by the lens imposes a stronger limit on the resolution of the camera than does the 3 μm grain size.

FIGURE 4.11 A paraxial ray emanating from 0 is reflected from the spherical mirror to 0'.

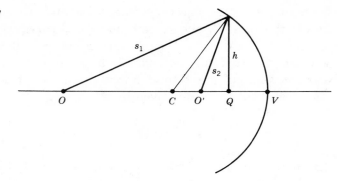

4.4.5 Reflection by a Spherical Mirror

The condition for a focus applied to the situation shown in Figure 4.11 gives

$$s_1 + s_2 = OV + O'V$$

If we use approximations (4.36) and (4.37), we get

$$\frac{1}{OV} + \frac{1}{O'V} = \frac{2}{R} \qquad (4.41)$$

R is taken positive if and only if the center of curvature is on the side from which the light is incident. As in Section 4.4.3, positive values of OV and/or $O'V$ refer to real object and/or image; negative values imply that the object and/or image is virtual. The magnification is given by (4.40).

PROBLEM 4.4.5   A cook has a shiny, spherically shaped spoon. On looking into the concave side he sees his inverted image 4 cm from the spoon. Without changing the distance between himself and the spoon he turns the spoon over and sees an erect image of himself 3 cm from the spoon. What is the radius of curvature of the spoon?

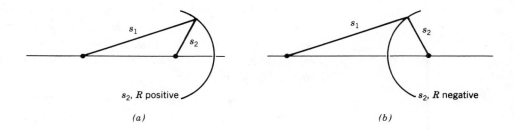

The signs of R and s_2 are reversed in the two cases. The mirror equation gives

$$\frac{1}{s_1} + \frac{1}{4} = \frac{2}{R}$$

$$\frac{1}{s_1} - \frac{1}{3} = \frac{2}{-R}$$

The difference of these two equations gives us R:

$$\frac{1}{4} + \frac{1}{3} = \frac{4}{R} = \frac{7}{12}$$

$$R = \frac{48}{7} \text{ cm.}$$

REVIEW PROBLEMS

4.1. An opaque screen in the $x-y$ plane is perforated by three circular holes, each of diameter a, with their centers at points whose (x, y) coordinates are $(d, 0)$, $(-d/2, -d\sqrt{3}/2)$, $(-d/2, +d\sqrt{3}/2)$. The screen is illuminated from below by light of wavelength λ from a distant source on the $-z$ axis. Find the distribution of diffracted intensity above the $x-y$ plane, as a function of the polar coordinates θ, ϕ.

Answer: $I(\theta, \phi) \sim \left[1 + 4\cos^2\left(\frac{\pi\sqrt{3}}{\lambda} \sin\theta \sin\phi \right) \right.$

$$+ 4\cos\left(\frac{\pi\sqrt{3}}{\lambda} \sin\theta \sin\phi \right)$$

$$\left. \times \cos\left(\frac{3\pi}{\lambda} \sin\theta \cos\phi \right) \right]$$

$$\times \left[\frac{J_1\left(\frac{2\pi a}{\lambda} \sin\theta \right)}{\sin\left(\frac{2\pi a}{\lambda} \sin\theta \right)} \right]^2 .$$

4.2. Plane parallel 10-cm radar waves emerge from a horn antenna whose opening is a rectangle that is 2 m wide and 1 m high. What are the angular width and height of the central maximum in the beam?

Answer: angular width $= 2 \times \arcsin\left(\frac{1}{20} \right)$

angular height $= 2 \times \arcsin\left(\frac{1}{10} \right)$.

4.3. A pinhole camera is made by cutting a small circular hole of diameter a in one side of a closed rectangular box, and projecting an image on the opposite side of the box, a distance L away. The camera is to be used with light of wavelength λ. What is the optimum pinhole size, optimum in the sense that it produces the smallest possible image for a point source of light far away from the camera? *Answer:* $a \approx \sqrt{1.22\lambda L}$.

4.4. A parallel beam with $\lambda = 5890$ Å is incident normal to a diffraction grating with 40 lines/mm. How many orders are visible? What is the dispersive power $\frac{d\alpha}{d\lambda}$ in the 4th order? *Answer:* 42; 1600 rad/cm.

4.5. Light from a source slit falls on two narrow slits 1 mm apart and 100 mm from the source. The fringes are observed on a screen 1 m away. Only the band between 4800 and 5200 Å is used.

(a) What is the fringe separation? *Answer:* angular separation $\sim 5 \times 10^{-4}$ rad, linear separation $\sim .5$ mm.

(b) How may fringes are clearly visible? *Answer:* about 13.

(c) How wide can the source slit be without seriously reducing the fringe visibility? *Answer:* angular width $\ll 5 \times 10^{-4}$ rad, so linear width $< 2 \times 10^{-2}$ mm.

4.6. Newton's rings are observed when a plano-convex lens is placed convex side down on a flat glass plate and illuminated from above with monochromatic light. The first bright ring has radius 1 mm.

(a) If the radius of the convex surface is 4 m, what is the wave length of the light? *Answer:* 5000 Å

(b) If the space between the lens and the flat plate is filled with water ($n = 4/3$), what will be the radius of the first bright ring? *Answer:* .87 mm.

4.7. Interference effects such as beats can easily be observed with two sound sources, such as two tuning forks or two loud speakers driven by oscillators. However, interference effects cannot be demonstrated with light from two different sources. What is the reason for this difference?

4.8. A slab of birefringent crystal of thickness T has its lower face in the x–y plane. Light propagating in the $+\hat{z}$ direction through the crystal has refractive index n_1 if its **E** vector points in the $\hat{x}$ direction and refractive index n_2 if its **E** vector points in the $\hat{y}$ direction. Suppose that linearly polarized light of wavelength λ is incident on this crystal from below, with its **E** vector making an angle of 45° with the $\hat{x}$ axis. Show that the light emerging from the upper crystal face is elliptically polarized, and find the ratio of the semi-axes of the ellipse. *Answer*: $\tan\left(\dfrac{\pi(n_2 - n_1)}{\lambda} T \right)$.

4.9. A light ray is incident at angle ψ to the normal of a plane-parallel glass plate of refractive index n and thickness t. Find the lateral displacement d of the ray if $n \simeq 1$. *Answer*: $d = t \sin\left(\psi\left(1 - \dfrac{1}{n} \right) \right)$.

4.10. The concave mirror of a reflecting telescope is 50 cm in diameter and has focal length 250 cm.

   (a) The moon as seen from the earth subtends an angle of 30 min. Find the diameter of the moon's image formed by the mirror. *Answer*: 2.18 cm.

   (b) Find the diameter of the image of a star in this instrument. (Use $\lambda = 5500$ Å) *Answer*: 3.4×10^{-4} cm.

4.11. A large concave mirror has a radius of 4 feet. A man stands 40 ft. in front of the mirror.

   (a) Where is his image? *Answer*: $2\dfrac{2}{19}$ ft in front of the mirror.

   (b) If the man walks toward the mirror at 3 mph, what is the speed of his image, and what is its percentage rate of change of size? *Answer*: $\dfrac{3}{361}$ mph away from the mirror, 11.6% per s.

4.12. Show that the lateral magnification produced by the spherical refracting surface shown in Figure 4.8 is given by

$$m = -\frac{n_1 \cdot O'V}{n_2 \cdot OV}$$

where the minus sign signifies that the image is inverted relative to the object.

4.13. The lens shown is made of glass of refractive index $n = 4/3$. An object is located at position 0, 27 cm from the left-hand vertex. Where is the image? *Answer*: 60/13 cm to the right of the right-hand vertex.

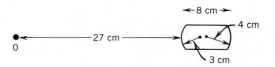

4.14. The first focal point of a lens is defined as the location of the object point on the lens axis whose image is at infinity. Similarly, the second focal point of a lens is the image point of an infinitely distant object. Locate the first and second focal points of the lens considered in Problem 13. *Answer*: 18/5 cm to the left of the first vertex, and 12/5 cm to the right of the second vertex, respectively.

4.15. Prove that the distance between the object and its image formed by a thin converging lens can never be less than four times the focal length of the lens.

CHAPTER 5

QUANTUM MECHANICS

Every section of this chapter on quantum mechanics contains references to our previous work in mechanics, electricity and magnetism, and optics. Thus, although quantum mechanics gives us a completely new way of understanding physical reality, its concepts and methods are closely related to those we first learn in classical physics. When confronting a problem in quantum mechanics, it is usually a profitable first step to visualize the corresponding classical system, if there is one. The experienced practitioner of quantum mechanics knows just how far classical reasoning can be trusted, and where nonclassical effects will appear. Unfortunately, this kind of knowledge is not easily summarized, because the classical limit of quantum mechanics is a complicated one, with many aspects. Nevertheless, the ability to use classical and quantum concepts simultaneously, and reliably, is a valuable one, and every student should attempt to develop it.

5.1 CLASSICAL MECHANICS VERSUS QUANTUM MECHANICS

How can we tell whether it is important to take quantum mechanics into account when we are analyzing the motion of a physical system? A simple and useful criterion can be expressed in terms of Planck's constant h, or in terms of the related constant $\hbar$, defined by $\hbar \equiv h/2\pi \approx 1.05 \times 10^{-27}$ erg s:

> From the parameters that characterize the motion being considered, extract a representative physical quantity, A, whose dimensions are *action* ($M \cdot L^2/T =$ energy $\times$ time = momentum $\times$ distance). If $A \approx \hbar$, it is important to take quantum effects into account; if $A \gg \hbar$, quantum effects may be ignored. It will never happen that $A \ll \hbar$.

We consider two examples of the application of this criterion: motion of the earth around the sun and motion of an electron around a proton. In the former case, suppose the earth and sun are moving at a constant separation R. If the gravitational attraction provides the centripetal acceleration, we have

$$G\frac{m_E m_S}{R^2} = \frac{m_E m_S}{m_E + m_S}\frac{v^2}{R}$$

$$v = \sqrt{\frac{G(m_E + m_S)}{R}}$$

$$A = pR = m_E m_S\sqrt{\frac{GR}{m_E + m_S}}$$

If $A \approx \hbar$, this implies

$$R \approx \left(\frac{\hbar}{m_E m_S}\right)^2\frac{m_E + m_S}{G} \approx 10^{-37}\,\text{m}$$

Thus, any value of R that is realistic in a discussion of planetary orbits will lead to an A much greater than $\hbar$. We can ignore quantum effects when we do celestial mechanics. A similar analysis applied to the motion of an electron around a proton gives

$$\frac{e^2}{R} = \frac{m_e m_p}{m_e + m_p}\frac{v^2}{R}$$

$$A = pR = \sqrt{\frac{m_e m_p}{m_e + m_p}e^2 R}$$

If $A \approx \hbar$,

$$R = \frac{\hbar^2}{\left(\dfrac{m_e m_p}{m_e + m_p}\right)e^2} = 0.53 \times 10^{-8}\,\text{cm}$$

$$= 0.53\,\text{Å} \tag{5.1}$$

This is in fact the order of magnitude of the electron-proton separation in the ground state of hydrogen.[1] For this state, therefore, $A \approx \hbar$ and quantum effects are important. They are not important for an electron-proton state with, say, $R \approx 10{,}000 \times 0.53\,\text{Å}$, which would lead to an A value of $\sqrt{10{,}000}\,\hbar \approx 100\hbar$.

The classical energy of an atom with an electron moving in the orbit whose radius is given by (5.1) would be

$$E = \frac{m_e m_p}{m_e + m_p}\frac{v^2}{2} - \frac{e^2}{R} = \frac{e^2}{2R} - \frac{e^2}{R}$$

$$= -\frac{\left(\dfrac{m_e m_p}{m_e + m_p}\right)e^4}{2\hbar^2} \approx -13.6\,\text{eV} \tag{5.2}$$

The estimates in (5.1) and (5.2) suggest that in systems dominated by the Coulomb interaction of electrons and protons, which essentially include all of

---

[1] This quantity is called the "Bohr radius."

atomic and molecular physics, the important lengths will be of the order of Å, and the important energies will be of the order of eV. With this information alone, we can make many useful order-of-magnitude estimates.

PROBLEM 5.1.1 How much energy is released when 1 kg of nitroglycerine is detonated? (The gram-molecular weight of nitroglycerine is 227.)

Since this is a chemical explosive, involving rearrangement of atomic and molecular structure, we will make the rough estimate that 1 eV is liberated for each molecule of nitroglycerine. A kilogram contains

$$1000 \, g \times \frac{1 \, mol}{227 \, g} \times 6 \times 10^{23} \frac{molecules}{mol} = 2.6 \times 10^{24} \, molecules$$

Thus, the energy liberated is approximately

$$2.6 \times 10^{24} \, eV = 2.6 \times 10^{24} \, eV \times 1.6 \times 10^{-12} \frac{erg}{eV}$$
$$= 4 \times 10^{12} \, erg = 4 \times 10^{5} \, J$$

PROBLEM 5.1.2 What force is needed to pull apart a rod of copper, 1 cm^2 in cross-section?

Consider two facing planes of copper ions, perpendicular to the axis of the rod. If we take the ionic spacing to be 1 Å $= 10^{-8}$ cm, then each plane will contain 10^{16} ions. Let us guess that the force required to pull two adjacent ions apart is 1 eV/Å. Thus, the total force needed to separate the planes is

$$10^{16} \times \frac{1 \, eV}{Å} = 10^{16} \times \frac{1.6 \times 10^{-12} \, erg}{10^{-8} \, cm} = 1.6 \times 10^{12} \, dyne$$
$$\approx 6 \times 10^{6} \, lb$$

In fact this number greatly exceeds the actual tensile strength of copper ($\sim 2 \times 10^{9}$ dyne). A real copper crystal contains many defects, which substantially decrease its tensile strength.[2]

An equivalent way of stating the $A \approx \hbar$ criterion is that quantum effects are important when

$$R \approx \frac{\hbar}{p} \quad (\equiv \lambda) \tag{5.3}$$

As before, R and p are distances and momenta that characterize the motion being considered. The length $\hbar/p = \lambda$ is called the *de Broglie wavelength*.

PROBLEM 5.1.3 Suppose a particle of charge $+Z_1 e$ and mass m moves with asymptotic speed v toward a stationary particle of charge $+Z_2 e$, mass M. Take the "characteristic length" of the orbit to be the distance of closest approach in a head-on collision. Find a criterion for deciding whether quantum effects are important.

---

[2] For a detailed discussion, see R. W. Lardner, *Mathematical Theory of Dislocations and Fracture*, University of Toronto Press, 1974, chapter 1. The discrepancy is explained by the presence of edge dislocations. The full value is realized for whiskers, which are near-perfect crystals.

The asymptotic kinetic energy is $mv^2/2$. At the distance R of closest approach, this will all be converted into potential energy,

$$\frac{mv^2}{2} = \frac{Z_1 Z_2 e^2}{R}$$

$$R = \frac{2Z_1 Z_2 e^2}{mv^2}$$

Thus, the problem can be treated by classical mechanics when

$$R = \frac{2Z_1 Z_2 e^2}{mv^2} \gg \lambda = \frac{\hbar}{mv}$$

$$2\frac{Z_1 Z_2 e^2}{\hbar v} \gg 1$$

The quantity $\eta \equiv Z_1 Z_2 e^2 / \hbar v$ is called the Sommerfeld parameter for the orbit. It plays an important role in problems of scattering of charged particles.

5.2 THE FORMALISM OF QUANTUM MECHANICS

A complete presentation of the formalism of quantum mechanics is beyond the scope of this review. We will limit our discussion to those aspects of the formalism needed for the applications given in the remainder of this chapter.

A *state* of a physical system is represented in quantum mechanics by a vector Ψ in an abstract "state space." This is a complex vector space of, in general, infinite dimension. Corresponding to any two state vectors Φ and Ψ of the system, a positive-definite scalar product, $\langle \Phi | \Psi \rangle = \langle \Psi | \Phi \rangle^*$ is defined. If c is any complex number, the state vectors Ψ and $c\Psi$ represent the same physical state of the system. A state vector Ψ is said to be *normalized* if $\langle \Psi | \Psi \rangle = 1$.

Physical observables are represented by Hermitian operators which act in the state space on the state vectors. If A is such an operator, then $A\Psi$ symbolizes the state vector obtained by acting with A on Ψ. The scalar product of Φ and $A\Psi$ is called a "matrix element",

$$\langle \Phi | A\Psi \rangle \equiv \langle \Phi | A | \Psi \rangle$$

The Hermitian character of A implies that

$$\langle \Phi | A\Psi \rangle = \langle A\Phi | \Psi \rangle = \langle \Psi | A\Phi \rangle^*$$

or, equivalently,

$$\langle \Phi | A | \Psi \rangle = \langle \Psi | A | \Phi \rangle^*$$

The "diagonal" matrix element $\langle \Psi | A | \Psi \rangle$ is called the "expectation value of the operator A for the state Ψ." If Ψ is a normalized state vector, then $\langle \Psi | A | \Psi \rangle$ is the average value of the physical observable represented by A, when measured in the state of the system represented by Ψ. In general, operators representing different physical observables do not commute with each other.

If A, Ψ, and a number of λ, satisfy the relation

$$A\Psi = \lambda\Psi \tag{5.4}$$

then Ψ is said to be an eigenstate of A, with eigenvalue λ. It is not difficult to show that the eigenvalues of a Hermitian operator are real. If (5.4) is obeyed, a measurement in the state represented by Ψ of the observable represented by A will certainly yield a value equal to λ. If Φ and Ψ are both normalized, and Ψ satisfies (5.4), then

$$P \equiv |\langle \Psi | \Phi \rangle|^2$$

is the *probability* that a measurement of A in the state Φ will yield a value equal to λ. For example, if A has a set of orthonormal eigenstates ψ_j,

$$A\psi_j = \lambda_j \psi_j$$

$$\langle \psi_j | \psi_k \rangle = \delta_{jk}$$

and ϕ is a linear combination of the ψ_j with numerical coefficients c_j,

$$\phi = \sum_j c_j \psi_j$$

then a measurement of the observable represented by A in the state represented by ϕ yields a result λ_k with probability

$$P_k = |\langle \psi_k | \phi \rangle|^2 = \left| \sum_j c_j \langle \psi_k | \psi_j \rangle \right|^2$$

$$= \left| \sum_j c_j \delta_{jk} \right|^2 = |c_k|^2$$

Explicit calculations in quantum mechanics are most easily performed in the "Schrödinger picture." The state Ψ is represented by the *wave function*, a function of the generalized coordinates[3] $q_1, \ldots, q_d$ and time t

$$\Psi = \Psi(q_1, \ldots, q_d, t) \tag{5.5}$$

The scalar product $\langle \Phi | \Psi \rangle$ is given by an integral over the generalized coordinates:

$$\langle \Phi | \Psi \rangle = \int dq_1, \ldots, dq_d \Phi^*(q_1, \ldots, q_d, t) \Psi(q_1, \ldots, q_d, t)$$

The operator corresponding to the physical observable q_j is simply multiplication of (5.5) by the variable q_j; the operator corresponding to the conjugate momentum p_j is $\hbar/i \cdot \partial/\partial q_j$. The rules of differentiation imply that

$$(p_j q_k - q_k p_j)\Psi(q_1, \ldots, q_d, t) = \frac{\hbar}{i}\left(\frac{\partial}{\partial q_j}(q_k \Psi) - q_k \frac{\partial}{\partial q_j}\Psi \right)$$

$$= \frac{\hbar}{i}\left(\frac{\partial q_k}{\partial q_j}\Psi + q_k \frac{\partial \Psi}{\partial q_j} - q_k \frac{\partial \Psi}{\partial q_j} \right)$$

$$= \frac{\hbar}{i}\delta_{kj}\Psi$$

This result applies to any state Ψ, so that we can write

$$p_j q_k - q_k p_j = \frac{\hbar}{i}\delta_{kj} \tag{5.6}$$

---

[3] The student should review the discussion of generalized coordinates and their conjugate momenta given in Section 1.9.

as an *operator* equation. The commutation relations (5.6) apply to any set of generalized coordinates and their conjugate momenta. For example, if the system consists of a single particle and (q_1, q_2, q_3) are taken to be its Cartesian coordinates (x, y, z), then (5.6) becomes

$$p_x x - x p_x = \frac{\hbar}{i}$$

$$p_x y - y p_x = 0, \qquad \text{etc.}$$

To determine the time dependence of Ψ, one starts with the classical Hamiltonian (1.9.47) expressed in terms of the q_j, p_j, and t. The p_j are replaced by $(\hbar/i)\partial/\partial q_j$, and the resulting *Hamiltonian operator* is used in the *time-dependent Schrödinger equation*:

$$H\left(q_1, \ldots, q_d, \frac{\hbar}{i}\frac{\partial}{\partial q_1}, \ldots, \frac{\hbar}{i}\frac{\partial}{\partial q_d}, t\right)\Psi(q_1, \ldots, q_d, t)$$

$$= i\hbar\frac{\partial}{\partial t}\Psi(q_1, \ldots, q_d, t) \tag{5.7}$$

If H does not depend explicitly on t, one can use the method of separation of variables to separate the q and t dependence of Ψ. This leads to solutions of (5.7) of the form

$$\Psi(q_1, \ldots, q_d, t) = e^{-(i/\hbar)Et}\Phi(q_1, \ldots, q_d) \tag{5.8a}$$

where Φ and E appear in the *time-independent Schrödinger equation*

$$H\left(q_1, \ldots, q_d, \frac{\hbar}{i}\frac{\partial}{\partial q_1}, \ldots, \frac{\hbar}{i}\frac{\partial}{\partial q_d}\right)\Phi(q_1, \ldots, q_d)$$

$$= E\Phi(q_1, \ldots, q_d) \tag{5.8b}$$

For example, the time-independent Schrödinger equation for the one-dimensional harmonic oscillator is obtained by making the substitution $p \to \hbar/i \cdot d/dx$ in the classical harmonic oscillator Hamiltonian (1.9.49):

$$\left[\frac{1}{2m}\left(\frac{\hbar}{i}\frac{d}{dx}\right)^2 + \frac{m\omega^2}{2}x^2\right]\phi(x) = E\phi(x)$$

$$\left[-\frac{\hbar^2}{2m}\frac{d^2}{dx^2} + \frac{m\omega^2}{2}x^2\right]\phi(x) = E\phi(x) \tag{5.9}$$

The normalized solutions of this equation are discussed in Section 5.4.4.

This prescription for finding the Hamiltonian operator to use in the Schrödinger equation can only be used when the system has a classical counterpart, from which we can obtain a classical Hamiltonian using (1.9.47). There are some quantum phenomena for which there are no true classical counterparts — for example, the intrinsic spin angular momentum of an electron. In these cases, other considerations must be used to infer the Schrödinger equation.

5.3 ANGULAR MOMENTUM

In both classical and quantum mechanics, angular momentum conservation is an important property of systems with rotational symmetry. According to (1.1b), if the external torque on a classical system has no component along some direction

$\hat{n}$, then the component along $\hat{n}$ of the angular momentum **L** of the system is conserved. The corresponding quantum mechanical statement is: if the Hamiltonian H of a system is invariant under rotations about $\hat{n}$, then the eigenstates of H can be chosen to be eigenstates of $\mathbf{L} \cdot \hat{n}$. Constructing eigenstates of $\mathbf{L} \cdot \hat{n}$ is then a useful first step in the search for eigenstates of H.

5.3.1 Commutation Relations

Classically, if the external torque has no component along several directions, say $\hat{n}_1$ and $\hat{n}_2$, then both $\mathbf{L} \cdot \hat{n}_1$ and $\mathbf{L} \cdot \hat{n}_2$ are conserved. However, in quantum mechanics $\mathbf{L} \cdot \hat{n}_1$ and $\mathbf{L} \cdot \hat{n}_2$ are operators that generally do not commute with each other. Thus, even though H may be invariant under rotations about both $\hat{n}_1$ and $\hat{n}_2$, it is not possible to find simultaneous eigenstates of $\hat{n}_1$ and $\hat{n}_2$. The uses of angular momentum in quantum mechanics are limited, therefore, by the noncommutativity of the different components of **L**. As we will see below, these operators are closely related to operators that actually rotate the state vectors. The commutation relations between the components of **L** are ultimately related to the multiplication rule for the 3-dimensional rotation group.

In quantum mechanics there are several different kinds of angular momentum operators:

1. Single-particle orbital angular momentum:

$$\mathbf{l} = \mathbf{r} \times \mathbf{p} = \frac{\hbar}{i} \mathbf{r} \times \nabla$$

2. Total orbital angular momentum of N particles:

$$\mathbf{L} = \sum_{i=1}^{N} \mathbf{l}_i$$

3. Single-particle intrinsic spin angular momentum **s**

4. Single-particle total angular momentum

$$\mathbf{j} = \mathbf{l} + \mathbf{s}$$

etc.

The commutation relations are of the same form in all these cases. Let us use J_x, J_y, J_z to label the Cartesian components of any of these kinds of angular momentum. Then

$$\left[J_x, J_y \right] \equiv J_x J_y - J_y J_x \equiv i\hbar J_z \tag{5.10a}$$

$$\left[J_z, J_x \right] \equiv J_z J_x - J_x J_z = i\hbar J_y \tag{5.10b}$$

$$\left[J_y, J_z \right] \equiv J_y J_z - J_z J_y = i\hbar J_x \tag{5.10c}$$

Since J_x, J_y, and J_z are physical observables, the operators that represent them are Hermitian.

A consequence of (5.10) is that the operator $\mathbf{J} \cdot \mathbf{J}$ commutes with all three operators J_α. Therefore, we can find simultaneous eigenstates of $\mathbf{J} \cdot \mathbf{J}$ and one of the J_α, say J_z. These eigenstates are conventionally labeled ψ_m^l, with the l and m indices related to the eigenvalues of $\mathbf{J} \cdot \mathbf{J}$ and J_z by

$$\mathbf{J} \cdot \mathbf{J} \psi_m^l = \hbar^2 l(l+1)\psi_m^l \qquad \left(l = 0, \frac{1}{2}, 1, \frac{3}{2}, 2, \dots \right) \tag{5.11a}$$

$$J_z \psi_m^l = \hbar m \psi_m^l \qquad (m = -l, -l+1, \dots, l) \tag{5.11b}$$

Moreover, if the ψ_m^l are normalized, their relative phases can be chosen so that

$$J_+\psi_m^l \equiv (J_x + iJ_y)\psi_m^l = \hbar\sqrt{(l-m)(l+m+1)}\,\psi_{m+1}^l \qquad (5.12a)$$

$$J_-\psi_m^l \equiv (J_x - iJ_y)\psi_m^l = \hbar\sqrt{(l+m)(l-m+1)}\,\psi_{m-1}^l \qquad (5.12b)$$

In the particular case of $l = 1/2$, (5.12) reduces to

$$J_+\psi_{1/2}^{1/2} = 0, \qquad J_-\psi_{-1/2}^{1/2} = 0 \qquad (5.13a)$$

$$J_+\psi_{-1/2}^{1/2} = \hbar\psi_{1/2}^{1/2}, \qquad J_-\psi_{1/2}^{1/2} = \hbar\psi_{-1/2}^{1/2} \qquad (5.13b)$$

PROBLEM 5.3.1   Use (5.11) and (5.13) to calculate the matrices of J_x, J_y, and J_z with respect to $\psi_m^{1/2}$ ($m = \pm 1/2$).

According to (5.11b),

$$\langle \psi_{m_1}^{1/2}|J_z|\psi_{m_2}^{1/2}\rangle = \frac{\hbar}{2}\delta_{m_1,m_2}$$

According to (5.13),

$$\langle \psi_{-1/2}^{1/2}|J_x + iJ_y|\psi_{1/2}^{1/2}\rangle = 0$$

$$\langle \psi_{-1/2}^{1/2}|J_x - iJ_y|\psi_{1/2}^{1/2}\rangle = \hbar$$

Adding and subtracting these two equations gives

$$\langle \psi_{-1/2}^{1/2}|J_x|\psi_{1/2}^{1/2}\rangle = \frac{\hbar}{2}$$

$$\langle \psi_{-1/2}^{1/2}|J_y|\psi_{1/2}^{1/2}\rangle = i\frac{\hbar}{2}$$

A similar argument can be made for $\langle \psi_{1/2}^{1/2}|J_{x,\,y}|\psi_{-1/2}^{1/2}\rangle$. The result can be written

$$\langle \psi_{m_1}^{1/2}|J_\alpha|\psi_{m_2}^{1/2}\rangle = \frac{\hbar}{2}[\sigma_\alpha]_{m_1,m_2} \qquad (5.14a)$$

where

$$\sigma_x = \begin{bmatrix} 0 & 1 \\ 1 & 0 \end{bmatrix}, \qquad \sigma_y \equiv \begin{bmatrix} 0 & -i \\ i & 0 \end{bmatrix}, \qquad \sigma_z \equiv \begin{bmatrix} 1 & 0 \\ 0 & -1 \end{bmatrix} \qquad (5.14b)$$

The three matrices (5.14b) are called the Pauli spin matrices. The first row and column of each correspond to $m = +1/2$, the second row and column to $-1/2$.

5.3.2 Orthogonality of Angular Momentum Eigenstates

We now prove the fundamental orthogonality theorem:

$$\langle \phi_{m_1}^{l_1}|\psi_{m_2}^{l_2}\rangle = \delta_{l_1,l_2}\delta_{m_1,m_2} \times \text{something independent of } m_1, m_2 \qquad (5.15)$$

Proof   Since the J_α (and hence $\mathbf{J} \cdot \mathbf{J}$) are Hermitian, we can deduce from (5.11a) that

$$\langle \phi_{m_1}^{l_1}|\mathbf{J} \cdot \mathbf{J}\psi_{m_2}^{l_2}\rangle = \hbar^2 l_2(l_2 + 1)\langle \phi_{m_1}^{l_1}|\psi_{m_2}^{l_2}\rangle$$

$$= \langle \mathbf{J} \cdot \mathbf{J}\phi_{m_1}^{l_1}|\psi_{m_2}^{l_2}\rangle = \hbar^2 l_1(l_1 + 1)\langle \phi_{m_1}^{l_1}|\psi_{m_2}^{l_2}\rangle$$

$$0 = [l_1(l_1 + 1) - l_2(l_2 + 1)]\langle \phi_{m_1}^{l_1}|\psi_{m_2}^{l_2}\rangle \qquad (5.16)$$

The l_1, l_2 values are restricted to those listed in (5.11a). Therefore, (5.16) implies that $\langle \phi_{m_1}^{l_1} | \psi_{m_2}^{l_2} \rangle = 0$ unless $l_1 = l_2$. A similar argument based on (5.11b) shows that $\langle \phi_{m_1}^{l_1} | \psi_{m_2}^{l_1} \rangle = 0$ unless $m_1 = m_2$. Next assume that $l_1 = l_2 = l$, $m_1 = m_2 = m$, and use (5.12) to show that

$$\langle \phi_{m+1}^l | \psi_{m+1}^l \rangle = \left\langle \frac{(J_x + iJ_y)\phi_m^l}{\hbar\sqrt{(l-m)(l+m+1)}} \middle| \frac{(J_x + iJ_y)\psi_m^l}{\hbar\sqrt{(l-m)(l+m+1)}} \right\rangle$$

$$= \frac{\langle \phi_m^l | (J_x - iJ_y)(J_x + iJ_y) | \psi_m^l \rangle}{\hbar^2 (l-m)(l+m+1)} \qquad (-l \le m \le l-1)$$

$$= \frac{\langle \phi_m^l | J_x^2 + J_y^2 - \hbar J_z | \psi_m^l \rangle}{\hbar^2 [l(l+1) - m(m+1)]}$$

$$= \frac{\langle \phi_m^l | \mathbf{J} \cdot \mathbf{J} - J_z^2 - \hbar J_z | \psi_m^l \rangle}{\hbar^2 [l(l+1) - m(m+1)]} = \langle \phi_m^l | \psi_m^l \rangle$$

Since we have shown that

$$\langle \phi_{m+1}^l | \psi_{m+1}^l \rangle = \langle \phi_m^l | \psi_m^l \rangle$$

for all $-l \le m \le l-1$, we have in effect proved the equality of the $\langle \phi_m^l | \psi_m^l \rangle$ for all m (for fixed l). The proof of (5.15) is complete.

5.3.3 Connection Between Angular Momentum Operators and Rotations

Now suppose that a vector $\mathbf{r}$ is rotated about a fixed unit vector $\hat{n}$ (Figure 5.1) The argument that is used to prove that $\mathbf{v} = \boldsymbol{\omega} \times \mathbf{r}$ can be modified to prove that

$$\frac{d\mathbf{r}(\alpha)}{d\alpha} = \hat{n} \times \mathbf{r}(\alpha)$$

If we apply the chain rule to differentiate $\psi(\mathbf{r}(\alpha))$ with respect to α, we get

$$\frac{d}{d\alpha}\psi(\mathbf{r}(\alpha)) = \frac{d\mathbf{r}(\alpha)}{d\alpha} \cdot \nabla\psi(\mathbf{r}(\alpha))$$

$$= \hat{n} \times \mathbf{r} \cdot \nabla\psi(\mathbf{r})$$

$$= \hat{n} \cdot \mathbf{r} \times \nabla\psi(\mathbf{r})$$

$$= \frac{i}{\hbar}\hat{n} \cdot \mathbf{l}\psi(\mathbf{r}(\alpha)) \qquad (5.17a)$$

FIGURE 5.1 The plane of the circle is perpendicular to $\hat{n}$. As α increases, the tip of $\mathbf{r}(\alpha)$ moves around the circle, starting at $\mathbf{r}(0)$.

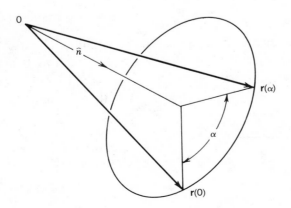

The solution of this differential equation for $\psi(\mathbf{r}(\alpha))$ is

$$\psi\big(\mathbf{r}(\alpha)\big) = e^{(i/\hbar)\alpha\hat{n}\cdot\mathbf{l}}\psi\big(\mathbf{r}(0)\big) \tag{5.17b}$$

where the exponential operator in (5.17b) is defined in terms of its power series expansion

$$e^{(i/\hbar)\alpha\hat{n}\cdot\mathbf{l}} = \sum_{m=0}^{\infty} \frac{1}{m!}\left(\frac{i}{\hbar}\alpha\hat{n}\cdot\mathbf{l}\right)^m \tag{5.18}$$

We can interpret (5.17b) as saying that the operator $\exp((i/\hbar)\alpha\hat{n}\cdot\mathbf{l})$ rotates the function $\psi(\mathbf{r})$ about the axis $\hat{n}$, through the angle $-\alpha$.[4]

Equation (5.17a) implies that if $l\psi = 0$, then ψ is unchanged by any rotation. Thus, states ψ_m^l with $l = m = 0$ are spherically symmetric.

A wave function $\psi(\mathbf{r})$ is appropriate only for a spinless particle. For a particle with intrinsic spin (say an electron or proton) $\mathbf{r}$ must be supplemented by a "spinor index" σ, which is also affected by rotation. This is accomplished by adding an intrinsic spin angular momentum operator $\mathbf{s}$ to the orbital angular momentum operator $\mathbf{l}$ in (5.17). Thus $\exp((i/\hbar)\alpha\hat{n}\cdot\mathbf{j})$ rotates the wave function of a particle with intrinsic spin. For an N-particle wave function, the corresponding operator is $\exp(i/\hbar\alpha\hat{n}\cdot\sum_{k=1}^{N}\mathbf{j}_k) = \exp(i/\hbar\alpha\hat{n}\cdot\mathbf{J})$.

Suppose that an operator V is spherically symmetric. Then

$$e^{(i/\hbar)\alpha\hat{n}\cdot\mathbf{J}}V\psi = Ve^{(i/\hbar)\alpha\hat{n}\cdot\mathbf{J}}\psi \qquad (\text{any } \psi, \hat{n}, \alpha)$$

$$e^{(i/\hbar)\alpha\hat{n}\cdot\mathbf{J}}V = Ve^{(i/\hbar)\alpha\hat{n}\cdot\mathbf{J}} \qquad (\text{any } \hat{n}, \alpha)$$

Differentiate this equation with respect to α and set α equal to zero, to get

$$\hat{n}\cdot\mathbf{J}V = V\hat{n}\cdot\mathbf{J} \qquad (\text{any } \hat{n})$$

If we now set $\hat{n} = \hat{x}, \hat{y}, \hat{z}$ in turn, we get

$$J_\alpha V = VJ_\alpha \qquad (\alpha = x, y, z) \tag{5.19}$$

Thus, a spherically symmetric operator commutes with all three components of the angular momentum operator. From this it follows that

$$J_z V\psi_{m_2}^{l_2} = VJ_z\psi_{m_2}^{l_2} = \hbar m_2 V\psi_{m_2}^{l_2}$$

$$\big(J_x \pm iJ_y\big)V\psi_{m_2}^{l_2} = V\big(J_x \pm iJ_y\big)\psi_{m_2}^{l_2}$$

$$= \hbar\sqrt{(l_2 \mp m_2)(l_2 \pm m_2 + 1)}\,V\psi_{m_2}^{l_2}$$

We see that $V\psi_{m_2}^{l_2}$ has the same angular momentum properties as $\psi_{m_2}^{l_2}$ and, therefore, the orthogonality theorem (5.15) holds, with $V\psi_{m_2}^{l_2}$ replacing $\psi_{m_2}^{l_2}$,

$$\big\langle \phi_{m_1}^{l_1}|V\psi_{m_2}^{l_2}\big\rangle \equiv \big\langle \phi_{m_1}^{l_1}|V|\psi_{m_2}^{l_2}\big\rangle$$

$$= \delta_{l_1,l_2}\delta_{m_1,m_2} \times \text{something independent of } m_1, m_2 \tag{5.20}$$

If V is invariant with respect to rotation about $\hat{z}$ only, then we can only conclude that V commutes with J_z, and (5.20) is replaced by the weaker condition

$$\big\langle \phi_{m_1}^{l_1}|V|\psi_{m_2}^{l_2}\big\rangle = 0, \qquad \text{unless } m_1 = m_2 \tag{5.21}$$

---

[4] For example, use (5.17b) with $\hat{n} = \hat{z}$, $\alpha = \pi/2$, and $\mathbf{r}(0) = r\hat{x}$. Then $\mathbf{r}(\alpha) = r\hat{y}$, and (5.17b) says that ψ evaluated at $r\hat{y}$ equals $e^{(i/\hbar)(\pi/2)l_z}\psi$ evaluated at $r\hat{x}$. Thus, $e^{(i/\hbar)(\pi/2)l_z}\psi$ can be regarded as ψ rotated through $-\pi/2$ about $\hat{z}$.

The Hamiltonian of a system in free space is spherically symmetric, so that (5.20) can be applied with $V = H$. Thus, if our objective is to find a set of states that diagonalizes the Hamiltonian, it is advantageous to work with angular momentum eigenstates.

PROBLEM 5.3.2 An electron is in a state with z-component of spin angular momentum $\hbar/2$. An observation designed to measure the component of spin angular momentum along an arbitrary direction $\hat{n}$ is made. What is the probability of observing a component of spin angular momentum $\hbar/2$ along $\hat{n}$?

This problem refers to two sets of states, $\chi_m^{1/2}$ and $\psi_m^{1/2}$ ($m = \pm 1/2$), such that

$$
\left.
\begin{aligned}
s_z \chi_m^{1/2} &= \hbar m \chi_m^{1/2} \\
\hat{n} \cdot \mathbf{s}\, \psi_m^{1/2} &= \hbar m \psi_m^{1/2}
\end{aligned}
\right\} \quad m = \pm \frac{1}{2}
\tag{5.22a}
$$
$$
\tag{5.22b}
$$

The answer to the problem is $P \equiv |\langle \psi_{1/2}^{1/2} | \chi_{1/2}^{1/2} \rangle|^2$. To determine this quantity we take the scalar product of the $m = 1/2$ component of (5.22b) with $\chi_{1/2}^{1/2}$:

$$
\langle \chi_{1/2}^{1/2} | \hat{n} \cdot \mathbf{s} | \psi_{1/2}^{1/2} \rangle = \frac{\hbar}{2} \langle \chi_{1/2}^{1/2} | \psi_{1/2}^{1/2} \rangle = \langle \hat{n} \cdot \mathbf{s} \chi_{1/2}^{1/2} | \psi_{1/2}^{1/2} \rangle
\tag{5.23}
$$

The dot product $\hat{n} \cdot \mathbf{s}$ can be written

$$
\hat{n} \cdot \mathbf{s} = n_x s_x + n_y s_y + n_z s_z
$$
$$
= \frac{n_x + i n_y}{2}\left(s_x - i s_y \right) + \frac{(n_x - i n_y)}{2}\left(s_x + i s_y \right) + n_z s_z
$$

Thus, (5.13) and (5.11b) imply that

$$
\hat{n} \cdot \mathbf{s} \chi_{1/2}^{1/2} = \hbar \frac{n_x + i n_y}{2} \chi_{-1/2}^{1/2} + n_z \frac{\hbar}{2} \chi_{1/2}^{1/2}
$$

and (5.23) becomes

$$
\frac{\hbar}{2} \langle \chi_{1/2}^{1/2} | \psi_{1/2}^{1/2} \rangle = \frac{\hbar}{2} \left[\left(n_x - i n_y \right) \langle \chi_{-1/2}^{1/2} | \psi_{1/2}^{1/2} \rangle + n_z \langle \chi_{1/2}^{1/2} | \psi_{1/2}^{1/2} \rangle \right]
$$
$$
\left(n_x - i n_y \right) \langle \chi_{-1/2}^{1/2} | \psi_{1/2}^{1/2} \rangle = \left(1 - n_z \right) \langle \chi_{1/2}^{1/2} | \psi_{1/2}^{1/2} \rangle
\tag{5.24}
$$

If the $\chi_m^{1/2}$ and $\psi_m^{1/2}$ are normalized, then

$$
|\langle \chi_{-1/2}^{1/2} | \psi_{1/2}^{1/2} \rangle|^2 + |\langle \chi_{1/2}^{1/2} | \psi_{1/2}^{1/2} \rangle|^2 = 1
$$

This equation and (5.24) can be satisfied by

$$
\langle \chi_{-1/2}^{1/2} | \psi_{1/2}^{1/2} \rangle = \frac{(1 - n_z)}{\sqrt{|n_x - i n_y|^2 + (1 - n_z)^2}} = \frac{(1 - n_z)}{\sqrt{2(1 - n_z)}}
$$

$$
\langle \chi_{1/2}^{1/2} | \psi_{1/2}^{1/2} \rangle = \frac{(n_x - i n_y)}{\sqrt{|n_x - i n_y|^2 + (1 - n_z)^2}} = \frac{n_x - i n_y}{\sqrt{2(1 - n_z)}}
$$

Thus,

$$
P = |\langle \chi_{1/2}^{1/2} | \psi_{1/2}^{1/2} \rangle|^2 = \frac{n_x^2 + n_y^2}{2(1 - n_z)}
$$
$$
= \frac{1 - n_z^2}{2(1 - n_z)} = \frac{1 + n_z}{2} = \frac{1 + \cos\theta}{2} = \cos^2\frac{\theta}{2}
\tag{5.25}
$$

where θ is the angle between $\hat{n}$ and $\hat{z}$. Note that this implies that if a measurement of s_z yields $\hbar/2$, a subsequent measurement of s_x yields $+\hbar/2$ with a probability $P = 1/2$ (and $-\hbar/2$ with $P = 1/2$).

PROBLEM 5.3.3 The Pauli spin matrices (5.14b) obey $\sigma_\alpha\sigma_\beta = \delta_{\alpha\beta}\mathbb{1} + i\varepsilon_{\alpha\beta\gamma}\sigma_\gamma$. Here $\mathbb{1}$ is a 2×2 unit matrix, $\varepsilon_{\alpha\beta\gamma}$ is the Levi-Civita symbol, and summation over repeated indices is implied. Use this result to calculate the matrix element

$$\langle \psi_{m_1}^{1/2}|e^{(i/\hbar)\alpha\hat{v}\cdot\mathbf{s}}|\psi_{m_2}^{1/2}\rangle$$

of the rotation operator.

Let us use the power series (5.18). For the $m = 2$ term, we have

$$(\hat{v} \cdot \mathbf{s})^2 = (v_\alpha s_\alpha)(v_\beta s_\beta) = v_\alpha v_\beta s_\alpha s_\beta = \left(\frac{\hbar}{2}\right)^2 v_\alpha v_\beta \sigma_\alpha \sigma_\beta$$

$$= \left(\frac{\hbar}{2}\right)^2 v_\alpha v_\beta \left[\delta_{\alpha\beta}\mathbb{1} + i\varepsilon_{\alpha\beta\gamma}\sigma_\gamma\right]$$

$$= \left(\frac{\hbar}{2}\right)^2 v_\alpha v_\alpha \mathbb{1} = \left(\frac{\hbar}{2}\right)^2 \mathbb{1} \tag{5.26}$$

Here we have used the fact that $\hat{v}$ is a unit vector, so that $v_\alpha v_\alpha (= v_x^2 + v_y^2 + v_z^2) = 1$. Repeated application of (5.26) gives

$$\left.\begin{array}{l}(\hat{v} \cdot \mathbf{s})^{2m} = \left(\dfrac{\hbar}{2}\right)^m \mathbb{1} \\[12pt] (\hat{v} \cdot \mathbf{s})^{2m+1} = \left(\dfrac{\hbar}{2}\right)^{2m+1}(\mathbf{v} \cdot \boldsymbol{\sigma})\end{array}\right\} \quad m = 0, 1, 2, \ldots$$

If these terms are used in (5.18), we get

$$\langle \psi_{m_1}^{1/2}|e^{(i/\hbar)\alpha\hat{v}\cdot\mathbf{s}}|\psi_{m_2}^{1/2}\rangle = \sum_{m=0}^{\infty}\frac{1}{(2m)!}\left(\frac{i\alpha}{2}\right)^{2m}\mathbb{1}$$

$$+ \sum_{m=0}^{\infty}\frac{1}{(2m+1)!}\left(\frac{i\alpha}{2}\right)^{2m+1}\hat{v}\cdot\boldsymbol{\sigma}$$

$$= \cos\frac{\alpha}{2}\mathbb{1} + i\sin\frac{\alpha}{2}\hat{v}\cdot\boldsymbol{\sigma}$$

$$= \begin{bmatrix} \cos\dfrac{\alpha}{2} + iv_z\sin\dfrac{\alpha}{2} & i\sin\dfrac{\alpha}{2}\times(v_x - iv_y) \\[12pt] i\sin\dfrac{\alpha}{2}\times(v_x + iv_y) & \cos\dfrac{\alpha}{2} - iv_z\sin\dfrac{\alpha}{2} \end{bmatrix}$$

This result can be used to answer the previous problem. If we put $\hat{v}$ in the x–y plane and set $\alpha = \theta$, the resulting rotation would send $\hat{z}$ into a vector $\hat{n}$ making an angle θ with $\hat{z}$. What we previously called $\langle \chi_{1/2}^{1/2}|\psi_{1/2}^{1/2}\rangle$ would now be

$$\langle \psi_{1/2}^{1/2}|e^{(i/\hbar)\theta(v_x s_x + v_y s_y)}|\psi_{1/2}^{1/2}\rangle = \cos\frac{\theta}{2} + iv_z\sin\frac{\theta}{2} = \cos\frac{\theta}{2}$$

leading to $P = \cos^2\dfrac{\theta}{2}$, in agreement with (5.25).

5.3.4 Parity

We have seen that when a Hamiltonian is invariant under rotations, it is convenient to work with states ψ_m^l that are simultaneous eigenstates of $\mathbf{J} \cdot \mathbf{J}$ and J_z. Now suppose that the Hamiltonian is invariant under *inversion*

$$x, y, z \rightarrow -x, -y, -z$$

Let I represent the effect of inversion on a wave function

$$I\psi(\mathbf{r}_1, \ldots, \mathbf{r}_n) \equiv \psi(-\mathbf{r}_1, \ldots, -\mathbf{r}_n)$$

The operator I can have eigenvalues[5] ± 1. Let us label the corresponding eigenstates with the symbol π (the *parity*):

$$I\psi^\pi = \pi\psi^\pi \qquad (\pi = \pm 1)$$

Derivations similar to those used to obtain (5.16) and (5.20) show that if an operator is invariant with respect to inversion, its matrix elements between states of opposite parity vanish.

A rotation is a homogeneous linear transformation of x, y, z, and so every rotation commutes with the operation of inversion. The result (5.19) then shows that

$$J_\alpha I = I J_\alpha \qquad (\alpha = x, y, z)$$

This means that it is possible to find simultaneous eigenstates of $\mathbf{J} \cdot \mathbf{J}$, J_z and I;

$$I\psi_m^{\pi l} = \pi\psi_m^{\pi l} \qquad (\pi = \pm 1)$$

$$\mathbf{J} \cdot \mathbf{J}\psi_m^{\pi l} = \hbar^2 l(l+1)\psi_m^{\pi l}$$

$$J_z\psi_m^{\pi l} = \hbar m\psi_m^{\pi l}$$

A rotationally invariant, inversion invariant Hamiltonian is diagonal in l, m, and π.

5.3.5 Spherical Harmonics

The spherical harmonics $Y_m^l(\theta, \phi)$ are single-valued functions of position on the unit sphere that are eigenstates of l_z and $\mathbf{l} \cdot \mathbf{l}$:

$$l_z Y_m^l(\theta, \phi) = \hbar m Y_m^l(\theta, \phi) \tag{5.27a}$$

$$(l_x + il_y)Y_m^l(\theta, \phi) = \hbar\sqrt{(l \mp m)(l \pm m + 1)} \, Y_{m \pm 1}^l(\theta, \phi) \tag{5.27b}$$

$$\mathbf{l} \cdot \mathbf{l} Y_m^l(\theta, \phi) = \hbar^2 l(l+1)Y_m^l(\theta, \phi) \tag{5.27c}$$

Because of (5.27), the orthogonality theorem (5.15) can be applied. The normalization of the Y_m^l is chosen so that the "factor independent of m" in (5.15) is equal to unity:

$$\langle Y_{m_1}^{l_1} | Y_{m_2}^{l_2} \rangle = \int \sin\theta \, d\theta \, d\phi \, Y_{m_1}^{l_1*}(\theta, \phi) Y_{m_2}^{l_2}(\theta, \phi)$$

$$= \delta_{l_1, l_2} \delta_{m_1, m_2} \tag{5.28}$$

Any set of single-spinless-particle eigenstates $\psi_m^l(\mathbf{r})$ have their (θ, ϕ) dependence

---

[5] This is equivalent to the assertion that any function ψ can be written as the sum of functions $1/2[\psi + I\psi]$ and $1/2[\psi - I\psi]$, which are, respectively, even and odd under inversion.

given by the Y_m^l:

$$\psi_m^l(\mathbf{r}) = u(r)Y_m^l(\theta,\phi) \tag{5.29}$$

If (5.15) is to hold, the radial function $u(r)$ must be independent of m.

If we apply (5.17b) to $Y_m^l(\theta,\phi)$ with $\hat{n} = \hat{z}$, we get

$$Y_m^l(\theta,\alpha) = e^{(i/\hbar)\alpha l_z}Y_m^l(\theta,0) = e^{im\alpha}Y_m^l(\theta,0) \tag{5.30}$$

Since $Y_m^l(\theta,\phi)$ is a single-valued function of position on the unit sphere, it must be independent of ϕ when $\theta = 0$. The only way that (5.30) can be consistent with this requirement is to have $Y_m^l(\theta = 0,\phi)$ vanish unless $m = 0$. For the normalization used in (5.28), this leads to

$$Y_m^l(\theta = 0,\phi) = \delta_{m,0}\sqrt{\frac{2l+1}{4\pi}} \tag{5.31}$$

According to (5.30), the $Y_0^l(\theta,\phi)$ functions are independent of ϕ. Sometimes the Y_m^l are defined in such a way that a factor i^l appears in (5.31).

The Legendre polynomial $P_l(\cos\theta)$ can be defined in terms of $Y_0^l(\theta,\phi)$ by

$$Y_0^l(\theta,0) = \sqrt{\frac{2l+1}{4\pi}}\, P_l(\cos\theta) \tag{5.32}$$

Thus, $P_l(\cos 0) = P_l(1) = 1$. The orthogonality theorem (5.28) applied to Y_0^l can be written in terms of the P_l as

$$\int_{-1}^{1} P_{l_1}(x)P_{l_2}(x)\,dx = \frac{2}{2l_1+1}\delta_{l_1,l_2} \tag{5.33}$$

The parity of $Y_m^l(\theta,\phi)$ is $(-1)^l$:

$$Y_m^l(\pi-\theta,\phi+\pi) = (-1)^l Y_m^l(\theta,\phi)$$

$$Y_m^l(-\hat{r}) = (-1)^l Y_m^l(\hat{r}) \tag{5.34}$$

In atomic and nuclear spectroscopy the letters $s, p, d, f, g, \ldots$ are used to label single-particle states with orbital angular momenta $0, 1, 2, 3, 4, \ldots$.

5.3.6 Laplacian in Spherical Polar Coordinates

From the identity

$$l_x^2 = \frac{\hbar}{i}\left(y\frac{\partial}{\partial z} - z\frac{\partial}{\partial y}\right)\frac{\hbar}{i}\left(y\frac{\partial}{\partial z} - z\frac{\partial}{\partial y}\right)$$

$$= -\hbar^2\left[y^2\frac{\partial^2}{\partial z^2} + z^2\frac{\partial^2}{\partial y^2} - 2yz\frac{\partial^2}{\partial y\partial z} - y\frac{\partial}{\partial y} - z\frac{\partial}{\partial z}\right]$$

and from the corresponding identities for l_y^2 and l_z^2, we can derive the following expression for $\mathbf{l}\cdot\mathbf{l}$:

$$\mathbf{l}\cdot\mathbf{l} = l_x^2 + l_y^2 + l_z^2$$

$$= -\hbar^2\left[r^2\nabla^2 - 2r_\alpha\frac{\partial}{\partial r_\alpha} - r_\alpha r_\beta\frac{\partial^2}{\partial r_\alpha\partial r_\beta}\right] \tag{5.35}$$

In (5.35) we have used the notation r_α to stand for x, y, z, and we imply summation over repeated indices. Now consider the effect of $r_\alpha \cdot \partial/\partial r_\alpha$ on a monomial $x^{n_x} y^{n_y} z^{n_z}$:

$$r_\alpha \frac{\partial}{\partial r_\alpha} x^{n_x} y^{n_y} z^{n_z} = \left(x \frac{\partial}{\partial x} + y \frac{\partial}{\partial y} + z \frac{\partial}{\partial z} \right) x^{n_x} y^{n_y} z^{n_z}$$

$$= (n_x + n_y + n_z) x^{n_x} y^{n_y} z^{n_z}$$

If we write $x^{n_x} y^{n_y} z^{n_z}$ in spherical polar coordinates, we get an expression of the form $r^{n_x + n_y + n_z} f(\theta, \phi)$, where $f(\theta, \phi)$ does not depend on r. Since

$$r \frac{\partial}{\partial r} x^{n_x} y^{n_y} z^{n_z} = r \frac{\partial}{\partial r} \left[r^{n_x + n_y + n_z} f(\theta, \phi) \right]$$

$$= (n_x + n_y + n_z) r^{n_x + n_y + n_z} f(\theta, \phi)$$

$$= (n_x + n_y + n_z) x^{n_x} y^{n_y} z^{n_z}$$

we see that $r_\alpha \cdot \partial/\partial r_\alpha$ and $r \cdot \partial/\partial r$ have the same effect on any monomial and, therefore, on any polynomial in x, y, z:

$$r_\alpha \frac{\partial}{\partial r_\alpha} = r \frac{\partial}{\partial r} \tag{5.36a}$$

A similar argument shows that

$$r_\alpha r_\beta \frac{\partial^2}{\partial r_\alpha \partial r_\beta} = r^2 \frac{\partial^2}{\partial r^2} \tag{5.36b}$$

Using (5.36a) and (5.36b) in (5.35), we find

$$\mathbf{l} \cdot \mathbf{l} = -\hbar^2 \left[r^2 \nabla^2 - 2r \frac{\partial}{\partial r} - r^2 \frac{\partial^2}{\partial r^2} \right]$$

$$\nabla^2 = \frac{\partial^2}{\partial r^2} + \frac{2}{r} \frac{\partial}{\partial r} - \frac{1}{\hbar^2} \frac{\mathbf{l} \cdot \mathbf{l}}{r^2} \tag{5.37a}$$

Thus, the effect of ∇^2 on a function $w(r) Y_m^l(\theta, \phi)$ is

$$\nabla^2 w(r) Y_m^l(\theta, \phi) = \left[\frac{\partial^2}{\partial r^2} + \frac{2}{r} \frac{\partial}{\partial r} - \frac{l(l+1)}{r^2} \right] w(r) Y_m^l(\theta, \phi) \tag{5.37b}$$

$$= \left[\frac{1}{r^2} \frac{\partial}{\partial r} r^2 \frac{\partial}{\partial r} - \frac{l(l+1)}{r^2} \right] w(r) Y_m^l(\theta, \phi) \tag{5.37c}$$

$$= \frac{1}{r} \left[\frac{\partial^2}{\partial r^2} - \frac{l(l+1)}{r^2} \right] r w(r) Y_m^l(\theta, \phi) \tag{5.37d}$$

If $w(r) Y_m^l(\theta, \phi)$ satisfies Laplace's equation (3.17a) then $w(r)$ must satisfy

$$\left[\frac{d^2}{dr^2} - \frac{l(l+1)}{r^2} \right] r w(r) = 0$$

whose general solution is

$$w(r) = ar^l + br^{-(l+1)} \qquad (a, b \text{ constant})$$

Thus, any solution $\psi(\mathbf{r})$ of Laplace's equation has the expansion

$$\psi(\mathbf{r}) = \sum_{l,m} \left[a_{l,m} r^l + b_{l,m} r^{-(l+1)} \right] Y_m^l(\theta, \phi) \tag{5.38}$$

The constants $a_{l,m}$, $b_{l,m}$ in (5.38) are determined by the boundary conditions on $\psi(\mathbf{r})$ (see, for example, Section 3.4).

5.3.7 Addition of Angular Momenta

The *vector addition* of two angular momenta, l_1 and l_2, leads to a series of possible *total* angular momenta L:

$$L = |l_1 - l_2|, |l_1 - l_2| + 1, \ldots, l_1 + l_2 \tag{5.39}$$

L, l_1, and l_2 are said to satisfy the *triangle relations*, since, if (5.39) is true, L, l_1, and l_2 could be the lengths of the sides of a triangle. If l_1 and l_2 are single-particle orbital angular momenta, the two-particle states with total angular momenta given by (5.39) would all have the same parity, $(-1)^{l_1+l_2}$. Many-particle states with total orbital angular momenta $L = 0, 1, 2, 3, 4, \ldots$ are designated by the capital letters, $S, P, D, F, G, \ldots$.

PROBLEM 5.3.4   The nucleus ^{6}Li has a 0^+ state. It is energetically possible for this state to decay into a deuteron (a 1^+ state) and an α-particle (a 0^+ state), yet this decay does not occur. Why not?

Suppose that the 0^+ state of ^{6}Li has decayed into a deuteron and an α-particle. The total angular momentum of zero could then be regarded as arising from the vector coupling of three angular momenta:

$$(l_1 = 0 \text{ from the } \alpha\text{-particle}) \times (l_2 = 1 \text{ from the deuteron})$$
$$\times (l \text{ from relative motion}).$$

According to the triangle relation (5.39), the only way in which we can achieve the total angular momentum $L = 0$ is to have $l = 1$ for the relative orbital angular momentum. But then the total parity would be

$$(+1 \text{ from the even-parity } \alpha) \times (+1 \text{ from the even-parity deuteron})$$
$$\times \left((-1)^l \text{ from the relative motion} \right)$$

[c.f. (5.34)]. Hence, if $l = 1$, the (α + deuteron) state has odd or negative parity, and would not be populated by the decay of an even-parity ^{6}Li state. There is no way in which a 0^+ state of ^{6}Li can decay into an α-particle and a deuteron, consistent with the conservation of total angular momentum and parity.

PROBLEM 5.3.5   (a) Prove that the function

$$\psi(\hat{r}_1, \hat{r}_2) \equiv \sum_{m=-l}^{l} Y_m^l(\hat{r}_1) Y_m^{l*}(\hat{r}_2)$$

is spherically symmetric, i.e., it is unchanged by simultaneous rotation of $\hat{r}_1$ and $\hat{r}_2$.

From the definition $\mathbf{l} \equiv \hbar/i \cdot \mathbf{r} \times \nabla$, it follows that

$$l_z^* = -l_z \qquad (l_x \pm i l_y)^* = -(l_x \mp i l_y)$$

Thus,

$$L_z\psi(\hat{r}_1, \hat{r}_2) = \left[l_z(1) + l_z(2)\right]\psi(\hat{r}_1, \hat{r}_2)$$

$$= \sum_{m=-l}^{l} \left\{\left[l_z(1)Y_m^l(\hat{r}_1)\right]Y_m^{*l}(\hat{r}_2) - Y_m^l(\hat{r}_1)\left[l_z(2)Y_m^l(\hat{r}_2)\right]^*\right\}$$

$$= \hbar \sum_{m=-l}^{l} (m - m)Y_m^l(\hat{r}_1)Y_m^{*l}(\hat{r}_2)$$

$$= 0$$

$$\left(L_x \pm iL_y\right)\psi(\hat{r}_1, \hat{r}_2) = \sum_{m=-l}^{l} \left\{\left[\left(l_x(1) \pm il_y(1)\right)Y_m^l(\hat{r}_1)\right]Y_m^{*l}(\hat{r}_2)\right.$$

$$\left. - Y_m^l(\hat{r}_1)\left[\left(l_x(2) \mp il_y(2)\right)Y_m^l(\hat{r}_2)\right]^*\right\}$$

$$= \hbar \sum_{m=-l}^{l} \left\{\sqrt{(l \mp m)(l \pm m + 1)}\; Y_{m\pm 1}^l(\hat{r}_1)Y_m^{*l}(\hat{r}_2)\right.$$

$$\left. - \sqrt{(l \pm m)(l \mp m + 1)}\; Y_m^l(\hat{r}_1)Y_{m \mp 1}^{*l}(\hat{r}_2)\right\}$$

$$= 0$$

This proves that $L_\alpha\psi(\hat{r}_1, \hat{r}_2) = 0$ for $\alpha = x, y, z$. We have seen on p. 208 that this guarantees the rotational invariance of $\psi(\hat{r}_1, \hat{r}_2)$.

(b) Use this result to prove the *spherical harmonic addition theorem*

$$\sum_{m=-l}^{l} Y_m^l(\hat{r}_1)Y_m^{*l}(\hat{r}_2) = \frac{2l + 1}{4\pi}P_l(\cos\omega_{12}) \qquad (5.40)$$

Here ω_{12} is the angle between $\hat{r}_1$ and $\hat{r}_2$.

Let us subject both $\hat{r}_1$ and $\hat{r}_2$ to the rotation that sends $\hat{r}_1$ into the $\hat{z}$ axis and $\hat{r}_2$ into the x–z plane. Then the new value of θ_1 will be 0, and the new polar coordinates of $\hat{r}_2$ will be $(\omega_{12}, 0)$. The invariance of the sum in (5.40) implies that

$$\sum_{m=-l}^{l} Y_m^l(\hat{r}_1)Y_m^{*l}(\hat{r}_2) = \sum_{m=-l}^{l} Y_m^l(0, \phi_1)Y_m^{*l}(\omega_{12}, 0)$$

According to (5.31), only the $m = 0$ term in the right-hand side is non-zero, and (5.32) implies that the value of this term is

$$\sqrt{\frac{2l + 1}{4\pi}} \times \sqrt{\frac{2l + 1}{4\pi}}\; P_l(\cos\omega_{12})$$

which proves the validity of (5.40).

5.4 BOUND STATE SOLUTIONS OF THE ONE-PARTICLE SCHRÖDINGER EQUATION

5.4.1 One-Dimensional Square Well

The potential well is shown in Figure 5.2. The zero of the vertical energy scale is taken to be at the bottom of the well. The time-independent Schrödinger equation (5.8b) for a particle of mass m moving in this potential well is

$$-\frac{\hbar^2}{2m}\frac{d^2\psi(x)}{dx^2} + V(x)\psi(x) = E\psi(x) \qquad (5.41)$$

FIGURE 5.2 A one-dimensional square-well potential.

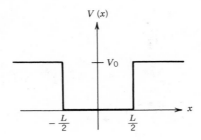

$V(x)$

V_0

x

$-\dfrac{L}{2}$ $\dfrac{L}{2}$

with

$$V(x) = 0, \quad \text{for } |x| < \frac{L}{2} \tag{5.42a}$$

$$= V_0, \quad \text{for } |x| > \frac{L}{2} \tag{5.42b}$$

If $\psi(x)$ is to describe a bound state, it must be normalizable,

$$\int_{-\infty}^{\infty} |\psi(x)|^2 \, dx = \text{a finite number} \tag{5.43}$$

We must satisfy (5.41) and (5.43) with a $\psi(x)$ which is continuous and has a continuous derivative.

We can easily find solutions of (5.41) valid for $|x| < L/2$ and for $|x| > L/2$, since $V(x)$ is constant in each region. We must combine the solutions valid in these two regions in such a way that the resulting $\psi(x)$ is continuous and has a continuous derivative at $x = \pm L/2$. If this is done, we get two types of bound-state solutions:

(a) even parity ($\psi(-x) = \psi(x)$),

$$\psi(x) = e^{-\kappa L/2}\cos(kx), \quad |x| \leq \frac{L}{2} \tag{5.44a}$$

$$= \cos\left(\frac{kL}{2}\right)e^{-\kappa|x|}, \quad |x| \geq \frac{L}{2} \tag{5.44b}$$

Here k and κ are defined by

$$k = \sqrt{\frac{2mE}{\hbar^2}} \tag{5.45a}$$

$$\kappa = \sqrt{\frac{2m}{\hbar^2}(V_0 - E)} \tag{5.45b}$$

The expressions given in (5.44a) and (5.44b) are obviously equal at $|x| = L/2$. Their derivatives will be equal if

$$\tan\left(\frac{kL}{2}\right) = \frac{\kappa}{k}$$

$$\tan\left(\sqrt{\frac{mL^2E}{2\hbar^2}}\right) = \sqrt{\frac{V_0}{E} - 1} \tag{5.46}$$

FIGURE 5.3 Graphical solution of (5.46) for $V_0 = 100$ MeV and $mL^2/(2\hbar^2) = 0.45$ MeV^{-1}. The solid line represents the left-hand side of (5.46) and the dashed line the right-hand side, both plotted as functions of E. From the three circled intersections we infer that there are three even-parity bound states, at energies of approximately 4.15, 36.67 and 94.48 MeV above the bottom of the well.

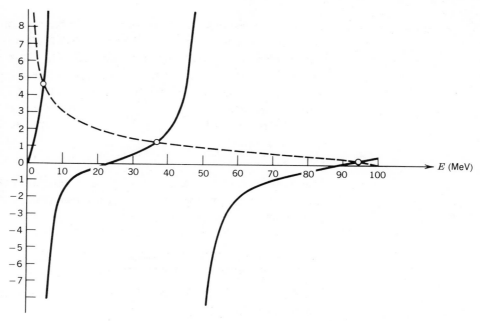

This is a transcendental equation for E, whose solutions may be obtained numerically or graphically (c.f. Figure 5.3). Once a solution of (5.46) has been found, (5.43) and (5.44) determine the corresponding normalizable wave function. It is easy to generalize from Figure 5.3 that the well will have n even-parity bound states if

$$(n-1)^2 \pi^2 < \frac{mL^2 V_0}{2\hbar^2} < n^2 \pi^2$$

(b) odd parity ($\psi(-x) = -\psi(x)$)

$$\psi(x) = e^{-\kappa L/2}\sin(kx), \qquad |x| \leq \frac{L}{2} \tag{5.47a}$$

$$= \sin\left(\frac{kL}{2}\right)e^{-\kappa x}, \qquad x \geq \frac{L}{2} \tag{5.47b}$$

$$= -\sin\left(\frac{kL}{2}\right)e^{+\kappa x}, \qquad x \leq -\frac{L}{2} \tag{5.47c}$$

Here k and κ are still defined by (5.46), but the condition for continuity of $\psi(x)$ at $|x| = L/2$ is now

$$\cot\left(\frac{kL}{2}\right) = -\frac{\kappa}{k}$$

$$\cot\left(\sqrt{\frac{mL^2 E}{2\hbar^2}}\right) = -\sqrt{\frac{V_0}{E} - 1} \tag{5.48}$$

The normalizable wave functions (5.44) and (5.47), which describe states in which the particle is *bound* in the well, are not the only physically useful eigenstates of the Schrödinger equation. In Section 5.9 we will see that there are also eigenstates describing *unbound* particles, i.e., particles which have enough energy to escape to $x = \pm\infty$. There are such eigenstates corresponding to every energy greater than the depth of the well. These are referred to as "continuum" solutions, since this part of the energy spectrum is a continuum. The continuum solutions are needed if we want to predict what will happen to a particle which comes in from $x = +\infty$ (or $x = -\infty$) and is scattered by the potential well.

The potential $V(x)$ defined in (5.42) is a particular example of a reflection-symmetric potential,

$$V(x) = V(-x) \tag{5.49}$$

The bound state eigenfunctions of any reflection-symmetric potential can be chosen to be parity eigenstates, that is to say, can be chosen to be either even or odd under the operation $x \rightleftharpoons -x$.

5.4.2 One-Dimensional Square Well with Infinitely High Walls

If we allow V_0 in (5.46) to become infinite, we see that the energies of the even parity states are given by

$$\sqrt{\frac{mL^2}{2\hbar^2} E_n} = \left(n + \frac{1}{2} \right) \pi \tag{5.50a}$$

$$E_n = \frac{2\hbar^2\pi^2}{mL^2} \left(n + \frac{1}{2} \right)^2 \qquad n = 0, 1, 2, \ldots \tag{5.50b}$$

$$k_n = \sqrt{\frac{2mE_n}{\hbar^2}} = \frac{(2n+1)\pi}{L} \tag{5.50c}$$

The corresponding interior wave function (5.44a) has spatial dependence given by

$$\cos(k_n x) = \cos\left([2n+1]\frac{\pi x}{L} \right) \tag{5.51}$$

Thus, the wave functions vanish at the walls of the well ($x = \pm L/2$). Similarly, the energies of the odd parity states are given by (5.48) to be

$$\frac{mL^2}{2\hbar^2} E_n = n\pi \tag{5.52a}$$

$$E_n = \frac{2\hbar^2\pi^2}{mL^2} n^2 \qquad n = 1, 2, 3, \ldots \tag{5.52b}$$

$$k_n = \frac{2n\pi}{L} \tag{5.52c}$$

with wave functions

$$\sin(k_n x) = \sin\left(\frac{2n\pi x}{L} \right) \tag{5.53}$$

which also vanish at the walls of the well. For many purposes it is useful to have wave functions normalized to unity over the width of the potential,

$$\int_{-L/2}^{L/2} |\psi(x)|^2 \, dx = 1 \tag{5.54}$$

This is accomplished by replacing (5.51) and (5.53) by

$$\psi_n(x) = \sqrt{\frac{2}{L}} \cos\left([2n + 1]\frac{\pi x}{L}\right), \qquad n = 0, 1, 2, \ldots \qquad (5.55a)$$

and

$$\psi_n(x) = \sqrt{\frac{2}{L}} \sin\left(\frac{2n\pi x}{L}\right), \qquad n = 1, 2, 3, \ldots \qquad (5.55b)$$

respectively.

PROBLEM 5.4.1   A particle of mass m is contained in a one-dimensional impenetrable box extending from $x = -L/2$ to $x = +L/2$. The particle is in its ground state. The walls of the box are now moved out symmetrically and instantaneously, to form a box extending from $x = -L$ to $x = +L$.

(a) Calculate the probability that, after this sudden expansion, the particle will be in the ground state.

Before (and immediately after) the sudden expansion, the wave function of the particle is

$$\psi_i(x) = \sqrt{\frac{2}{L}} \cos\left(\frac{\pi x}{L}\right), \qquad |x| \leq \frac{L}{2}$$

$$= 0, \qquad |x| \geq \frac{L}{2}$$

The ground state wave function in the expanded box is

$$\psi_f(x) = \sqrt{\frac{2}{2L}} \cos\left(\frac{\pi x}{2L}\right), \qquad |x| \leq L$$

Thus, the probability that the initial state of the particle will become the ground state of the expanded box is

$$\left| \int_{-L}^{L} \psi_f^*(x)\psi_i(x)\, dx \right|^2 = \left| \int_{-L/2}^{L/2} \left(\sqrt{\frac{1}{L}} \cos\left(\frac{\pi x}{2L}\right) \right) \left(\sqrt{\frac{2}{L}} \cos\left(\frac{\pi x}{L}\right) \right) dx \right|^2$$

$$= \left| \frac{2\sqrt{2}}{\pi} \int_{-\pi/4}^{\pi/4} \cos(\theta)\cos(2\theta)\, d\theta \right|^2 = \left(\frac{8}{3\pi}\right)^2 = 0.7205$$

(b) What is the probability that the initial state of the particle will become the first excited state of the expanded box?

In this case,

$$\psi_f(x) = \sqrt{\frac{1}{L}} \sin\left(\frac{\pi x}{L}\right), \qquad |x| \leq L$$

Then

$$\int_{-L}^{L} \psi_f^*(x)\psi_i(x)\, dx = \sqrt{\frac{2}{L}} \sqrt{\frac{1}{L}} \int_{-L}^{L} \sin\left(\frac{\pi x}{L}\right) \cos\left(\frac{\pi x}{L}\right) dx = 0$$

since the integrand is an odd function of x. Thus, the probability of ending up in the first excited state of the expanded box is zero.

(c) What is the expectation value of the energy of the particle after the rapid expansion? (Answer without calculation.)

During the rapid expansion, the particle cannot transfer energy to the walls. Thus, the expectation value of the energy is the same as it was before the expansion, namely:

$$E_0 = \frac{\hbar^2 \pi^2}{2mL^2} \qquad [\text{cf. (5.50b)}]$$

(d) Now consider a situation in which the particle is in the ground state of the original box, extending from $x = -L/2$ to $x = +L/2$. The walls of the box again move out to $x = \pm L$, but this time they move very slowly. What is the expectation value of the energy after this slow expansion?

If the box expands sufficiently slowly, the state of the particle will continuously adjust in such a way that it is always in the ground state. Thus, when the walls reach $x = \pm L$, the state will be an eigenstate of the expanded box with energy $E_0 = \hbar^2\pi^2/2m(2L)^2 = \hbar^2\pi^2/8mL^2$, and this will be the expectation value of the energy. The difference between this answer and the one obtained in (c) is due to the work done by the particle against the slowly moving walls.

In some cases it is useful to put the walls of the box at $x = 0$ and $x = L$. The well is then no longer symmetric, and the energy eigenstates will no longer be parity eigenstates. To find the energy eigenstates, we again require that the wave function vanish at the infinitely high walls of the well. The result can be written

$$\psi_n(x) = \sqrt{\frac{2}{L}} \sin\left(\frac{n\pi x}{L}\right) \left.\vphantom{\sqrt{\frac{2}{L}}}\right\} \qquad (5.56a)$$

$$E_n = \frac{\hbar^2}{2m}\left(\frac{n\pi}{L}\right)^2 = \frac{\hbar^2\pi^2}{2mL^2}n^2 \left.\vphantom{\frac{\hbar^2}{2m}}\right\} \qquad n = 1,2,3,\ldots \qquad (5.56b)$$

It is easily verified that the sequence of energies given by (5.56b) is the same as the sequence given by (5.50b) and (5.52b).

PROBLEM 5.4.2 Suppose that the initial positions of the walls of the box in the previous problem were at $x = 0$ and $x = L$. Now the box is suddenly expanded by having its right-hand wall move to $x = 2L$.

(a) What is the probability that, after this sudden expansion, the particle will be in the ground state?

Now we have

$$\psi_i(x) = \sqrt{\frac{2}{L}} \sin\left(\frac{\pi x}{L}\right), \qquad 0 \leq x \leq L$$

$$= 0, \qquad\qquad\qquad x \geq L$$

$$\psi_f(x) = \sqrt{\frac{2}{2L}} \sin\left(\frac{\pi x}{2L}\right), \qquad 0 \leq x \leq 2L$$

so the probability of ending up in the ground state is

$$\left|\int_0^{2L} \psi_f^*(x)\psi_i(x)\,dx\right|^2 = \left|\int_0^L \left(\sqrt{\frac{2}{L}}\sin\left(\frac{\pi x}{L}\right)\right)\left(\sqrt{\frac{1}{L}}\sin\left(\frac{\pi x}{2L}\right)\right)dx\right|^2$$

$$= \left|\frac{2\sqrt{2}}{\pi}\int_0^{\pi/2}\sin(2\theta)\sin(\theta)\,d\theta\right|^2 = \left(\frac{4\sqrt{2}}{3\pi}\right)^2 = 0.3603$$

(b) What is the probability of ending up in the first excited state of the expanded box?

$\psi_i(x)$ is the same as before, but now we use

$$\psi_f(x) = \sqrt{\frac{2}{2L}}\,\sin\left(\frac{2\pi x}{2L}\right)$$

$$\left|\int_0^{2L}\psi_f^*(x)\psi_i(x)\,dx\right|^2 = \left|\sqrt{\frac{2}{L}}\,\sqrt{\frac{1}{L}}\int_0^L\sin^2\left(\frac{\pi x}{L}\right)dx\right|^2$$

$$= \frac{1}{2}$$

Note that these results are different from those we found when both walls of the box expanded simultaneously.

5.4.3 One-Dimensional δ-Function Interaction

Suppose that the potential is an attractive δ-function at the origin, so that the Schrödinger equation is

$$-\frac{\hbar^2}{2m}\frac{d^2\psi}{dx^2}(x) - G\delta(x)\psi(x) = E\psi(x) \tag{5.57}$$

Here $\delta(x)$ is the Dirac δ-function, defined[6] to have the properties

$$\delta(x) = 0, \qquad \text{if } x \neq 0 \tag{5.58a}$$

$$\int_a^b \delta(x)\,dx = 1, \qquad \text{if } a < 0 \text{ and } b > 0 \tag{5.58b}$$

and G is a positive constant. When $x \neq 0$, (5.56) becomes

$$-\frac{\hbar^2}{2m}\frac{d^2\psi}{dx^2}(x) = E\psi(x)$$

whose even-parity normalized solutions can be written as

$$\psi(x) = e^{-\kappa x} \qquad (x \geq 0) \tag{5.59a}$$
$$= e^{+\kappa x} \qquad (x \leq 0) \tag{5.59b}$$

with

$$\kappa = \sqrt{\frac{2m}{\hbar^2} \times (-E)} \tag{5.59c}$$

To find κ, we integrate (5.57) from $x = -\varepsilon$ to $+\varepsilon$, and then allow ε to approach zero:

$$-\frac{\hbar^2}{2m}\int_{-\varepsilon}^{\varepsilon}\frac{d^2\psi(x)}{dx^2}\,dx - G\int_{-\varepsilon}^{\varepsilon}\delta(x)\psi(x)\,dx = E\int_{-\varepsilon}^{\varepsilon}\psi(x)\,dx$$

$$-\frac{\hbar^2}{2m}[\psi'(\varepsilon) - \psi'(-\varepsilon)] - G\psi(0) = E \times 2\varepsilon \times \bar{\psi}(0) \tag{5.60}$$

---

[6]This is the one-dimensional version of the three-dimensional δ-function defined in the footnote on p. 114.

The prime represents differentiation, and $\overline{\psi}(0)$ is the average value of $\psi(x)$ over the interval $-\varepsilon < x < \varepsilon$. If we allow ε to approach zero in (5.60), we find that

$$\lim_{\varepsilon \to 0} \left[\psi'(\varepsilon) - \psi'(-\varepsilon) \right] = -\frac{2mG}{\hbar^2} \psi(0) \tag{5.61}$$

Thus, the δ-function potential produces a discontinuity in ψ' at $x = 0$. If we use (5.59) in (5.61), we get

$$\kappa = \frac{mG}{\hbar^2}$$

$$E = -\frac{\hbar^2}{2m} \left(\frac{mG}{\hbar^2} \right)^2 = -\frac{mG^2}{\hbar^2} \tag{5.62}$$

We see that the attractive δ-potential has a single even-parity bound state. Note that a δ-function potential at the origin has no effect on the odd-parity states, since a particle in an odd-parity state never is at the origin. If $\psi(-x) = -\psi(x)$, then $\psi(0) = 0$ and the right-hand side of (5.61) vanishes even if $G > 0$.

PROBLEM 5.4.3   Find the energies of the bound states of the potential

$$V(x) = -G\delta(x), \qquad -a \le x \le a \tag{5.63a}$$

$$= \infty, \qquad\qquad |x| > a \tag{5.63b}$$

which consists of a δ-function in the middle of an infinite square well.

This problem is somewhat analogous to the point mass at the midpoint of a stretched string (Problem 1.14.5). For odd-parity states, the δ-function has no effect and so (5.52b) applies. Let us take the even-parity wave function to be

$$\psi(x) = \sin(k(x - a)), \qquad x > 0$$
$$= \sin(k(-x - a)), \qquad x < 0$$

This state is continuous at $x = 0$, and vanishes at the infinite walls at $x = \pm a$. If we apply the condition (5.61), we find

$$2k \cos(-ka) = -\frac{2mG}{\hbar^2} \sin(-ka)$$

$$\tan(ka) = \frac{\hbar^2}{mG} k = \frac{\hbar^2}{mGa} \times (ka)$$

This can be regarded as a transcendental equation for ka. There are infinitely many solutions, and the nth solution k_n yields an energy $E_n = \hbar^2 k_n^2 / 2m$ above the flat inside part of the well.

5.4.4 Harmonic Oscillator Potential

The Schrödinger equation for an oscillator with frequency $\omega = \sqrt{k/m}$ has been given in (5.9). The eigenvalues and normalized eigenfunctions are

$$E_n = \hbar\omega \times \left(n + \tfrac{1}{2} \right) \tag{5.64a}$$

$$\psi_n(x) = \left[\frac{m\omega}{\pi\hbar} \right]^{1/4} \frac{1}{\sqrt{2^n n!}} h_n\left(\sqrt{\frac{m\omega}{\hbar}} \, x \right) e^{-(m\omega/2\hbar)x^2} \qquad n = 0, 1, 2, \ldots \tag{5.64b}$$

Here h_n is a Hermite polynomial. Since $h_n(-y) = (-1)^n h_n(y)$, it follows from (5.64b) that the parity of ψ_n is $(-1)^n$. Note that the energy scale is determined by $\hbar\omega$, and the distance scale by $\sqrt{\hbar/m\omega}$.

PROBLEM 5.4.4   The Hamiltonian for two interacting spin-$\frac{1}{2}$ identical fermions in one dimension is

$$H = \frac{p_1^2}{2m} + \frac{p_2^2}{2m} + \frac{m\omega^2}{2}(x_2 - x_1)^2 \tag{5.65}$$

What is the energy spectrum and what are the corresponding eigenfunctions?

Since the interaction between the particles is a function of their separation, it will be convenient to work in relative and mass-center coordinates, x and X:

$$x = x_2 - x_1 \tag{5.66a}$$

$$X = \frac{x_1 + x_2}{2} \tag{5.66b}$$

The corresponding momenta are given by (1.76):

$$p = \tfrac{1}{2}(p_2 - p_1) \tag{5.67a}$$

$$P = p_1 + p_2 \tag{5.67b}$$

P is the total momentum of the two particles, as measured in the laboratory, and $\pm p$ are the momenta of the individual particles as measured by an observer moving with the mass center. If we express H in terms of these new coordinates and momenta, we get

$$H = \frac{P^2}{2M} + \left(\frac{p^2}{2\mu} + \frac{\mu(\sqrt{2}\,\omega)^2}{2} x^2 \right) \tag{5.68}$$

Here $M(= 2m)$ is the total mass and $\mu(= m/2)$ is the reduced mass (1.74b).

The Hamiltonian (5.68) is separable[7] with respect to the relative and mass-center degrees of freedom. An eigenfunction of (5.68) is

$$\psi_{K,n}(x, X) = e^{iKX} h_n\left(\sqrt{\frac{\mu\sqrt{2}\,\omega}{\hbar}}\, x \right) e^{-1/2(\mu(\sqrt{2}\,\omega)/\hbar)x^2} \tag{5.69a}$$

$$= e^{iK(x_1 + x_2)/2} h_n\left(\sqrt{\frac{m\omega}{\sqrt{2}\,\hbar}}\,(x_2 - x_1) \right) e^{-m\omega(x_2 - x_1)^2/2\sqrt{2}\,\hbar} \tag{5.69b}$$

and the corresponding energy eigenvalue is

$$E_{K,n} = \frac{\hbar^2 K^2}{2M} + \hbar \times (\sqrt{2}\,\omega) \times (n + \tfrac{1}{2}) \tag{5.70}$$

According to (5.66), interchanging x_1 and x_2 has the effect of replacing x by $-x$, but has no effect on X. These operations cause (5.69a) to be multiplied by $(-1)^n$. Thus, the even-n states are symmetric with respect to permutation of the positions of the two particles. If the particles are fermions, these symmetric

---

[7] Let $H = H_1 + H_2$, where H_1 and H_2 refer to different degrees of freedom. Then if $H_1\phi_1 = E_1\phi_1$ and $H_2\phi_2 = E_2\phi_2$, it follows that $H\phi_1\phi_2 = (E_1 + E_2)\phi_1\phi_2$. Thus, we can find eigenstates of the separable Hamiltonian by taking products of the eigenstates of the component Hamiltonians H_1 and H_2.

"orbital" states must be multiplied by an antisymmetric two-particle spin state (which corresponds to $S = 0$, $m_S = 0$). On the other hand, odd-n orbital states are antisymmetric with respect to permutation of the positions of the particles, so they must be multiplied by symmetric two-particle spin states ($S = 1$, $m_S = 0, \pm 1$).

5.4.5 Spherically Symmetric Potential

The Schrödinger equation for a particle moving in three-dimensional space is a partial differential equation in three variables

$$\left[-\frac{\hbar^2}{2\mu} \nabla^2 + U(\mathbf{r}) \right] \psi(\mathbf{r}) = E\psi(\mathbf{r}) \tag{5.71}$$

However, if the potential energy is spherically symmetric ($U(\mathbf{r}) = U(r)$), this equation separates in spherical polar coordinates. We seek solutions of the form

$$\psi_m^l(\mathbf{r}) = \frac{u_l(r)}{r} Y_m^l(\theta, \phi) \qquad \left\{ \begin{array}{l} l = 0, 1, 2, \\ m = -l, -l+1, \ldots, l \end{array} \right. \tag{5.72}$$

and use (5.37d) to calculate the effect of ∇^2. Then instead of having to confront the partial differential equation (5.71), we need only solve a set of uncoupled ordinary differential equations (one for each l):

$$\left.\begin{array}{c} \left[\dfrac{d^2}{dr^2} - \dfrac{l(l+1)}{r^2} - \dfrac{2\mu}{\hbar^2} U(r) + k^2 \right] u_l(r) = 0 \\[2ex] k = \sqrt{\dfrac{2\mu E}{\hbar^2}} \end{array}\right\} E > 0 \tag{5.73a}$$

or

$$\left.\begin{array}{c} \left[\dfrac{d^2}{dr^2} - \dfrac{l(l+1)}{r^2} - \dfrac{2\mu}{\hbar^2} U(r) - \kappa^2 \right] u_l(r) = 0 \\[2ex] \kappa = \sqrt{\dfrac{2\mu(-E)}{\hbar^2}} \end{array}\right\} E < 0 \tag{5.73b}$$

We suppose that $U(r)$ obeys the following conditions:

$$r^2 U(r) \underset{r \to 0}{\to} 0 \tag{5.74a}$$

$$r U(r) \underset{r \to \infty}{\to} 0 \tag{5.74b}$$

As a consequence of (5.74a), the behavior of $u_l(r)$ near the origin is either

$$u_l(r) \underset{r \to 0}{\to} \alpha r^{l+1} \qquad \text{(regular)} \tag{5.75a}$$

or

$$u_l(r) \underset{r \to 0}{\to} \alpha r^{-l} \qquad \text{(irregular)} \tag{5.75b}$$

We are usually constrained to use the regular solution; otherwise $\psi_m^l(\mathbf{r})$ from (5.72) would be singular at the origin. As a consequence of (5.74b) the behavior of $u_l(r)$ near $r = \infty$ is either

$$u_l(r) \underset{r \to \infty}{\to} \beta_{\pm} e^{\pm ikr} \qquad (E > 0) \tag{5.76a}$$

or

$$U_l(r) \underset{r \to \infty}{\to} \beta_{\pm} e^{\pm \kappa r} \qquad (E < 0) \tag{5.76b}$$

If we are seeking a bound-state solution of (5.73), it must be normalizable,

$$\int_0^\infty [u_l(r)]^2 \, dr \quad \text{is finite,} \tag{5.77}$$

and the only one of the four possible asymptotic forms (5.76) consistent with this is $\beta_- e^{-\kappa r}$. Thus, E must be negative, and it must be chosen so that the solution of (5.73b) that is regular as $r \to 0$ behaves at $r = \infty$ like $e^{-\kappa r}$. This will occur at a discrete set of eigenenergies E_{nl}. The subscript n is called the principal quantum number, and is used to label the different eigenfunctions of (5.73b) with the same value of l.

In Section 5.10, when we discuss scattering, we will make use of the positive energy solutions of (5.73a).

PROBLEM 5.4.5   Find the condition that must be satisfied by the spherically symmetric square well potential

$$U(r) = -V_0 \quad (V_0 > 0) \quad (r < a) \tag{5.78a}$$
$$= 0 \quad\quad\quad\quad (r > a) \tag{5.78b}$$

if it is just barely deep enough to contain one bound state.

The centrifugal potential term $\hbar^2 l(l+1)/r^2$ in (5.73b) makes the total effective radial potential deepest in the $l = 0$ states. Thus, if there is to be only one bound state, it will be an $l = 0$ state. The radial Schrödinger equation (5.73b) becomes

$$\left[\frac{d^2}{dr^2} + \frac{2\mu V_0}{\hbar^2} - \kappa^2 \right] u_0(r) = 0 \quad (r < a) \tag{5.79a}$$

$$\left[\frac{d^2}{dr^2} - \kappa^2 \right] u_0(r) = 0 \quad (r > a) \tag{5.79b}$$

We require that $u_0(r)$ vanish at $r = 0$ and be a decaying exponential at $r = \infty$. Thus the acceptable solutions of (5.78) are proportional to

$$u_0(r) = \alpha \sin(kr) \quad (r < a) \tag{5.80a}$$

$$k = \sqrt{\frac{2\mu V_0}{\hbar^2} - \kappa^2}$$

$$u_0(r) = \beta e^{-\kappa r} \quad (r > a) \tag{5.80b}$$

The requirements of continuity of $u_0(r)$ and $u_0'(r)$ at $r = a$ can be satisfied by equating the interior and exterior logarithmic derivatives u_0'/u_0 calculated at $r = a$ from (5.80),

$$k \cot(ka) = -\kappa$$

$$\tan\left(\sqrt{\frac{2\mu a^2}{\hbar^2}(V_0 + E)} \right) = -\sqrt{\frac{V_0 + E}{-E}} \quad (-V_0 < E < 0) \tag{5.81}$$

This is a transcendental equation for E whose solutions give the energies of possible $l = 0$ bound states. If one of these solutions is $E \approx 0$, then (5.81) implies that

$$\sqrt{\frac{2\mu a^2}{\hbar^2} V_0} = (2n + 1)\frac{\pi}{2}, \quad n = 0, 1, 2, \ldots$$

Thus, the smallest value of $V_0 a^2$ that admits an $l = 0$ bound state is $V_0 a^2 = (\hbar\pi/2)^2/2\mu$.

5.4.6 Coulomb Potential $U(r) = -Ze^2/r$

In this case the bound-state eigenfunctions and eigenvalues are given by

$$\frac{u_{nl}(r)}{r} = \left\{ \left(\frac{2}{na} \right)^3 \frac{(n-l-1)!}{2n[(n+l)!]^3} \right\}^{1/2} e^{-(r/na)} \left(\frac{2r}{na} \right)^l L_{n+1}^{2l+1} \left(\frac{2r}{na} \right) \quad (5.82a)$$

$$E_n = -\frac{Ze^2}{2a} \times \frac{1}{n^2}; \quad n = 1, 2, \ldots \quad (5.82b)$$

$$l = 0, 1, 2, \ldots, n-1$$

with $L_{n+l}^{2l+1}(x)$ an associated Laguerre polynomial, and the constant a defined by

$$a \equiv \frac{\hbar^2}{Z\mu e^2} \quad (5.83)$$

In the case of an electron moving in the field of a fixed proton, μ is the electron mass m_e, and the distance and energy scales are given by

$$a_0 = \frac{\hbar^2}{m_e e^2} = 0.53 \text{ Å} \qquad [\text{cf. (5.1)}] \quad (5.84a)$$

$$\frac{e^2}{2a_0} = \frac{m_e e^4}{2\hbar^2} = 13.6 \text{ eV} \qquad [\text{cf. (5.2)}] \quad (5.84b)$$

If we wish to apply this analysis to the hydrogen atom, then μ in (5.83) must be the *reduced* electron-proton mass $m_p m_e/(m_p + m_e)$.

If an electromagnetic transition occurs between two levels separated by energy ΔE, then the frequency ν, and wavelength, λ, of the emitted photons are given by the Bohr relation

$$\Delta E = h\nu = \frac{hc}{\lambda} \approx \frac{12,400 \text{ eV Å}}{\lambda (\text{Å})}$$

$$\approx \frac{1.24 \text{ eV micron}}{\lambda (\text{microns})} \quad (5.85)$$

PROBLEM 5.4.5   How many photons does a radio station emit in one second when broadcasting with a power of 100 kilowatts in the 300-meter band?

Each photon has energy

$$E = \frac{hc}{\lambda} = \frac{12,400 \text{ eV Å}}{300 \times 10^{10} \text{ Å}} = \frac{1.24}{3} \times 10^{-8} \text{ eV}$$

Thus, photons are emitted at the rate of

$$10^5 \frac{J}{s} \times \frac{1 \text{ eV}}{1.6 \times 10^{-19} \text{ J}} \times \frac{1 \text{ photon}}{\dfrac{1.24}{3} \times 10^{-8} \text{ eV}} \approx 1.5 \times 10^{32} \text{ photons/s}$$

PROBLEM 5.4.6   The $n = 3$ to $n = 2$ transition in hydrogen gives rise to the "Hα" spectral line. The light from a discharge containing hydrogen and helium is found to have a line 2.86 Å away from the Hα line. This is attributed to a transition in singly-ionized helium atoms (He$^+$).

(a) Find the principal quantum numbers of the levels involved in He$^+$.

According to (5.82b) and (5.84b), the energy of the Hα transition is

$$E_3 - E_2 = \frac{\mu_H e^4}{2\hbar^2}\left(\frac{1}{2^2} - \frac{1}{3^2}\right) \tag{5.86a}$$

For a transition between n' and n'' levels in He^+,

$$E_{n''} - E_{n'} = \frac{4\mu_{He^+} e^4}{2\hbar^2}\left(\frac{1}{(n')^2} - \frac{1}{(n'')^2}\right) \tag{5.86b}$$

Since μ_H and μ_{He^+} are nearly equal, the energy differences in (5.86a) and (5.86b) will be nearly equal if

$$\frac{1}{2^2} - \frac{1}{3^2} = 4\times\left(\frac{1}{(n')^2} - \frac{1}{(n'')^2}\right) = \left(\frac{2}{n'}\right)^2 - \left(\frac{2}{n''}\right)^2$$

which is obviously satisfied if $n' = 4$ and $n'' = 6$.

(b) Now use these data and the fact that the proton-electron mass ratio is 1836, to calculate the Rydberg constant for an infinitely heavy nucleus,

$$R_\infty = \frac{m_e e^4}{2\hbar^2}$$

Equations (5.86a) and (5.86b) imply that

$$E_3 - E_2 = \frac{5}{36}\frac{e^4}{2\hbar^2}\frac{m_p m_e}{m_p + m_e}$$

$$E_6 - E_4 = \frac{5}{36}\frac{e^4}{2\hbar^2}\frac{4m_p m_e}{4m_p + m_e}$$

Thus,

$$\frac{\lambda_H - \lambda_{He^+}}{hc} = \frac{1}{E_3 - E_2} - \frac{1}{E_6 - E_4} = \frac{36}{5}\cdot\frac{2\hbar^2}{e^4}\left[\frac{1}{m_e} + \frac{1}{m_p} - \frac{1}{m_e} - \frac{1}{4m_p}\right]$$

$$= \frac{36}{5}\cdot\frac{2\hbar^2}{e^4}\cdot\frac{3}{4m_p} = \frac{36}{5}\cdot\frac{2\hbar^2}{m_e e^4}\cdot\frac{3}{4}\times\frac{1}{1836}$$

$$\frac{2.68\ \text{Å}}{12{,}400\ \text{eV Å}} = \frac{36}{5}\cdot\frac{3}{4}\cdot\frac{1}{1836}\cdot\frac{1}{R_\infty}; \qquad R_\infty = 13.6\ \text{eV}$$

PROBLEM 5.4.7   The radial wave function of an electron in the ground state of a one-electron atom is $2(Z/a_0)^{3/2}e^{-Zr/a_0}$, where Z is the charge of the nucleus. Suppose that through beta decay the nuclear charge is instantaneously increased by one. What is the probability that the electron is in the ground state of the new atom?

Immediately after the decay, the electron radial wave function is

$$u(r) = 2\left(\frac{Z}{a_0}\right)^{3/2} e^{-(Zr/a_0)}$$

whereas the ground-state radial wave function is

$$u_{gs}(r) = 2\left(\frac{Z+1}{a_0}\right)^{3/2} e^{-((Z+1)r/a_0)}$$

Thus, the probability that the new atom is in its ground state is

$$
\left| \int_0^\infty u(r) u_{gs}(r) r^2 \, dr \right|^2 = 16 \left[\frac{Z(Z+1)}{a_0^2} \right]^3 \left[\int_0^\infty e^{-(2Z+1)r/a_0} r^2 \, dr \right]^2
$$

$$
= 16 \left[\frac{Z(Z+1)}{a_0^2} \right]^3 \left[\left(\frac{a_0}{2Z+1} \right)^3 \cdot 2 \right]^2 = \left[\frac{Z(Z+1)}{Z(Z+1) + \frac{1}{4}} \right]^3
$$

PROBLEM 5.4.8   Atoms are made up of nuclei, consisting of protons and neutrons with masses $Mc^2 \approx 940$ Mev, and electrons with mass $m_e c^2 \approx 0.5$ Mev. Suppose that we lived in a hypothetical world in which the electron mass was $m_e c^2 = 100$ Mev, but the proton and neutron masses were unchanged. We also pretend that, in this hypothetical world, there is no weak interaction like the one in our world that results in β-decay $N \to P + e^- + \bar{\nu}$, $P \to N + e^+ + \nu$.

(a) In our world, materials tend to have densities in the range 1–10 g/cm³. What is the range of densities in the hypothetical world? (Ignore reduced-mass effects.)

In the new world, nuclei still contribute most of the mass of an atom. However, according to (5.84), atomic dimensions in the hypothetical world will be smaller by a factor of about (0.5 Mev/100 Mev), so material densities will be greater by a factor of about $(100/0.5)^3 = 8 \times 10^6$. Thus, we expect densities to be of the order of 10^7–10^8 g/cm³ in the hypothetical world.

(b) Chemical reactions occur at temperatures of the order of 1000 K. What are the corresponding temperatures in the hypothetical world?

According to (5.82b) and (5.83), atomic energies will be about 200 times greater in the hypothetical world. This will lead to reaction temperatures of the order of 200×1000 K $= 2 \times 10^5$ K.

(c) Would the Planck radiation law be changed in the hypothetical world and, if so, how?

Since the Planck radiation law does not involve the electron mass, it would be the same in the hypothetical world as in ours.

(d) Show that if the beta decay interaction were allowed, the hypothetical world would have no atoms.

Since $M_p + M_e > M_n$ in the hypothetical world, all protons would capture electrons and become neutrons: $p + e^- \to N + \nu$.

PROBLEM 5.4.9   The Lyman-α transition in atomic hydrogen has a wavelength of 1215 Å and a transition rate of 0.6×10^9 s^{-1}. Estimate the minimum $\Delta\lambda/\lambda$.

According to the Heisenberg uncertainty principle, if a state has lifetime τ, its energy is uncertain by an amount

$$
\Delta E \approx \frac{\hbar}{\tau}
$$

Since $\lambda = hc/E$, a small uncertainty ΔE implies that

$$
\frac{\Delta\lambda}{\lambda} = \Delta(\log \lambda) = \Delta(-\log E) = -\frac{\Delta E}{E}
$$

$$
\approx -\frac{1}{E} \frac{h/2\pi}{\tau} = -\frac{\lambda}{2\pi c \tau}
$$

$$
\left| \frac{\Delta\lambda}{\lambda} \right| = \frac{1215 \text{ Å}}{2\pi \cdot 3 \times 10^{18} \text{ Å/s}} \times (0.6 \times 10^9 \text{ s}^{-1}) \approx 4 \times 10^{-8}
$$

5.5 DENSITY OF STATES

The use of Fourier series is based on the observation that a large class of functions can be expanded over the interval $-L/2 \leq x \leq L/2$ in terms of the denumerable set of functions

$$\phi_n(x) \equiv \sqrt{\frac{1}{L}} \, e^{i(2\pi/L)nx} \qquad (n = 0, \pm 1, \pm 2, \ldots) \qquad (5.87)$$

We say, somewhat loosely, that the $\phi_n(x)$ of (5.87) form a *complete* set of functions over the interval $-L/2 < x < L/2$. Similarly, the functions

$$\phi_{n_x n_y n_z}(x, y, z) \equiv \sqrt{\frac{1}{L^3}} \, e^{i(2\pi/L)(n_x x + n_y y + n_z z)} \qquad (5.88a)$$

$$(n_x, n_y, n_z = 0, \pm 1, \pm 2, \ldots) \qquad (5.88b)$$

form a complete set of functions in a cube of volume $V = L^3$ centered at the origin. The functions $\phi_{n_x n_y n_z}$ defined in (5.88) are single-particle linear momentum eigenstates:

$$\mathbf{p}\phi_{n_x n_y n_z} = \frac{\hbar}{i}\nabla\phi_{n_x n_y n_z} = \hbar\mathbf{k}\phi_{n_x n_y n_z} \qquad (5.89a)$$

with

$$k_x = \frac{2\pi}{L}n_x, \qquad k_y = \frac{2\pi}{L}n_y, \qquad k_z = \frac{2\pi}{l}n_z \qquad (5.89b)$$

It is convenient to regard the three numbers k_x, k_y, k_z as Cartesian coordinates of a point in "k-space". Each function (5.89) of our complete set corresponds to a point on a cubic lattice in k-space. The cubes have sides of length $2\pi/L$, and volume $(2\pi/L)^3$. Thus, we can say that each of our functions $\phi_{n_x n_y n_z}$ has associated with it a k-space volume of $(2\pi/L)^3$, and the number of functions (5.89) whose **k** vector lies within d^3k is equal to

$$dN = \frac{d^3k}{\left(\dfrac{2\pi}{L}\right)^3} = \frac{V}{(2\pi)^3}\,d^3k \qquad (5.90a)$$

We can also introduce spherical polar coordinates in k-space. Since $d^3k = k^2\,dk\,d\Omega_k$, we conclude that the number of functions (5.88) with k within dk of k_0, and $\hat{k}$ within solid angle $d\Omega_k$ of $\hat{k}_0$, is

$$dN = \frac{V}{(2\pi)^3}k_0^2\,dk\,d\Omega_k \qquad (5.90b)$$

If we need the number of functions (5.88) with k within dk of k_0, irrespective of the direction of $\hat{k}$, we can integrate (5.90b) over $\hat{k}$ to get

$$dN = \frac{V}{(2\pi)^3}k_0^2\,dk\int d\Omega_k = \frac{V}{2\pi^2}k_0^2\,dk \qquad (5.90c)$$

It is also useful to express (5.90a) in terms of $\mathbf{p} = \hbar\mathbf{k}$:

$$dN = \frac{V}{(2\pi)^3}\frac{d^3p}{\hbar^3} = \frac{V}{(2\pi\hbar)^3}\,d^3p = \frac{V\,d^3p}{h^3} \qquad (5.91)$$

Since $V\,d^3p$ is a volume in phase space, we can interpret (5.91) by saying that each function of our complete set (5.88) has associated with it a phase space volume of h^3.

So far we have counted only the number of *spatial* states in various momentum intervals. If the particle has intrinsic spin, there will be several internal states associated with each spatial state. If the number of these internal states is g, then (5.90) should be replaced by

$$\frac{dN}{d^3k} = g\frac{V}{(2\pi)^3} \tag{5.92a}$$

$$\frac{dN}{dk\,d\Omega_k} = g\frac{V}{(2\pi)^3}k_0^2 \tag{5.92b}$$

$$\frac{dN}{dk} = g\frac{Vk_0^2}{2\pi^2} \tag{5.92c}$$

We sometimes need expressions for the number of single-particle states per unit (kinetic) energy interval. These can be obtained by multiplying the expressions (5.92) by dk/dE. For example, we can have:

1. nonrelativistic spinless particle ($g = 1$):

$$E = \frac{p^2}{2m} = \frac{\hbar^2 k^2}{2m}, \qquad k^2\frac{dk}{dE} = \frac{1}{2}\left(\frac{2m}{\hbar^2}\right)^{3/2}E^{1/2}$$

$$\frac{dN}{dE} = \frac{V}{4\pi^2}\left(\frac{2m}{\hbar^2}\right)^{3/2}E_0^{1/2} \tag{5.93a}$$

2. nonrelativistic electron ($g = 2 \times \frac{1}{2} + 1 = 2$):

$$\frac{dN}{dE} = 2\cdot\frac{V}{4\pi^2}\left(\frac{2m}{\hbar^2}\right)^{3/2}E_0^{1/2} \tag{5.93b}$$

3. Neutrinos, photons, and ultrarelativistic electrons ($g = 2$):

$$E = \sqrt{c^2 p^2 + m^2 c^4} - mc^2 \approx cp = \hbar ck$$

(exact for photons and neutrinos with $m = 0$; approximate for electrons if $p \gg mc$)

$$k^2\frac{dk}{dE} = \frac{E^2}{(\hbar c)^3}$$

$$\frac{dN}{dE} = 2\cdot\frac{V}{2\pi^2}\frac{E_0^2}{(\hbar c)^3} \tag{5.93c}$$

Electrons and neutrinos require $g = 2$ because they have intrinsic spin $\hbar/2$. Photons of momentum $\hbar\mathbf{k}$ have two independent states of polarization, so they also require that we use $g = 2$ (see Section 5.12).

These derivations have counted the number of independent states per unit momentum or energy interval in a cube of volume V. It can be shown[8] that the expressions (5.90) and (5.93) are correct for a volume of any shape, as long as the phase space volume is large enough to hold many states.

---

[8] W. Ledermann, *Proc. Roy. Soc. (London)* A182, 362 (1944).

5.6 APPROXIMATION METHODS

5.6.1 Rayleigh-Ritz Variational Principle

Let E_0 be the lowest eigenvalue of H. If ψ is any state of the system (not necessarily an eigenstate), then

$$\frac{\langle \psi|H|\psi \rangle}{\langle \psi|\psi \rangle} \geq E_0 \tag{5.94}$$

The equality holds only if ψ is an eigenstate ψ_0 corresponding to the eigenvalue E_0. Moreover, the ratio in (5.94) is stationary with respect to small variations about ψ_0. This means that if ψ differs from ψ_0 by a small amount $\lambda\phi$, then (5.94) will differ from E_0 by a number of order λ^2. Thus, if ψ is a fairly good approximation to ψ_0, (5.94) will probably yield an excellent approximation to E_0.

If ψ contains parameters α, we can get the best approximation to E_0 by calculating

$$E(\alpha) \equiv \frac{\langle \psi(\alpha)|H|\psi(\alpha) \rangle}{\langle \psi(\alpha)|\psi(\alpha) \rangle} \tag{5.95}$$

and minimizing it with respect to α. This minimum value of $E(\alpha)$ is still, of course, greater than or equal to E_0. Unfortunately, this method is unable to tell us how much greater $[E(\alpha)]_{\min}$ is than E_0.

PROBLEM 5.6.1   It is known that the stable ion H^- exists (two electrons bound to a proton). The simplest approximation to its ground state consists of two electrons moving in a $1s$ ($n = 1$, $l = 0$) orbit. Estimate the ground state energy of H^- using this approximate state as a trial function, and see if you can prove the stability of the H^--ion.

Useful information:

$$\psi_{1s}(\mathbf{r}) = \left(\frac{1}{\pi a_0^3} \right)^{1/2} e^{-(r/a_0)} \qquad \text{(normalized H-atom ground state)} \tag{5.96a}$$

$$\left\langle \psi_{1s}(\mathbf{r}) \Big| \frac{1}{r} \Big| \psi_{1s}(\mathbf{r}) \right\rangle = \frac{1}{a_0} \tag{5.96b}$$

$$\left\langle \psi_{1s}(\mathbf{r}) \big| \nabla^2 \big| \psi_{1s}(\mathbf{r}) \right\rangle = -\frac{1}{a_0^2} \tag{5.96c}$$

$$\int d^3r \, d^3r' \, [\psi_{1s}(\mathbf{r})]^2 [\psi_{1s}(\mathbf{r}')]^2 \times \frac{1}{|\mathbf{r} - \mathbf{r}'|} = \frac{5}{8a_0} \tag{5.96d}$$

Use

$$\psi(\mathbf{r}_1, \mathbf{r}_2; a) = \frac{1}{\pi a^3} e^{-(r_1 + r_2)/a} \tag{5.97}$$

as a variational wave function, with a as the variational parameter. This state is normalized for any value of a.

The Hamiltonian is

$$H = \frac{p_1^2}{2m} + \frac{p_2^2}{2m} - \frac{e^2}{r_1} - \frac{e^2}{r_2} + \frac{e^2}{|\mathbf{r}_1 - \mathbf{r}_2|} \tag{5.98}$$

According to (5.96b and c),

$$\left\langle \psi(\mathbf{r}_1, \mathbf{r}_2; a) \left| \frac{e^2}{r_1} + \frac{e^2}{r_2} \right| \psi(\mathbf{r}_1, \mathbf{r}_2; a) \right\rangle = \frac{2e^2}{a}$$

$$\left\langle \psi(\mathbf{r}_1, \mathbf{r}_2; a) \left| \frac{p_1^2}{2m} + \frac{p_2^2}{2m} \right| \psi(\mathbf{r}_1, \mathbf{r}_2; a) \right\rangle = 2 \cdot \frac{\hbar^2}{2m} \cdot \frac{1}{a^2}$$

If we also use (5.96d), we can calculate that

$$E(a) = \frac{\left\langle \psi(\mathbf{r}_1, \mathbf{r}_2; a) | H | \psi(\mathbf{r}_1, \mathbf{r}_2; a) \right\rangle}{\left\langle \psi(\mathbf{r}_1, \mathbf{r}_2; a) | \psi(\mathbf{r}_1, \mathbf{r}_2; a) \right\rangle}$$

$$= \frac{\hbar^2}{ma^2} - \frac{2e^2}{a} + \frac{5e^2}{8a}$$

If we minimize this expression with respect to a, we get

$$0 = \frac{d}{da} E(a) = \frac{d}{da} \left[\frac{\hbar^2}{ma^2} - \frac{11e^2}{8a} \right]$$

$$= -\frac{2\hbar^2}{ma^3} + \frac{11e^2}{8a^2}$$

Thus,

$$a = \frac{16}{11} \frac{\hbar^2}{me^2} \tag{5.99a}$$

$$E\left(\frac{16}{11} \frac{\hbar^2}{me^2} \right) = -\frac{121}{256} \frac{me^4}{\hbar^2} \geq E_0 \tag{5.99b}$$

Note that (5.99a) is larger than the single-electron value of $\hbar^2/me^2$. This can be regarded as a consequence of the mutual repulsion of the two electrons. Alternatively, we can say that each electron partially screens the nuclear charge, leading to an effective Z in (5.83) of $11/16$.

According to (5.84b), the energy of a hydrogen atom in its ground state, with a second electron at rest infinitely far away, is $-1/2 \cdot me^4/\hbar^2 = -128/256 \cdot me^4/\hbar^2$. The H^- ion will be stable only if its energy is less than $-128/256 \cdot me^4/\hbar^2$, and, unfortunately, the inequality we have found in (5.99b) is not strong enough to tell us whether this is so.

We can get a better variational wave function if we allow the two electrons to be in different single-particle states, with different values of a. This leads[9] to a variational upper bound of $-0.5133 me^4/\hbar^2$, which is enough to prove the stability of H^-.

5.6.2 Rayleigh-Schrödinger Perturbation Theory

This method can be applied when the Hamiltonian can be put in the form

$$H = H_0 + \lambda V \tag{5.100}$$

H_0 must be sufficiently simple for us to be able to find exact eigenvalues ε_i and

---

[9]H. Shull and P. O. Lowdin, *J. Chem. Phys.* 25, 1035 (1956).

eigenfunctions ϕ_i,

$$H_0\phi_i = \varepsilon_i\phi_i \tag{5.101a}$$

$$\langle \phi_i | \phi_j \rangle = \delta_{ij} \tag{5.101b}$$

and λV must be small (in a sense to be defined below).

Let E_i and ψ_i be exact eigenvalues and eigenfunctions of H, such that, as we allow λ to approach zero,

$$\psi_i \underset{\lambda \to 0}{\to} \phi_i$$

$$E_i \underset{\lambda \to 0}{\to} \varepsilon_i$$

Then

$$E_i = \varepsilon_i + \lambda\langle \phi_i | V | \phi_i \rangle + O(\lambda^2) \tag{5.102a}$$

$$\psi_i = \phi_i + \lambda \sum_{\substack{j \\ (j \neq i)}} \phi_j \frac{\langle \phi_j | V | \phi_i \rangle}{\varepsilon_i - \varepsilon_j} + O(\lambda^2) \tag{5.102b}$$

In general, the convergence of the series (5.102) will be rapid if the off-diagonal matrix elements $\langle \phi_j | \lambda V | \phi_i \rangle$ are small compared to the corresponding unperturbed energy differences.

This method fails if some of the ε_i are equal. Suppose that ε_i is a d-fold degenerate eigenvalue, i.e., there are d states $\phi_{i\alpha}$ such that

$$H_0\phi_{i\alpha} = \varepsilon_i\phi_{i\alpha} \qquad (\alpha = 1, 2, \ldots, d)$$

Then the first step in the analysis is to find the eigenvalues and eigenvectors of the matrix $M_{\alpha\beta} \equiv \langle \phi_{i\alpha} | \lambda V | \phi_{i\beta} \rangle$. The eigenvalues of this matrix provide the first-order corrections to the degenerate unperturbed eigenvalue ε_i.

PROBLEM 5.6.2 Treat the electron-electron interaction in the helium atom as a perturbation, and calculate the ground state energy to first order.

We take

$$H_0 = \frac{p_1^2}{2m} - \frac{2e^2}{r_1} + \frac{p_2^2}{2m} - \frac{2e^2}{r_2}$$

$$\lambda V = + \frac{e^2}{|\mathbf{r}_1 - \mathbf{r}_2|}$$

The spatial part of the lowest eigenstate of H_0 is[10]

$$\Phi_0 = \phi_{1s}(\mathbf{r}_1)\phi_{1s}(\mathbf{r}_2) = \frac{1}{\pi\left(\frac{a_0}{2}\right)^3} e^{-2(r_1 + r_2)/a_0}$$

[cf. (5.96a), keeping in mind that the nuclear charge here is $2e$]. The corresponding unperturbed eigenvalue is $2 \times (-4me^4/2\hbar^2) = -4me^4/\hbar^2$. The first-order correction to the energy is then

$$\langle \Phi_0 | \frac{e^2}{|\mathbf{r}_1 - \mathbf{r}_2|} | \Phi_0 \rangle = \frac{e^2}{\left[\pi\left(\frac{a_0}{2}\right)^3\right]^2} \int \frac{e^{-(4/a_0)(r_1 + r_2)}}{|\mathbf{r}_1 - \mathbf{r}_2|} d^3r_1 \, d^3r_2 \tag{5.103}$$

----

[10]Cf. the footnote on p. 223.

To evaluate the integral we apply the spherical harmonic addition theorem (5.40) to a multipole expansion of the interaction:

$$\frac{1}{|\mathbf{r}_1 - \mathbf{r}_2|} = \sum_{l=0}^{\infty} \frac{r_<^l}{r_>^{l+1}} P_l(\cos \omega_{12})$$

$$= \sum_{l=0}^{\infty} \frac{r_<^l}{r_>^{l+1}} \frac{4\pi}{2l+1} \sum_{m=-l}^{l} Y_m^l(\hat{r}_1) Y_m^{l*}(\hat{r}_2) \qquad (5.104)$$

Because the factor $\exp(-4(r_1 + r_2)/a_0)$ is invariant under separate rotations of $\mathbf{r}_1$ and $\mathbf{r}_2$, only the $l = 0$ term in (5.104) survives the angular integration in (5.103). The result is

$$\left\langle \Phi_0 \left| \frac{e^2}{|\mathbf{r}_1 - \mathbf{r}_2|} \right| \Phi_0 \right\rangle = \frac{(4\pi)^2 e^2}{\left[\pi \left(\frac{a_0}{2} \right)^3 \right]^2} \int_{r_1=0}^{\infty} r_1^2\, dr_1 \int_{r_2=0}^{\infty} r_2^2\, dr_2\, e^{-(4/a_0)(r_2+r_2)} \times \frac{1}{r_>}$$

$$= \frac{1024}{a_0^6} \cdot 2e^2 \int_{r_1=0}^{\infty} r_1^2\, dr_1 e^{-(4/a_0)r_1} \int_{r_2=r_1}^{\infty} r_2^2\, dr_2\, e^{-(4/a_0)r_2} \times \frac{1}{r_2}$$

$$= \frac{5}{4} \frac{e^2}{a_0} = \frac{5}{4} \frac{me^4}{\hbar^2}$$

Thus, the approximate helium ground state energy is

$$-4 \frac{me^4}{\hbar^2} + \frac{5}{4} \frac{me^4}{\hbar^2} = -\frac{11}{4} \frac{me^4}{\hbar^2}$$

correct to first order in the electron-electron interaction.

PROBLEM 5.6.3   Estimate the energy difference between the singlet and triplet states of the $(1s2s)$ configuration in helium. The $2s$ single-particle state in helium is

$$\psi_{2s}(\mathbf{r}) = \frac{1}{\sqrt{4\pi}} \left(\frac{1}{a_0} \right)^{3/2} \left(2 - \frac{2r}{a_0} \right) e^{-(r/a_0)}$$

The 2-electron wave function must be antisymmetric with respect to interchange of position and spin variables. Since a singlet ($S = 0$) spin state is antisymmetric with respect to interchange of the electron spin variables, a singlet spin state must be combined with a position wave function that is symmetric with respect to the interchange of the electron position variables. The reverse is true for the triplet ($S = 1$) state.

Singlet:

$$\psi_S(\mathbf{r}_1, \mathbf{r}_2, \sigma_1, \sigma_2) = \frac{\psi_{1s}(\mathbf{r}_1)\psi_{2s}(\mathbf{r}_2) + \psi_{1s}(\mathbf{r}_2)\psi_{2s}(\mathbf{r}_1)}{\sqrt{2}} \chi_{m_s=0}^{S=0}(\sigma_1, \sigma_2)$$

Triplet:

$$\psi_T(\mathbf{r}_1, \mathbf{r}_2, \sigma_1, \sigma_2) = \frac{\psi_{1s}(\mathbf{r}_1)\psi_{2s}(\mathbf{r}_2) - \psi_{1s}(\mathbf{r}_2)\psi_{2s}(\mathbf{r}_1)}{\sqrt{2}} \chi_{m_s=0}^{S=1}(\sigma_1, \sigma_2)$$

The factors $1/\sqrt{2}$ are needed for normalization. The unperturbed energies of these two states will be equal. The first order energy corrections will differ by

$$\langle \psi_S | \frac{e^2}{|\mathbf{r}_1 - \mathbf{r}_2|} | \psi_S \rangle - \langle \psi_T | \frac{e^2}{|\mathbf{r}_1 - \mathbf{r}_2|} | \psi_T \rangle$$

$$= 2 \langle \psi_{1s}(\mathbf{r}_1) \psi_{2s}(\mathbf{r}_2) | \frac{e^2}{|\mathbf{r}_1 - \mathbf{r}_2|} | \psi_{1s}(\mathbf{r}_2) \psi_{2s}(\mathbf{r}_1) \rangle$$

$$= 2 \cdot \left(\frac{8}{\pi a_0^3} \right) \cdot \left(\frac{1}{4 \pi a_0^3} \right) e^2 \int \left(2 - \frac{2r_1}{a_0} \right) \left(2 - \frac{2r_2}{a_0} \right) \frac{e^{-3(r_1 + r_2)/a_0}}{|\mathbf{r}_1 - \mathbf{r}_2|} \, d^3 r_1 \, d^3 r_2$$

$$= \frac{64}{729} \frac{e^2}{a_0} \tag{5.105}$$

The method of evaluating the six-fold integral is similar to that used for the evaluation of (5.103) in the previous problem. Note that the singlet and triplet states have different energies, although we have not included any spin-dependent forces in our Hamiltonian. The energy difference (5.105) is the result of the greater effectiveness of the Coulomb repulsion between the electrons in the position-symmetric state, in which the electrons are closer together on the average. The spin angular momentum of the state enters the problem indirectly, via the connection between position and spin permutation symmetry mandated by Fermi statistics.

PROBLEM 5.6.4   Consider the system described by the Hamiltonian

$$H = \frac{p^2}{2m} + \frac{m\omega^2}{2\alpha} \left(1 - e^{-\alpha x^2} \right)$$

(a) Calculate an approximate value for the ground state energy using first-order perturbation theory, perturbing about the solution for

$$H_0 = \frac{p^2}{2m} + \frac{m\omega^2}{2} x^2$$

whose unnormalized ground-state eigenfunction is $e^{-(m\omega/2\hbar)x^2}$, and whose ground state energy is $\hbar\omega/2$.

The perturbation is $(m\omega^2/2)((1 - e^{-\alpha x^2})/\alpha - x^2) \equiv \lambda V$. The normalized unperturbed eigenstate is $(m\omega/\hbar\pi)^{1/4} e^{-(m\omega/2\hbar)x^2} \equiv \phi_0$. Thus, the first-order correction to the energy is

$$\langle \phi_0 | \lambda V | \phi_0 \rangle = \left(\frac{m\omega}{\hbar\pi} \right)^{1/2} \int_{-\infty}^{\infty} e^{-(m\omega/\hbar)x^2} \frac{m\omega^2}{2} \left[\frac{1 - e^{-\alpha x^2}}{\alpha} - x^2 \right] dx$$

Integrals of this sort are conveniently evaluated using the general formula given at the bottom of p. 282. The result is

$$\frac{m\omega^2}{2\alpha} \left[1 - \frac{1}{\sqrt{1 + \dfrac{\alpha\hbar}{m\omega}}} \right] - \frac{\hbar\omega}{4}$$

If we add this to the unperturbed energy $\hbar\omega/2$, we get

$$E \approx \hbar\omega \left[\frac{1}{4} + \frac{m\omega}{2\hbar\alpha} \left(1 - \frac{1}{\sqrt{1 + \dfrac{\alpha\hbar}{m\omega}}} \right) \right] \qquad \text{(first order)} \tag{5.106a}$$

(b) Calculate an approximate value for the ground state energy using the Rayleigh-Ritz variational principle, and compare the result with (5.106a)

We take our trial wave function to be $\phi(\beta) = e^{-(\beta x^2/2)}$ and calculate

$$E(\beta) = \frac{\langle \phi(\beta)|H|\phi(\beta)\rangle}{\langle \phi(\beta)|\phi(\beta)\rangle}$$

$$= \frac{\displaystyle\int_{-\infty}^{\infty} e^{-\beta x^2/2}\left[-\frac{\hbar^2}{2m}\frac{d^2}{dx^2} + \frac{m\omega^2}{2\alpha}\left(1 - e^{-\alpha x^2}\right)\right]e^{-\beta x^2/2}\,dx}{\displaystyle\int_{-\infty}^{\infty} e^{-\beta x^2}\,dx}$$

$$= \hbar\omega\left[\frac{1}{4}\left(\frac{\beta\hbar}{m\omega}\right) + \frac{m\omega}{2\hbar\alpha}\left(1 - \frac{1}{\sqrt{1+\dfrac{\alpha}{\beta}}}\right)\right] \qquad \text{(variational)}$$

(5.106b)

To find β, we solve the equation

$$0 = \frac{dE(\beta)}{d\beta} = \frac{\hbar^2}{4m} - \frac{m\omega^2}{4\beta^2\left(1+\dfrac{\alpha}{\beta}\right)^{3/2}}$$

$$\beta\left(1 + \frac{\alpha}{\beta}\right)^{3/4} = \frac{m\omega}{\hbar} \tag{5.107}$$

This is equivalent to a quatric equation for β. If $\alpha \ll m\omega/\hbar$, an approximate solution to (5.107) is $\beta = m\omega/\hbar$. If this is substituted into the variational estimate (5.106b), the result agrees with the first order perturbation theory estimate (5.106a). If we substituted into (5.106b) the value of β that is the exact solution of (5.107), we would get a lower, and more accurate estimate of the ground state energy. For example, if $\alpha = .2m\omega/\hbar$, the solution to (5.107) is approximately $\beta = .85401(m\omega/\hbar)$. This leads to a variational estimate of $.46316\hbar\omega$, as opposed to the perturbation theory estimate of $.46782\hbar\omega$.

5.6.3 Fermi's Golden Rule for Transitions to Continuum Final States

Here the Hamiltonian has the form

$$H = H_0 + We^{-i\omega t} \tag{5.108}$$

H_0 and W are constant operators and ω is a constant frequency. Let ϕ_0 and ϕ_1 be eigenstates of H_0,

$$H_0\phi_0 = \varepsilon_0\phi_0$$
$$H_0\phi_1 = \varepsilon_1\phi_1$$

ε_0 is in the discrete part of the spectrum, but ε_1 is in the continuum. Then the transition probability per unit time, from the discrete state ϕ_0 to the continuum state ϕ_1, is

$$\Gamma_{0\to 1} = \frac{2\pi}{\hbar}|\langle\phi_1|W|\phi_0\rangle|^2\rho(\varepsilon_1) \tag{5.109}$$

subject to the condition that $\varepsilon_1 = \varepsilon_0 + \hbar\omega$. Here $\rho(\varepsilon_1)$ is the density of final states at ε_1. This is calculated (cf. Section 5.5) in a box of volume V, and ϕ_1 must be normalized in the same box:

$$\int_V |\phi_1|^2 \, d\tau = 1$$

If this is done, $\Gamma_{0 \to 1}$ given by (5.109) will be independent of V.

PROBLEM 5.6.5   Nuclei sometimes decay from excited states to the ground state by *internal conversion*, a process in which an atomic electron is emitted instead of a photon. Let the initial and final nuclear states have wave functions $\Phi_I(\mathbf{r}_1, \mathbf{r}_2, \ldots, \mathbf{r}_Z)$ and $\Phi_F(\mathbf{r}_1, \mathbf{r}_2, \ldots, \mathbf{r}_Z)$ respectively, where $\mathbf{r}_i$ $(i = 1, 2, \ldots, Z)$ describes the protons.[11] The perturbation giving rise to the transition is just the proton-electron interaction

$$W = -\sum_{i=1}^{Z} \frac{e^2}{|\mathbf{r} - \mathbf{r}_i|}$$

where $\mathbf{r}$ is the electron coordinate.

(a) Write down the matrix element for the process in lowest-order perturbation theory, assuming that the electron is initially in a state characterized by the quantum numbers (n, l, m), and that its energy, after it is emitted, is large enough so that its final state may be described by a plane wave. Neglect spin.

The initial state is $\phi_0 \equiv \psi_m^{nl}(\mathbf{r}) \Phi_I(\mathbf{r}_1, \mathbf{r}_2, \ldots, \mathbf{r}_Z)$. The final state is $\phi_1 \equiv 1/\sqrt{V} \, e^{i\mathbf{k} \cdot \mathbf{r}} \Phi_F(\mathbf{r}_1, \mathbf{r}_2, \ldots, \mathbf{r}_Z)$. Thus, the matrix element needed in (5.109) is

$$\langle \phi_1 | W | \phi_0 \rangle$$

$$= -\frac{1}{\sqrt{V}} \int d^3r \, e^{-i\mathbf{k} \cdot \mathbf{r}} \Big\langle \Phi_F(\mathbf{r}_1, \mathbf{r}_2, \ldots, \mathbf{r}_Z) \Big| \sum_{i=1}^{Z} \frac{e^2}{|\mathbf{r} - \mathbf{r}_i|} \Big| \Phi_I(\mathbf{r}_1, \mathbf{r}_2, \ldots, \mathbf{r}_Z) \Big\rangle \psi_m^{nl}(\mathbf{r})$$

(b) Write down an expression for the internal conversion rate.

The density of states for electrons of energy ε_1 is given by (5.93a) for each spin state. Since our interaction is independent of time, we use (5.108) with $\omega = 0$, so $\varepsilon_1 + E_F = \varepsilon_0 + E_I$. Then (5.109) becomes

$$\Gamma_{0 \to 1} = \frac{2\pi}{\hbar} \frac{1}{4\pi^2} \left(\frac{2m}{\hbar^2} \right)^{3/2} \varepsilon_1^{3/2} \times$$

$$\times \left| \int d^3r \, e^{-i\mathbf{k} \cdot \mathbf{r}} \langle \Phi_F | \sum_{i=1}^{Z} \frac{e^2}{|\mathbf{r} - \mathbf{r}_i|} | \Phi_I \rangle \psi_m^{nl}(\mathbf{r}) \right|^2$$

with

$$k = \sqrt{\frac{2m}{\hbar^2} \varepsilon_1} = \sqrt{\frac{2m}{\hbar^2} (\varepsilon_0 + E_I - E_F)}$$

(c) For light nuclei, the nuclear radius is much smaller than the Bohr radius for the given Z, and we can use the expansion

$$\frac{1}{|\mathbf{r} - \mathbf{r}_i|} \approx \frac{1}{r} + \frac{\mathbf{r} \cdot \mathbf{r}_i}{r^3} \tag{5.110}$$

(cf. 3.18a). Use this approximation to express the transition rate in terms of the

---

[11]We make no explicit reference to the neutrons in the nucleus since they play no direct role in this process.

dipole matrix element

$$\mathbf{d} \equiv \left\langle \Phi_F \Big| \sum_{i=1}^{Z} \mathbf{r}_i \Big| \Phi_I \right\rangle$$

Because $1/r$ does not involve the nuclear coordinates, it cannot connect orthogonal nuclear states. Thus,

$$\left\langle \Phi_F \Big| \frac{1}{r} \Big| \Phi_I \right\rangle = 0$$

and the first surviving term in our multipole expansion (5.110) is

$$\left\langle \Phi_F \Big| \sum_{i=1}^{Z} \frac{e^2}{|\mathbf{r} - \mathbf{r}_i|} \Big| \Phi_I \right\rangle \approx \frac{e^2 \mathbf{r}}{r^3} \cdot \left\langle \Phi_F \Big| \sum_{i=1}^{Z} \mathbf{r}_i \Big| \Phi_I \right\rangle$$

$$= \frac{e^2 \mathbf{r} \cdot \mathbf{d}}{r^3}$$

This leads to

$$\Gamma_{0 \to 1} = \frac{e^4}{2\pi\hbar} \left(\frac{2m}{\hbar^2} \right)^{3/2} \varepsilon_1^{1/2}$$

$$\times \left| \mathbf{d} \cdot \int d^3 r \, e^{-i\mathbf{k}\cdot\mathbf{r}} \frac{\mathbf{r}}{r^3} \psi_m^{nl}(\mathbf{r}) \right|^2$$

5.7 NONRELATIVISTIC CHARGED PARTICLE IN AN ELECTROMAGNETIC FIELD

In (3.64) we have given the Hamiltonian of a nonrelativistic charged particle moving in a specified electromagnetic field. Let us apply this Hamiltonian to the problem of an electron of charge $-e$ moving in the Coulomb field of a stationary nucleus of charge $+Ze$, in the presence of an externally applied uniform magnetic field $\mathbf{B} = B\hat{n}$. The potentials appropriate to this situation are

$$\phi(\mathbf{r}, t) = \frac{Ze}{r} \tag{5.111a}$$

$$\mathbf{A}(\mathbf{r}, t) = \frac{B}{2} \hat{n} \times \mathbf{r} \tag{5.111b}$$

(cf. Problem 3.11.1). In all situations attainable in the laboratory the dominant force on the electron is the Coulomb force due to the nucleus, and the external magnetic field is a weak perturbation. In this case, $|eA/c| \ll p$, and (3.64) can be approximated by

$$H \approx \frac{p^2}{2m} - e\phi(\mathbf{r}, t) + \frac{e}{2mc}(\mathbf{p} \cdot \mathbf{A} + \mathbf{A} \cdot \mathbf{p})$$

$$= \frac{p^2}{2m} - e\phi(\mathbf{r}, t) + \frac{eB}{4mc}(\mathbf{p} \cdot \hat{n} \times \mathbf{r} + \hat{n} \times \mathbf{r} \cdot \mathbf{p})$$

$$= \frac{p^2}{2m} - e\phi(\mathbf{r}, t) + \frac{eB}{2mc} \hat{n} \cdot \mathbf{r} \times \mathbf{p}$$

$$= \frac{p^2}{2m} - e\phi(\mathbf{r}, t) + \frac{e\mathbf{B}}{2mc} \cdot \mathbf{l} \tag{5.112}$$

Note that $\mathbf{p} \cdot n \times \mathbf{r}$ contains no $p_x x$, $p_y y$, or $p_z z$ products, so $\mathbf{p} \cdot (\hat{n} \times \mathbf{r}) =$

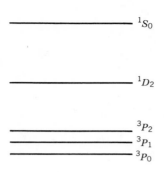

FIGURE 5.4 Schematic representation of the lowest energy levels of ^{12}C. The drawing is not to scale. The actual splittings between the $^{3}P_{J}$ levels are of the order of 0.002 eV, and are due to the spin-orbit interaction. The ^{1}S-^{3}P-^{1}D splittings are of the order of 1 eV, and are due to the Coulomb repulsion between the electrons.

$(\hat{n} \times \mathbf{r}) \cdot \mathbf{p}$. Equation (5.112) is often written in terms of a magnetic moment operator

$$\boldsymbol{\mu} \equiv \frac{-e}{2mc}\mathbf{l} \tag{5.113a}$$

$$H \equiv \frac{p^2}{2m} - e\phi(\mathbf{r}, t) - \boldsymbol{\mu} \cdot \mathbf{B} \tag{5.113b}$$

[cf. (3.57c, d)]. For a particle with intrinsic spin $\mathbf{s}$, (5.113a) is replaced by

$$\boldsymbol{\mu} = \frac{-e}{2mc}[\mathbf{l} + g\mathbf{s}] \tag{5.114}$$

It is a consequence of Dirac's theory of the electron that g in (5.114) must be taken to be 2. Thus, $\boldsymbol{\mu}$ is not proportional to the total angular momentum operator $\mathbf{j}\, (= \mathbf{l} + \mathbf{s})$. This leads to the "anomalous Zeeman effect."

PROBLEM 5.7.1 The lowest energy levels of the carbon atom ($Z = 6$) are shown in Figure 5.4 How would these levels be changed if a weak uniform magnetic field $\mathbf{B}$ were applied to the atom?

Suppose that the $\mathbf{B}$ field points along the $\hat{z}$ axis. Then it adds to the Hamiltonian a perturbation

$$\lambda V = \frac{eB}{2mc}\left[L_z + 2S_z\right] = \frac{eB}{2mc}\left[J_z + S_z\right] \tag{5.115}$$

where L_z, S_z, and J_z are, respectively, the z components of the electronic orbital, spin, and total angular momenta. We must calculate the expectation value of (5.115) in the free-atom eigenstates in order to obtain the first-order effect of the magnetic field. The free-atom eigenstates have specified values of L, S, J, and J_z, but not of S_z. The Wigner-Eckart theorem can be used to show[12] that

$$\langle S_z \rangle_{LSJM} = 0, \qquad \text{if } J = 0 \tag{5.116a}$$

$$= \frac{\langle \mathbf{S} \cdot \mathbf{J} \rangle_{LSJM}}{\langle \mathbf{J} \cdot \mathbf{J} \rangle_{LSJM}} \langle J_z \rangle_M = \frac{\hbar M \langle \mathbf{S} \cdot \mathbf{J} \rangle_{LSJM}}{\hbar^2 J(J + 1)} \qquad (J \neq 0) \tag{5.116b}$$

---

[12] The essential point is that (S_x, S_y, S_z) have the same rotational transformation properties as (J_x, J_y, J_z), so the matrix elements $\langle \psi_{M_1}^{J_1} | S_\alpha | \psi_{M_2}^{J_2} \rangle$ and $\langle \psi_{M_1}^{J_1} | J_\alpha | \psi_{M_2}^{J_2} \rangle$ are proportional, with the proportionality factor independent of M_1, M_2, α. See, for example, *Group Theory and Its Application to the Quantum Mechanics of Atomic Spectra*, E. P. Wigner, Academic Press, New York, 1959, p. 273.

But we can rewrite $\mathbf{S} \cdot \mathbf{J}$ as

$$\mathbf{S} \cdot \mathbf{J} = \frac{-(\mathbf{J} - \mathbf{S}) \cdot (\mathbf{J} - \mathbf{S}) + \mathbf{J} \cdot \mathbf{J} + \mathbf{S} \cdot \mathbf{S}}{2}$$

$$= \frac{-\mathbf{L} \cdot \mathbf{L} + \mathbf{J} \cdot \mathbf{J} + \mathbf{S} \cdot \mathbf{S}}{2}$$

Thus,

$$\langle \mathbf{S} \cdot \mathbf{J} \rangle_{LSJM} = \frac{\hbar^2}{2}[-L(L+1) + J(J+1) + S(S+1)]$$

and (5.116b) becomes

$$\langle S_z \rangle_{LSJM} = \frac{\hbar M}{2}\left[1 + \frac{S(S+1) - L(L+1)}{J(J+1)}\right] \qquad (5.116c)$$

If (5.116a) and (5.116c) are used to calculate the expectation value of (5.115), the result is

$$\Delta E \equiv \langle \lambda V \rangle_{LSJM} = 0, \qquad \text{if } J = 0$$

$$= \frac{e\hbar}{2mc}BM\left[\frac{3}{2} + \frac{S(S+1) - L(L+1)}{2J(J+1)}\right] \qquad (J \neq 0)$$

The application of this formula to the lowlying states of the carbon atom is given in the following table:

State	L	S	J	ΔE
$^1S_0$	0	0	0	0
$^1D_2$	2	0	2	$\dfrac{e\hbar}{2mc} \cdot BM$
$^3P_2$	1	1	2	$\dfrac{e\hbar}{2mc} \cdot \dfrac{3}{2}BM$
$^3P_1$	1	1	1	$\dfrac{e\hbar}{2mc} \cdot \dfrac{3}{2}BM$
$^3P_0$	1	1	0	0

The quantity $e\hbar/2mc$ is called a *Bohr magneton*.

PROBLEM 5.7.2 The hyperfine splitting in the hydrogen atom can be discussed in terms of a perturbation $A\mathbf{I} \cdot \mathbf{J}$ in the Hamiltonian, where $\mathbf{I}$ and $\mathbf{J}$ are the angular momentum operators of the proton and electron, respectively, and A is a constant.

(a) Calculate the splitting produced in the $1s$-level of atomic hydrogen by this perturbation.

The total angular momentum of the atom is $\mathbf{I} + \mathbf{J} = \mathbf{F}$. We write

$$\mathbf{I} \cdot \mathbf{J} = \frac{(\mathbf{I} + \mathbf{J}) \cdot (\mathbf{I} + \mathbf{J}) - \mathbf{I} \cdot \mathbf{I} - \mathbf{J} \cdot \mathbf{J}}{2} = \frac{\mathbf{F} \cdot \mathbf{F} - \mathbf{I} \cdot \mathbf{I} - \mathbf{J} \cdot \mathbf{J}}{2}$$

Then the expectation value of the perturbation in the state characterized by quantum numbers I, J, F is

$$\langle A\mathbf{I} \cdot \mathbf{J} \rangle_{IJF} = \frac{A\hbar^2}{2}[F(F+1) - I(I-1) - J(J+1)]$$

For the hydrogen atom, $I = J = \frac{1}{2}$, and $F = 0$ or 1. Thus, we have

$$\langle A\mathbf{I} \cdot \mathbf{J}\rangle_{(1/2)(1/2)F} = \frac{A\hbar^2}{2}\left[F(F+1) - \frac{3}{2}\right] = \begin{cases} \dfrac{A\hbar^2}{4} & \text{for } F = 1 \\[2mm] -\dfrac{3A\hbar^2}{4} & \text{for } F = 0 \end{cases} \qquad (5.117)$$

and the splitting between the $F = 0$ and 1 levels is $A\hbar^2$.

(b) Now suppose that a uniform magnetic field $\mathbf{B}$ is applied to the above-described hydrogen atom. This can be accounted for by the addition of perturbing terms

$$\frac{e}{2mc}\left[g_I \mathbf{I} \cdot \mathbf{B} + g_J \mathbf{J} \cdot \mathbf{B}\right]$$

Here g_I and g_J are gyromagnetic factors. Calculate the splitting produced in the $1s$ level of atomic hydrogen by this external field and by the hyperfine interaction $A\mathbf{I} \cdot \mathbf{J}$, acting simultaneously.

Let us choose the z direction to coincide with that of the $\mathbf{B}$ field, so $\mathbf{B} = B\hat{z}$. The total perturbation is

$$\lambda V = A\mathbf{I} \cdot \mathbf{J} + \frac{eB}{2mc}\left[g_I I_z + g_J J_z\right] \qquad (5.118)$$

We choose our unperturbed states to be $\psi_{m_p}^{1/2}\phi_{m_e}^{1/2}$, where

$$I_z\psi_{m_p}^{1/2} = \hbar m_p \psi_{m_p}^{1/2}, \qquad m_p = \pm\frac{1}{2}$$

$$J_z\phi_{m_e}^{1/2} = \hbar m_e \phi_{m_e}^{1/2}, \qquad m_e = \pm\frac{1}{2}$$

Since the unperturbed Hamiltonian is degenerate with respect to these four states $\psi_{m_p}^{1/2}\phi_{m_e}^{1/2}$, we must diagonalize the 4×4 matrix of λV with respect to these states (cf. Section 5.6.2). We have

$$\langle \psi_{m_p}^{1/2}\phi_{m_e}^{1/2}|g_I I_z + g_J J_z|\psi_{m_p'}^{1/2}\phi_{m_e'}^{1/2}\rangle = \delta_{m_p m_p'}\delta_{m_e m_e'}\hbar\left[g_I m_p + g_J m_e\right] \qquad (5.119a)$$

$$\langle \psi_{m_p}^{1/2}\phi_{m_e}^{1/2}|A\mathbf{I} \cdot \mathbf{J}|\psi_{m_p'}^{1/2}\phi_{m_e'}^{1/2}\rangle$$

$$= A\langle \psi_{m_p}^{1/2}\phi_{m_e}^{1/2}|I_zJ_z + \frac{(I_x + iI_y)(J_x - iJ_y) + (I_x - iI_y)(J_x + iJ_y)}{2}|\psi_{m_p'}^{1/2}\phi_{m_e'}^{1/2}\rangle \qquad (5.119b)$$

Application of (5.11b) and (5.13) to these $J = I = \frac{1}{2}$ proton and electron states gives the following matrix of the perturbation (5.118) with respect to the unperturbed states:

m_p	m_e	m_p': $\frac{1}{2}$ m_e': $\frac{1}{2}$	$\frac{1}{2}$ $-\frac{1}{2}$	$-\frac{1}{2}$ $\frac{1}{2}$	$-\frac{1}{2}$ $-\frac{1}{2}$
$\frac{1}{2}$	$\frac{1}{2}$	$\dfrac{eB\hbar}{2mc}\left[\dfrac{g_I + g_J}{2}\right] + \dfrac{A\hbar^2}{4}$	0	0	0
$\frac{1}{2}$	$-\frac{1}{2}$	0	$\dfrac{eB\hbar}{2mc}\left[\dfrac{g_I - g_J}{2}\right] - \dfrac{A\hbar^2}{4}$	$\dfrac{A\hbar^2}{2}$	0
$-\frac{1}{2}$	$\frac{1}{2}$	0	$\dfrac{A\hbar^2}{2}$	$\dfrac{eB\hbar}{2mc}\left[\dfrac{-g_I + g_J}{2}\right] - \dfrac{A\hbar^2}{4}$	0
$-\frac{1}{2}$	$-\frac{1}{2}$	0	0	0	$\dfrac{eB\hbar}{2mc}\left[\dfrac{-g_I - g_J}{2}\right] + \dfrac{A\hbar^2}{4}$

$$(5.120)$$

Notice that the only off-diagonal matrix elements are those with the same value of $m_e + m_p$. This is because the perturbation is invariant under rotations around the $\hat{z}$ axis, so that it can only connect eigenstates of F_z ($= I_z + J_z$) with the same eigenvalue [cf. (5.21)].

Two of the eigenvalues of the matrix (5.120) can be written down immediately:

$$\lambda_1 = \frac{eB\hbar}{2mc}\left[\frac{g_I + g_J}{2}\right] + \frac{A\hbar^2}{4} \tag{5.121a}$$

$$\lambda_2 = \frac{eB\hbar}{2mc}\left[\frac{-g_I - g_J}{2}\right] + \frac{A\hbar^2}{4} \tag{5.121b}$$

The other two are found by solving the secular equation

$$\begin{vmatrix} \dfrac{eB\hbar}{2mc}\left[\dfrac{g_I - g_J}{2}\right] - \dfrac{A\hbar^2}{4} - \lambda & \dfrac{A\hbar^2}{2} \\ \dfrac{A\hbar^2}{2} & \dfrac{eB\hbar}{2mc}\left[\dfrac{-g_I + g_J}{2}\right] - \dfrac{A\hbar^2}{4} - \lambda \end{vmatrix} = 0$$

$$\lambda^2 + \frac{A\hbar^2}{2}\lambda + \left[\left(\frac{A\hbar^2}{4}\right)^2 - \left(\frac{eB\hbar}{2mc}\left[\frac{g_I - g_j}{2}\right]\right)^2 - \left(\frac{A\hbar^2}{2}\right)^2\right] = 0$$

whose roots are

$$\lambda_3 = \frac{1}{2}\left\{-\frac{A\hbar^2}{2} + \sqrt{(A\hbar^2)^2 + \left[\frac{eB\hbar}{2mc}(g_I - g_J)\right]^2}\right\} \tag{5.121c}$$

$$\lambda_4 = \frac{1}{2}\left\{-\frac{A\hbar^2}{2} - \sqrt{(A\hbar^2)^2 + \left[\frac{eB\hbar}{2mc}(g_I - g_J)\right]^2}\right\} \tag{5.121d}$$

The four eigenvalues (5.121) are the first-order corrections to the energy due to the perturbation (5.118). If we allow B to become zero in (5.121), we recover the result (5.117), which applied when only the $A\mathbf{I} \cdot \mathbf{J}$ term was present. On the other hand, if we set $A = 0$ in (5.121), the result is

$$\lambda(m_p, m_e) = \frac{e\hbar}{2mc}B\left[g_I m_p + g_J m_e\right] \qquad \left(m_p, m_e = \pm\tfrac{1}{2}\right)$$

In this limit, the electron and proton spins are not coupled by the hyperfine interaction, and respond independently to the external magnetic field.

5.8   TIME-DEPENDENT PROBLEMS

Equation (5.8a) shows a solution of the time-dependent Schrödinger equation expressed as a product of a time-dependent phase and a time-independent eigenstate of the time-independent Schrödinger equation (5.8b). Such solutions are possible whenever the Hamiltonian does not depend explicitly on time. The probability density associated with the solution (5.8a) is

$$P = |\Psi(q_1, \ldots, q_d, t)|^2 = |\Phi(q_1, \ldots, q_d)|^2 \tag{5.122}$$

and, thus, is independent of time. However, suppose that we construct a "wave packet" by adding together terms such as (5.8a), corresponding to different energy eigenvalues E,

$$\Psi(q_1, \ldots, q_d, t) = e^{-(i/\hbar)E_1 t}\Phi_1(q_1, \ldots, q_d) + e^{-(i/\hbar)E_2 t}\Phi_2(q_1, \ldots, q_d) \tag{5.123}$$

Since (5.7) is linear in Ψ, the sum (5.123) is a solution of (5.7) if each contributing term is a solution. But the probability density associated with $\Psi(q_1, \ldots, q_d, t)$ of (5.123) is

$$P = |\Psi(q_1, \ldots, q_d, t)|^2$$
$$= |\Phi_1(q_1, \ldots, q_d)|^2 + |\Phi_2(q_1, \ldots, q_d)|^2$$
$$+ 2Re\left[\Phi_1^*(q_1, \ldots, q_d)\Phi_2(q_1, \ldots, q_d)e^{(i/\hbar)(E_1-E_2)t}\right] \tag{5.124}$$

which is time-dependent.

PROBLEM 5.8.1   A particle of mass m moves in one dimension in a square well with walls of infinite height a distance L apart. The particle is known to be in a state consisting of an equal admixture of the two lowest energy eigenstates of the system. Find the probability as a function of time that the particle will be found in the right-hand half of the well.

The wave function of the system is

$$\psi(x, t) = \frac{1}{\sqrt{2}}[\psi_0(x, t) + \psi_1(x, t)] \tag{5.125}$$

with

$$\psi_0(x, t) = \sqrt{\frac{2}{L}}\cos\left(\frac{\pi x}{L}\right)e^{-(i/\hbar)E_0 t}, \qquad E_0 = \frac{\hbar^2\pi^2}{2mL^2} \tag{5.126a}$$

$$\psi_1(x, t) = \sqrt{\frac{2}{L}}\sin\left(\frac{2\pi x}{L}\right)e^{-(i/\hbar)E_1 t}, \qquad E_1 = \frac{2\hbar^2\pi^2}{mL^2} \tag{5.126b}$$

[cf. (5.50)–(5.53)]. Thus, (5.124) becomes

$$P(x, t) = \frac{1}{L}\left[\cos^2\left(\frac{\pi x}{L}\right) + \sin^2\left(\frac{2\pi x}{L}\right) + 2\cos\left(\frac{\pi x}{L}\right)\sin\left(\frac{2\pi x}{L}\right)\cos\left(\frac{E_1 - E_0}{\hbar}t\right)\right] \tag{5.127}$$

and the probability that the particle is between $x = 0$ and $x = L/2$ is

$$\int_0^{L/2} P(x, t)\, dx = \frac{1}{L}\left[\frac{1}{2}\cdot\frac{L}{2} + \frac{1}{2}\cdot\frac{L}{2} - 4 \times \cos\left(\frac{E_1 - E_0}{\hbar}t\right) \times \frac{L}{3\pi}\right]$$
$$= \frac{1}{2} + \frac{4}{3\pi}\cos\left(\frac{3\hbar^2\pi^2}{2mL^2}t\right)$$

PROBLEM 5.8.2   What are the expectation values of H and x in the state (5.125)?

Since ψ_0 and ψ_1 of (5.125) are eigenstates of H corresponding to different eigenvalues, they are orthogonal. Thus,

$$
\begin{aligned}
\langle \psi(x, t)|H|\psi(x, t)\rangle &= \frac{1}{2}\langle \psi_0 + \psi_1|H|\psi_0 + \psi_1\rangle \\
&= \frac{1}{2}\left[\langle \psi_0|H|\psi_0\rangle + \langle \psi_1|H|\psi_1\rangle\right] \\
&= \frac{1}{2}\left[\frac{\hbar^2\pi^2}{2mL^2} + \frac{2\hbar^2\pi^2}{mL^2}\right] = \frac{5}{4}\frac{\hbar^2\pi^2}{mL^2}
\end{aligned}
$$

The expectation value of x is

$$
\begin{aligned}
\langle \psi(x, t)|x|\psi(x, t)\rangle &= \int_{-L/2}^{L/2} |\psi(x, t)|^2 x\, dx \\
&= \frac{1}{2}\cdot\frac{2}{L}\int_{-L/2}^{L/2}\left[\cos^2\left(\frac{\pi x}{L}\right) + \sin^2\left(\frac{2\pi x}{L}\right)\right. \\
&\qquad\left. + 2\cos\left(\frac{E_1 - E_0}{\hbar}t\right)\cos\left(\frac{\pi x}{L}\right)\sin\left(\frac{2\pi x}{L}\right)\right]x\, dx \\
&= \frac{2}{L}\cos\left(\frac{E_1 - E_0}{\hbar}t\right)\int_{-L/2}^{L/2}\cos\left(\frac{\pi x}{L}\right)\sin\left(\frac{2\pi x}{L}\right)x\, dx \\
&= \frac{16L}{9\pi^2}\cos\left(\frac{3}{2}\frac{\hbar^2\pi^2}{mL^2}t\right)
\end{aligned}
$$

It is sometimes convenient to calculate the time dependence of an expectation value $\langle \psi|V|\psi\rangle$ by using the equation

$$
\frac{d}{dt}\langle \psi|V|\psi\rangle = \langle \psi|\frac{\partial V}{\partial t}|\psi\rangle + \frac{1}{i\hbar}\langle \psi|HV - VH|\psi\rangle \tag{5.128}
$$

This can easily be derived by using the Schrödinger equation (5.7) to express the time dependence of ψ.

PROBLEM 5.8.3   An electron is immersed in a homogeneous magnetic field along the z axis. At time $t = 0$, the electron is in a state with component of spin angular momentum $\hbar/2$ along the x axis. Find the expectation values of the x, y, and z components of the spin angular momentum at any later time.

Set $V = S_\alpha$ in (5.128). Since the operator S_α does not depend explicitly on t, (5.128) becomes

$$
\frac{d}{dt}\langle \psi|S_\alpha|\psi\rangle = \frac{1}{i\hbar}\langle \psi|HS_\alpha - S_\alpha H|\psi\rangle
$$

The only part of H that does not commute with S_α is the spin-dependent part $(eB/mc)\cdot S_z$ [cf. (5.115)]. Thus,

$$
\frac{d}{dt}\langle \psi|S_\alpha|\psi\rangle = \frac{eB}{i\hbar mc}\langle \psi|S_z S_\alpha - S_\alpha S_z|\psi\rangle \tag{5.129}
$$

The components of **S** obey the usual angular momentum commutation relations

(5.10). Setting $\alpha = x, y, z$ in (5.129), we get

$$\frac{d}{dt}\langle \psi | S_x | \psi \rangle = \frac{eB}{mc}\langle \psi | S_y | \psi \rangle \tag{5.130a}$$

$$\frac{d}{dt}\langle \psi | S_y | \psi \rangle = -\frac{eB}{mc}\langle \psi | S_x | \psi \rangle \tag{5.130b}$$

$$\frac{d}{dt}\langle \psi | S_z | \psi \rangle = 0 \tag{5.130c}$$

A second time derivative of (5.130a) and (5.130b) gives

$$\frac{d^2}{dt^2}\langle \psi | S_\alpha | \psi \rangle = -\left(\frac{eB}{mc}\right)^2 \langle \psi | S_\alpha | \psi \rangle \qquad (\alpha = x, y) \tag{5.131}$$

The solution of (5.131) with the specified initial conditions is

$$\langle \psi | S_x | \psi \rangle = \frac{\hbar}{2}\cos\left(\frac{eB}{mc}t\right)$$

$$\langle \psi | S_y | \psi \rangle = -\frac{\hbar}{2}\sin\left(\frac{eB}{mc}t\right)$$

Finally, (5.130c) implies that $\langle \psi | S_z | \psi \rangle$ is independent of t, so it continues to have its initial value of zero. It is seen that $\langle \psi | \mathbf{S} | \psi \rangle$ has the time dependence of a vector that rotates clockwise around the z axis with circular frequency $\omega_B = eB/mc$ (cf. 3.55).

5.9 SCATTERING IN ONE DIMENSION BY A LOCALIZED POTENTIAL

The first step in the analysis of a scattering problem is the construction of a solution of the Schrödinger equation in the external region, outside the range of the potential. This will generally involve incident, reflected, and transmitted waves (cf. Problems 1.14.2, 1.14.3, and 3.16.2). The reflection and transmission coefficients are determined by the requirements that ψ and its derivatives be continuous everywhere.

PROBLEM 5.9.1   A particle with $E > 0$ is incident from the left on the potential step shown in Figure 5.5. Find the transmission and reflection coefficients corresponding to the step in the potential at $x = 0$.

The experimental conditions require incident and reflected waves for $x < 0$, and a transmitted wave for $x > 0$. Thus, we look for a solution of the form

$$\psi(x) = e^{ikx} + re^{-ikx}, \qquad k = \sqrt{\frac{2mE}{\hbar^2}} \qquad (x < 0) \tag{5.132a}$$

$$= te^{iKx}, \qquad K = \sqrt{\frac{2m(E + b)}{\hbar^2}} \qquad (x > 0) \tag{5.132b}$$

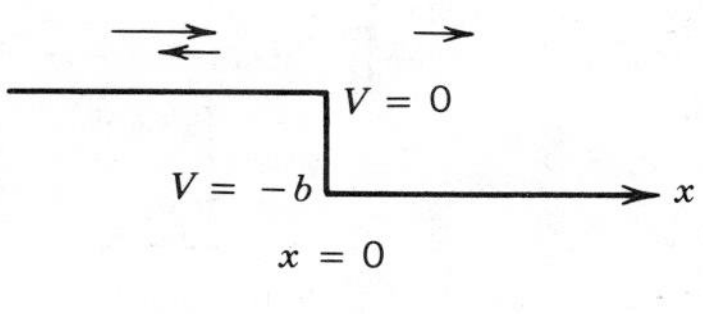

FIGURE 5.5 A one-dimensional step, with particles incident from the high-potential side. The arrows represent the incident, reflected, and transmitted flux.

If we require that $\psi(x)$ and $\psi'(x)$ be continuous at $x = 0$, we get two simultaneous equations for r and t:

$$1 + r = t$$
$$ik(1 - r) = iKt$$

The solution is

$$r = \frac{k - K}{k + K} \tag{5.133a}$$

$$t = \frac{2k}{k + K} \tag{5.133b}$$

The particle flux, $\mathbf{j}$, is calculated from

$$\mathbf{j} = \frac{\hbar}{2mi}\left[\psi^*\nabla\psi - \psi\nabla\psi^*\right] \tag{5.134}$$

For $x < 0$, the flux is $(\hbar k/m)[1 - |r|^2]$; for $x > 0$ the flux is $(\hbar K/m)|t|^2$. Thus, the reflection coefficient R and the transmission coefficient T are given by

$$R = \frac{\text{reflected flux}}{\text{incident flux}} = \frac{\frac{\hbar k}{m}|r|^2}{\frac{\hbar k}{m}} = |r|^2 = \left(\frac{k - K}{k + K}\right)^2 \tag{5.135a}$$

$$T = \frac{\text{transmitted flux}}{\text{incident flux}} = \frac{\frac{\hbar K}{m}|t|^2}{\frac{\hbar k}{m}} = \frac{K}{k}|t|^2 = \frac{4kK}{(k + K)^2} \tag{5.135b}$$

Note that these expressions satisfy the flux conservation condition

$$R + T = 1 \tag{5.136}$$

Of course, if this were a problem in classical physics, all the flux would be transmitted. We see from (5.135a) that quantum mechanics tells us there will always be some reflected flux, as long as there is a discontinuity in the potential. In fact, if that discontinuity is large compared to the incident particle energy, essentially all the incident flux will be reflected.

PROBLEM 5.9.2   A stream of particles of kinetic energy $E < V_0$ and mass m is incident from $x = -\infty$ on a finite rectangular barrier

$$V(x) = V_0 > 0, \qquad 0 < x < a$$
$$= 0, \qquad \text{otherwise}$$

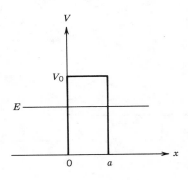

Find the transmission coefficient.

The experimental conditions require incident and reflected waves for $x < 0$,

$$\psi(x) = e^{ikx} + re^{-ikx}, \qquad k = \sqrt{\frac{2mE}{\hbar^2}} \qquad (x < 0) \qquad (5.137a)$$

and transmitted waves for $x > a$,

$$\psi(x) = te^{ikx} \qquad (x > a) \qquad (5.137b)$$

The transmittance T is simply $|t|^2$, since k has the same value for the incident and transmitted waves. The solution of the Schrödinger equation in the region $0 < x < a$ will be of the form

$$\psi(x) = \alpha e^{\kappa x} + \beta e^{-\kappa x}, \qquad \kappa = \sqrt{\frac{2m(V_0 - E)}{\hbar^2}} \qquad (0 < x < a) \quad (5.137c)$$

The requirements of continuity of $\psi(x)$ and $\psi'(x)$ at $x = 0$ and $x = a$ yield four simultaneous equations for the four unknowns, r, t, α, β:

$$1 + r = \alpha + \beta$$

$$ik(1 - r) = \kappa(\alpha - \beta)$$

$$\alpha e^{\kappa a} + \beta e^{-\kappa a} = te^{ika}$$

$$\kappa(\alpha e^{\kappa a} - \beta e^{-\kappa a}) = ikte^{ika}$$

The solution of these equations gives

$$\begin{aligned}
t &= \frac{4i\kappa k e^{-ika}}{(\kappa + ik)^2 e^{-\kappa a} - (\kappa - ik)^2 e^{\kappa a}} \\
&= \frac{2i\kappa k e^{-ika}}{(k^2 - \kappa^2)\sinh(\kappa a) + 2i\kappa k \cosh(\kappa a)}
\end{aligned} \qquad (5.138)$$

Finally we have

$$\begin{aligned}
T = |t|^2 &= \frac{4\kappa^2 k^2}{(k^2 - \kappa^2)^2 \sinh^2(\kappa a) + 4\kappa^2 k^2 \cosh^2(\kappa a)} \\
&= \frac{E(V_0 - E)}{\left[\dfrac{V_0}{2}\sinh\left(\dfrac{\sqrt{2m(V_0 - E)}\, a}{\hbar}\right)\right]^2 + E(V_0 - E)}
\end{aligned} \qquad (5.139)$$

In this case, classical particles would all be reflected by a barrier whose height was greater than their kinetic energy. According to (5.139), quantum mechanics tells us that a particle can tunnel through the barrier, with a probability that is controlled by the parameter $\hbar/(\sqrt{2m(V_0 - E)}\, a)$. If this parameter is small, (5.139) approaches

$$T \approx \frac{16E(V_0 - E)}{V_0^2} e^{-2\sqrt{2m(V_0 - E)}\, a/\hbar} \qquad \left(\hbar \ll \sqrt{2m(V_0 - E)}\, a\right) \quad (5.140)$$

Thus, the penetrability is decreased by increasing the mass of the particle or the width or height of the barrier. Note that the presence of the exponential factor in (5.140) implies that rather small changes in the energy of the particle or in the properties of the well can have a large effect on the penetrability. Processes such as α-decay, cold emission of electrons from metals, and nuclear reactions below the Coulomb barrier have rates that are limited by penetrability factors similar to (5.140). These rates are observed to change rapidly when small changes occur in the parameters corresponding to V_0, E, m, and a.

We can also apply the above analysis to the situation in which $E > V_0$. We need only replace κ in (5.137c), (5.138), and (5.139) by iK, where

$$K = \frac{\sqrt{2m(E - V_0)}}{\hbar^2} \qquad (E > V_0) \qquad (5.137d)$$

Now the particle has positive kinetic energy in the region $0 \le x \le a$. The transmittance becomes

$$t = \frac{2Kke^{-ika}}{-i(K^2 + k^2)\sin(Ka) + 2Kk\cos(Ka)}$$

$$T = |t|^2 = \frac{4K^2k^2}{(K^2 + k^2)^2 \sin^2(Ka) + 4K^2k^2\cos^2(Ka)}$$

$$= \frac{E(E - V_0)}{\left[\frac{V_0}{2}\sin\left(\frac{\sqrt{2m(E - V_0)}\,a}{\hbar}\right)\right]^2 + E(E - V_0)} \qquad (5.141)$$

Here we see that $T \to 1$ as E becomes much greater than the height of the barrier V_0. Note that $T = 1$ whenever E is such that

$$\frac{\sqrt{2m(E - V_0)}}{\hbar^2}a = Ka = n\pi \qquad (n = 1, 2, 3, \dots)$$

These discrete values of E, at which the barrier becomes perfectly transparent, are said to be the energies of transmission resonances.

The $E > V_0$ formulae we have just presented are also applicable to the situation in which $E > 0$, $V_0 < 0$. In this case we have scattering by an attractive square well. The corresponding wave functions (5.137) are the continuum (positive energy) solutions of the Schrödinger equation referred to on page 218. They exist for every $E > 0$.

5.10 SCATTERING BY A THREE-DIMENSIONAL LOCALIZED POTENTIAL

In the stationary-state treatment of scattering, we seek a solution of the Schrödinger equation that has asymptotic form

$$\psi(\mathbf{r}) \underset{r \to \infty}{\to} e^{ikz} + f(\theta, \phi)\frac{e^{ikr}}{r} \qquad (5.142)$$

The function $f(\theta, \phi)$ is called the scattering amplitude. It is related to the *differential cross-section* by

$$\frac{d\sigma}{d\Omega}(\theta, \phi) = \frac{\dfrac{\text{flux}}{\text{unit solid angle}}\ \text{scattered in direction } (\theta, \phi)}{\dfrac{\text{incident flux}}{\text{unit area}}} = |f(\theta, \phi)|^2$$

$$(5.143)$$

Here (θ, ϕ) are spherical polar coordinates chosen so that the beam is incident from $z = -\infty$, traveling in the $+z$ direction.

If the potential is spherically symmetric, centered at the origin, it is convenient to expand (5.142) in terms of angular momentum eigenstates

$$\psi(\mathbf{r}) = \sum_{l=0}^{\infty} (2l + 1) \frac{u_l(r)}{r} P_l(\cos\theta) \tag{5.144}$$

This converts the problem of solving a partial differential equation for $\psi(\mathbf{r})$ into the simpler problem of solving the ordinary differential equation (5.73a) to find the radial functions $u_l(r)$.

We seek the solution of (5.73a) that has the following properties:

$$u_l(r) \underset{r \to 0}{\to} \alpha r^{l+1} \qquad (\alpha \text{ constant}) \tag{5.145a}$$

$$\underset{r \gg \frac{l}{k}, R}{\to} \frac{-i^l}{2ik} \left[e^{-i(kr-(l\pi/2))} - S_l e^{i(kr-(l\pi/2))} \right] \tag{5.145b}$$

The coefficient $-(-1)^l/2ik$ of e^{-ikr} in (5.145b) is chosen for later convenience. We are free to make this choice since (5.73a) is homogeneous. However, once we have chosen this normalization for e^{-ikr}, the coefficient of e^{ikr} is specified by our solution $u_l(r)$. In (5.145b) this coefficient is given in terms of a parameter S_l. If the potential $U(r)$ is real and spherically symmetric, the incoming and outgoing fluxes must be equal for each partial (l) wave. This implies that S_l has unit modulus, so it can be written in terms of a real phase shift δ_l, as follows,

$$S_l = e^{2i\delta_l} \tag{5.145c}$$

Once we have found (5.145b), we can obtain the scattering amplitude $f(\theta, \phi)$ by subtracting from (5.144) the partial wave expansion of the plane wave

$$e^{ikz} = \sum_{l=0}^{\infty} i^l (2l + 1) j_l(kr) P_l(\cos\theta) \tag{5.146a}$$

$$j_l(kr) \underset{r \gg \frac{l}{k}}{\to} \frac{\sin\left(kr - \frac{l\pi}{2}\right)}{kr} \tag{5.146b}$$

We get

$$\psi(\mathbf{r}) - e^{ikz} \underset{r \to \infty}{\to} \frac{e^{ikr}}{r} f(\theta, \phi)$$

$$\underset{r \to \infty}{\to} \frac{e^{ikr}}{r} \sum_{l=0}^{\infty} (2l + 1) \frac{S_l - 1}{2ik} P_l(\cos\theta) \tag{5.146c}$$

Thus, at large r, $\psi(\mathbf{r}) - e^{ikz}$ contains only outgoing waves. This is a consequence of the choice we made in (5.145b) for the coefficient of the incoming wave part of $u_l(r)$. If we compare (5.146c) to (5.142) we get an expression for the scattering amplitude:

$$f(\theta, \phi) = \frac{1}{2ik} \sum_{l=0}^{\infty} (2l + 1) [S_l - 1] P_l(\cos\theta)$$

$$= \frac{1}{k} \sum_{l=0}^{\infty} (2l + 1) e^{i\delta_l} \sin\delta_l P_l(\cos\theta) \tag{5.147a}$$

$$\frac{d\sigma}{d\Omega}(\theta, \phi) = \frac{1}{k^2} \left| \sum_{l=0}^{\infty} (2l + 1) e^{i\delta_l} \sin\delta_l P_l(\cos\theta) \right|^2 \tag{5.147b}$$

The total scattering cross-section is

$$\sigma = \int_{\substack{\theta=0\to\pi \\ \phi=0\to2\pi}} \frac{d\sigma}{d\Omega} \sin\theta \, d\theta \, d\phi = \frac{4\pi}{k^2} \sum_{l=0}^{\infty} (2l+1)\sin^2\delta_l \qquad (5.147c)$$

The sums in (5.147) include every nonnegative integral l. Suppose, however, that the scattering potential has a finite range, R. Then if $l \gg kR$, the centrifugal potential term in (5.73a) will cause $u_l(r)$ to be very small when $r < R$. Thus, a particle with this orbital angular momentum will be very little affected by the presence of the potential. This means that $u_l(r)$ will be very close to the free-particle radial function (5.146b), so that $S_l \approx 1$ and $\delta_l = 0$. Terms such as this make very little contribution ot the partial-wave sums in (5.147). In practice, therefore, when we have a potential of finite range it is necessary to include only a finite number of partial waves (l-values). The classical interpretation of the neglected states with very high l is that they correspond to orbits with such large impact parameters that they never get close enough to the force center to feel a deflecting force, and so they are not scattered.

PROBLEM 5.10.1 Consider a potential of arbitrary shape that vanishes for $r \geq a$. Let $u(r)$ be the $l = 0$ radial wave function that vanishes at the origin, corresponding to the energy E, and let $f(E)$ be its logarithmic derivative at $r = a$:

$$f(E) \equiv \left[\frac{1}{u(r)} \frac{du(r)}{dr}\right]_{r=a}$$

(a) If there is a bound state at $E = -E_B$, what is the value of $f(-E_B)$? When $l = 0$ and $r > a$, (5.73b) becomes

$$\left[\frac{d^2}{dr^2} - \kappa^2\right]u(r) = 0 \qquad (r > a)$$

The normalizable solution is proportional to $e^{-\kappa r}$, whose logarithmic derivative at $r = a$ is

$$\left[\frac{1}{u}\frac{du}{dr}\right]_{r=a} = -\kappa = -\frac{\sqrt{2\mu(-E_B)}}{\hbar^2}$$

The required continuity of u and u' at $r = a$ implies that the logarithmic derivatives of the interior and exterior solutions agree at $r = a$. Thus,

$$f(-E_B) = -\frac{\sqrt{2\mu(-E_B)}}{\hbar^2}$$

(b) Assume that $f(E)$ is a slowly varying function of E, and that E_B is close to zero. Find an approximate expression for the scattering cross-section at low energy.

According to (5.145b), the exterior $l = 0$ positive-energy radial function is

$$u(r) = -\frac{1}{2ik}\left[e^{-ikr} - S_0 e^{ikr}\right]$$

whose logarithmic derivative at $r = a$ is

$$f(E) = \left[\frac{1}{u}\frac{du}{dr}\right]_{r=a} = -ik\frac{e^{-ika} + S_0 e^{ika}}{e^{-ika} - S_0 e^{ika}}$$

At low energy, $f(E) \approx f(-E_B)$, since both E and $-E_B$ are small, and we are told that $f(E)$ varies slowly with energy. Thus,

$$-ik\frac{e^{-ika} + S_0 e^{ika}}{e^{-ika} - S_0 e^{ika}} = -\kappa$$

$$S_0 = e^{-2ika}\frac{\kappa - ik}{\kappa + ik} = e^{-2i\left(ka + \arctan\frac{k}{\kappa}\right)}$$

and (5.145c) allows us to conclude that

$$\delta_0 = -\left(ka + \arctan\left(\frac{k}{\kappa}\right)\right)_{k \to 0} - k\left(a + \frac{1}{\kappa}\right)$$

The low-energy limit of σ is then given by (5.147c),

$$\sigma_{E \to 0} 4\pi\left(a + \frac{1}{\kappa}\right)^2 = 4\pi\left(a + \sqrt{\frac{\hbar^2}{-2\mu E_B}}\right)^2$$

(c) Apply this result to the calculation of σ for low-energy neutron-proton scattering. The neutron-proton system forms an $l = 0$ bound state with $E_B = -2.23$ MeV, the potential has a range $a \approx 10^{-13}$ cm, and the neutron and proton rest energies are both about 940 MeV.

We need

$$\sqrt{\frac{\hbar^2}{-2\mu E_B}} = \sqrt{\frac{\hbar^2 c^2}{-2(\mu c^2)E_B}} = \frac{197.3 \text{ MeV-fm}}{\sqrt{2 \times \dfrac{940}{2} \text{ MeV} \times 2.23 \text{ MeV}}} = 4.3 \text{ fm}$$

$$\sigma_{E \to 0} 4\pi \times (1.0 + 4.3)^2 \times 10^{-26} \text{ cm}^2$$

$$= 3.5 \times 10^{-24} \text{ cm}^2 = 3.5 \text{ barns}$$

Experiment gives a zero-energy neutron-proton total cross-section of 21 barns. Wigner noted that in the $E_B = -2.23$ MeV bound state, the neutron and proton are in a triplet state ($S = 1$), whereas the neutron-proton scattering involves both singlet ($S = 0$) and triplet spin states. Thus, if the neutron-proton force were different in singlet and triplet states, one could understand why a prediction based on the bound state could not account for the scattering cross-section. This was the first demonstration of the spin-dependence of nuclear forces.

5.11 BORN APPROXIMATION

The Schrödinger equation for a particle of mass μ, moving with positive energy in a localized potential field $U(\mathbf{r})$, can be written

$$(\nabla^2 + k^2)\psi(\mathbf{r}) = \frac{2\mu}{\hbar^2}U(\mathbf{r})\psi(\mathbf{r}) \tag{5.148}$$

An equivalent integral equation, incorporating the boundary conditions appropriate to scattering, is

$$\psi(\mathbf{r}) = e^{ikz} - \frac{\mu}{2\pi\hbar^2}\int\frac{e^{ik|\mathbf{r} - \mathbf{r}'|}}{|\mathbf{r} - \mathbf{r}'|}U(\mathbf{r}')\psi(\mathbf{r}')\,d^3r' \tag{5.149}$$

Any $\psi(\mathbf{r})$ satisfying this integral equation also satisfies the Schrödinger equation

(5.71). This can be proven by acting on (5.149) with $\nabla_r^2 + k^2$, and using the equation

$$\left(\nabla_r^2 + k^2\right)\frac{e^{ik|r-r'|}}{|r - r'|} = -4\pi\delta(r - r')$$

inside the integral on the right-hand side.

To extract the scattering amplitude from (5.149), we consider the limit of the integral as $r \to \infty$. Note that the presence of the factor $U(r')$ in the integrand ensures that r' will be finite wherever the integrand is nonzero. We have

$$|r - r'| = \sqrt{r^2 - 2r \cdot r' + (r')^2} = r\sqrt{1 - 2\frac{\hat{r} \cdot r'}{r} + \left(\frac{r'}{r}\right)^2}$$

$$\underset{\substack{r \to \infty \\ r' \text{ finite}}}{\to} r\left[1 - \frac{\hat{r} \cdot r'}{r}\right] = r - \hat{r} \cdot r'$$

Thus, we conclude from (5.149) that

$$\psi(r) \underset{r \to \infty}{\to} e^{ikz} + \frac{e^{ikr}}{r}\left[-\frac{\mu}{2\pi\hbar^2}\int e^{-ik\hat{r} \cdot r'}U(r')\psi(r')\, d^3r'\right]$$

and comparison with (5.142) shows that the scattering amplitude is

$$f(\theta, \phi) = -\frac{\mu}{2\pi\hbar^2}\int e^{-ik\hat{r} \cdot r'}U(r')\psi(r')\, d^3r' \tag{5.150}$$

In this expression, θ and ϕ are the polar coordinates of the fixed unit vector $\hat{r}$.

Equation (5.150) is exact, but it is not very useful as it stands, since we require knowledge of the exact wave function $\psi(r')$ in order to evaluate the integral. However, note that the presence of $U(r')$ in the integrand means that we only need to know $\psi(r')$ where $U(r')$ is nonzero, which may be a rather small region of configuration space. Let us assume that in this region $\psi(r')$ does not differ greatly from the incident wave, $e^{ikz'}$. Then (5.150) becomes

$$f(\theta, \phi) \approx -\frac{\mu}{2\pi\hbar^2}\int e^{-ik\hat{r} \cdot r'}U(r')e^{ikz'}\, d^3r'$$

$$= -\frac{\mu}{2\pi\hbar^2}\int e^{i(k_i - k_f) \cdot r'}U(r')\, d^3r'$$

$$= -\frac{\mu}{2\pi\hbar^2}\int e^{iq \cdot r'}U(r')\, d^3r' \tag{5.151}$$

In (5.151) we have introduced the initial and final propagation vectors, $k_i \equiv k\hat{z}$ and $k_f \equiv k\hat{r}$, and the "momentum transfer" $q \equiv k_i - k_f$ (actually $1/\hbar$ times the momentum transferred by the particle to the well). Equation (5.151) is called the Born approximation. Its consequence is that the scattering amplitude is proportional to the three-dimensional Fourier transform of the scattering potential. The reader should compare (5.151) to (4.3a), the basic equation of Fraunhofer scattering theory.

Note that the Born approximation does not require that $\psi(r')$ and $e^{ikz'}$ be equal everywhere. Obviously, when r' is large, $\psi(r')$ and $e^{ikz'}$ are quite different. All we require is that $\psi(r')$ and $e^{ikz'}$ be equal, or nearly equal, in the r' region where the potential $U(r')$ is strong. Many studies have been made of the circumstances under which the Born approximation can be expected to be valid. It often turns out that the Born approximation works better than we have any reason to expect.

PROBLEM 5.11.1   Find the Born approximation expression for the differential cross-section for scattering of a particle of mass μ from a crystal with atoms at lattice points

$$\mathbf{r}_{lmn} = l\mathbf{a} + m\mathbf{b} + n\mathbf{c} \qquad (\mathbf{a}, \mathbf{b}, \mathbf{c} \text{ constant})$$

$$(l = 0, \pm 1, \pm 2, \ldots, \pm L)$$
$$(m = 0, \pm 1, \pm 2, \ldots, \pm M)$$
$$(n = 0, \pm 1, \pm 2, \ldots, \pm N)$$

The interaction potential between the particle and the atom at $\mathbf{r}_{lmn}$ is $V(\mathbf{r} - \mathbf{r}_{lmn})$.
The total potential seen by the particle is

$$U(\mathbf{r}) = \sum_{lmn} V(\mathbf{r} - [l\mathbf{a} + m\mathbf{b} + n\mathbf{c}])$$

and so the Born approximation scattering amplitude is

$$
\begin{aligned}
f(\mathbf{q}) &= -\frac{\mu}{2\pi\hbar^2} \sum_{lmn} \int e^{i\mathbf{q}\cdot\mathbf{r}'} V(\mathbf{r}' - [l\mathbf{a} + m\mathbf{b} + n\mathbf{c}]) \, d^3r' \\
&= -\frac{\mu}{2\pi\hbar^2} \sum_{lmn} \int e^{i\mathbf{q}\cdot(\mathbf{r} + l\mathbf{a} + m\mathbf{b} + n\mathbf{c})} V(\mathbf{r}) \, d^3r \\
&= \left[\sum_{lmn} e^{i\mathbf{q}\cdot(l\mathbf{a} + m\mathbf{b} + n\mathbf{c})} \right] \times \frac{-\mu}{2\pi\hbar^2} \int e^{i\mathbf{q}\cdot\mathbf{r}} V(\mathbf{r}) \, d^3r
\end{aligned}
$$

The differential cross-section is

$$
\frac{d\sigma}{d\Omega}(\mathbf{q}) = |f(\mathbf{q})|^2
$$

$$
= \left| \sum_{lmn} e^{i\mathbf{q}\cdot(l\mathbf{a} + m\mathbf{b} + n\mathbf{c})} \right|^2 \times \left| \frac{-\mu}{2\pi\hbar^2} \int e^{i\mathbf{q}\cdot\mathbf{r}} V(\mathbf{r}) \, d^3r \right|^2 \qquad (5.152)
$$

The first factor on the right-hand side of (5.152) represents the effect of the interference between the scattering by different atoms. If $\mathbf{q}$ equals

$$
\mathbf{q}_{l', m', n'} \approx 2\pi \frac{n'\mathbf{a} \times \mathbf{b} + l'\mathbf{b} \times \mathbf{c} + m'\mathbf{c} \times \mathbf{a}}{\mathbf{a} \cdot \mathbf{b} \times \mathbf{c}} \qquad (l', m', n' \text{ integral}) \quad (5.153)
$$

then it is easily verified that

$$
\mathbf{q}_{l', m', n'} \cdot (l\mathbf{a} + m\mathbf{b} + n\mathbf{c}) = 2\pi \cdot (ll' + mm' + nn')
$$
$$
= 2\pi \times \text{an integer}
$$

This means that for $\mathbf{q} = \mathbf{q}_{l'\,m'\,n'}$, all the terms contribute constructively to the (l, m, n) sum in (5.152), and the scattering is very strong. The $\mathbf{q}_{l'\,m'\,n'}$ of (5.153) are said to form the lattice *reciprocal* to the original lattice $\mathbf{r}_{lmn}$. According to (5.152), this interference pattern is modulated by a factor associated with the scattering from an individual atom, the so-called *atomic form-factor*. We see that the general character of the scattering pattern is similar to the diffraction pattern produced by a series of identical holes (Section 4.2).

5.12 EMISSION OF ELECTROMAGNETIC RADIATION

In the quantum theory of radiation, the free electromagnetic field is analyzed into normal modes. Each mode can be characterized by a linear momentum $\hbar\mathbf{k}$, an energy $\hbar k/c = \hbar\omega$, and a state of polarization $\hat{\varepsilon}$. There are two independent polarization modes for each $\hbar\mathbf{k}$ (i.e., two orthogonal linear polarization modes, or left-hand-right-hand circular polarization modes). The possible energies associated with the $\mathbf{k}$, $\hat{\varepsilon}$ mode are $n\hbar\omega$, with $n = 0, 1, 2, \ldots$. The integer n is said to be the number of $\mathbf{k}$, $\hat{\varepsilon}$ *photons* present. In the free-field problem, the state of excitation of each mode (the number of photons) is constant. However, if charged particles are present, they interact with the radiation field. Then the photon number is not constant and we can have processes in which photons are created (emission) or annihilated (absorption). These processes are also associated with changes in the state of motion of the particles (transitions).

The radiation field, considered as a dynamical system, has an infinite number of degrees of freedom. This complication puts the above-described analysis beyond the scope of this book. We will present instead the conceptually simpler (but less rigorous) semi-classical theory of radiation. With the particle-state transition $\psi_a \rightarrow \psi_b$, there is associated a so-called *transition charge density* $\rho_{a \rightarrow b}(\mathbf{r}, t)$, defined by

$$\begin{aligned}
\rho_{a \rightarrow b}(\mathbf{r}, t) &\equiv \langle \psi_b | \sum_i q_i \delta(\mathbf{r} - \mathbf{r}_i) | \psi_a \rangle \\
&= \langle e^{-(i/\hbar)E_b t}\phi_b | \sum_i q_i \delta(\mathbf{r} - \mathbf{r}_i) | e^{-(i/\hbar)E_a t}\phi_a \rangle \\
&= e^{-i(E_a - E_b)t/\hbar} \langle \phi_b | \sum_i q_i \delta(\mathbf{r} - \mathbf{r}_i) | \phi_a \rangle \\
&= e^{-i\omega t} \langle \phi_b | \sum_i q_i \delta(\mathbf{r} - \mathbf{r}_i) | \phi_a \rangle
\end{aligned} \tag{5.154}$$

Thus, the frequency ω associated with the transition charge density is related to the energy difference $E_a - E_b$ by the Bohr relation

$$\hbar\omega = E_a - E_b \tag{5.155}$$

Associated with the time-dependent charge density (5.154), we have a time-dependent electric dipole moment

$$\begin{aligned}
\mathbf{P}_{a \rightarrow b}(t) &= \int \rho_{a \rightarrow b}(\mathbf{r}, t)\mathbf{r} \, dv \\
&= e^{-i\omega t} \langle \phi_b | \sum_i q_i \mathbf{r}_i | \phi_a \rangle
\end{aligned} \tag{5.156}$$

which is used in the classical expression (3.115) for the rate of emission of electromagnetic energy:

$$\frac{dP}{d\Omega} = \frac{\omega^4}{8\pi c^3} \left| \langle \phi_b | \sum_i q_i \mathbf{r}_i | \phi_a \rangle \cdot \hat{\varepsilon}^* \right|^2$$

This gives the energy flux per unit solid angle, measured by a detector sensitive only to radiation with polarization $\hat{\varepsilon}$. If we make the additional assumption that this energy is emitted as photons, each with energy $\hbar\omega$, the rate at which these photons will be recorded is

$$\frac{dN}{d\Omega \, dt} = \frac{\omega^3}{8\pi\hbar c^3} |\mathbf{D}_{ab} \cdot \hat{\varepsilon}^*|^2 \tag{5.157a}$$

where we have used $\mathbf{D}_{ab}$ to represent the electric dipole transition matrix element

$$\mathbf{D}_{ab} \equiv \langle \phi_b | \sum_i q_i \mathbf{r}_i | \phi_a \rangle \tag{5.157b}$$

We must remember that (5.157) is based on the electric dipole approximation implicit in (3.115). If $\mathbf{D}_{ab}$ in (5.157b) vanishes, there will be no electric dipole radiation. The transition can still occur (albeit more slowly) via higher terms in the multipole expansion of the radiation field. For the next terms in this expansion (magnetic dipole, electric quadrupole) the transition rates are smaller than electric dipole rates by a factor of order of magnitude $(L/\lambda)^2$, where L is the size of the radiation source and λ is the wavelength [cf. (3.111)]. For an atom emitting visible radiation, this factor is about $(1/5000)^2$.

Since electric dipole radiation is relatively strong, it is important to ascertain when it can or cannot occur. Let us suppose that the states ϕ_a and ϕ_b are eigenstates of a spherically symmetric, inversion-invariant Hamiltonian, so that they can be chosen to be angular momentum and parity eigenstates. Then

$$\mathbf{D}_{ab} = \langle \phi_{M_b}^{\pi_b J_b} | \sum q_i \mathbf{r}_i | \phi_{M_a}^{\pi_a J_a} \rangle$$

$$= \langle \phi_{M_b}^{\pi_b J_b} | \left(\sum q_i \mathbf{r}_i \right) \phi_{M_a}^{\pi_a J_a} \rangle \qquad (5.158)$$

Since $(\Sigma_i q_i \mathbf{r}_i)\phi_{M_a}^{\pi_a J_a}$ has parity $-\pi_a$, the scalar product (5.158) vanishes unless $-\pi_a = \pi_b$. This leads to the parity *selection rule*:

Electric dipole radiation can only occur between states of opposite parity

To get angular momentum selection rules, we express x_i, y_i, and z_i in terms of spherical harmonics

$$z_i = \sqrt{\frac{4\pi}{3}} \, r_i Y_0^1(\hat{r}_i) \qquad (5.159a)$$

$$x_i = \sqrt{\frac{2\pi}{3}} \, r_i \left[Y_{-1}^1(\hat{r}_i) - Y_1^1(\hat{r}_i) \right] \qquad (5.159b)$$

$$y_i = i\sqrt{\frac{2\pi}{3}} \, \hat{r}_i \left[Y_{-1}^1(\hat{r}_i) + Y_1^1(\hat{r}_i) \right] \qquad (5.159c)$$

Then we can rewrite (5.158) in the form

$$\mathbf{D}_{a \to b} = \sqrt{\frac{4\pi}{3}} \left\{ \hat{z} \langle \phi_{M_b}^{\pi_b J_b} | \sum_i q_i r_i Y_0^1(\hat{r}_i) \phi_{M_a}^{\pi_a J_a} \rangle \right.$$

$$+ \frac{\hat{x} + i\hat{y}}{\sqrt{2}} \langle \phi_{M_b}^{\pi_b J_b} | \sum_i q_i r_i Y_{-1}^1(\hat{r}_i) \phi_{M_a}^{\pi_a J_a} \rangle$$

$$\left. + \frac{\hat{x} - i\hat{y}}{\sqrt{2}} \langle \phi_{M_b}^{\pi_b J_b} | \sum_i q_i r_i Y_1^1(\hat{r}_i) \phi_{M_a}^{\pi_a J_a} \rangle \right\} \qquad (5.160)$$

The orthogonality theorem (5.15) cannot be applied directly to the matrix elements in (5.160), since their right-hand parts (kets) are not eigenstates of the total angular momentum. According to (5.39), they can be expanded in terms of states whose total angular momenta vary between $|1 - J_a|$ and $1 + J_a$. Then we can apply the orthogonality theorem to each term in this expansion. Unless J_b lies in the range $|1 - J_a|$ to $1 + J_a$, every application of (5.15) will result in zero. This leads to the total angular momentum selection rule

Electric dipole radiation is only possible if J_a, J_b and 1 satisfy the triangle relations (5.39).

The application of the orthogonality theorem to the M-quantum numbers in (5.160) is more immediate, since M is an additive quantum number:

$$\left[l_z(1) + l_z(2) \right] \psi_{m_1}^{l_1}(1)\psi_{m_2}^{l_2}(2) = \hbar(m_1 + m_2)\psi_{m_1}^{l_1}\psi_{m_2}^{l_2}$$

Thus, the product $\sum_i q_i r_i Y_m^1(\hat{r}_i)\phi_{M_a}^{\pi_a J_a}$ is an eigenstate of $\sum_i l_z(i)$ with eigenvalue $h(m + M_a)$, with $m = 0, \pm 1$. The orthogonality theorem (5.15) then implies that

Electric dipole radiation is only possible if $M_b = M_a$ or $M_a \pm 1$.

PROBLEM 5.12.1   Determine the angular distribution of the radiation emitted in the transition $\phi_M^{\pi_a J_a} \to \phi_M^{\pi_b J_b}$, where $\pi_a J_a \pi_b J_b$ are consistent with the electric dipole selection rules. Since $M_a = M_b = M$, the only part of (5.160) that survives is

$$\mathbf{D}_{ab} = \sqrt{\frac{4\pi}{3}}\,\hat{z}\langle\phi_M^{\pi_b J_b}|\sum_i q_i r_i Y_0^1(\hat{r}_i)|\phi_M^{\pi_a J_a}\rangle$$
$$= \hat{z}(\mathbf{D}_{ab})_z$$

Thus, (5.157a) becomes

$$\frac{dN}{d\Omega\,dt} = \frac{\omega^3}{8\pi\hbar c^3}\left|(\mathbf{D}_{ab})_z\,\hat{z}\cdot\hat{\varepsilon}^*\right|^2$$

Comparison with (3.115) shows that the angular distribution is the same as for the radiation emitted by a classical dipole oscillating along the $\hat{z}$ axis. In particular, the radiation is linearly polarized in the plane determined by the $\hat{z}$ axis and the direction of observation.

PROBLEM 5.12.2   Find the angular distribution of right-hand circularly polarized photons emitted during the transition $\phi_M^{\pi_a J_a} \to \phi_{M-1}^{\pi_b J_b}$.
Since $M_b = M_a - 1$, only the second term of (5.160) survives:

$$\mathbf{D}_{ab} = \sqrt{\frac{4\pi}{3}}\,\frac{\hat{x} + i\hat{y}}{\sqrt{2}}\langle\phi_{M-1}^{\pi_b J_b}|\sum_i q_i Y_{-1}^1(\hat{r}_i)|\phi_M^{\pi_a J_a}\rangle$$

This vector is proportional to $\mathbf{p}_1$ of Problem 3.19.4, where we considered the radiation from a dipole rotating in the x–y plane. Using the result of that analysis, we have

$$\hat{\varepsilon}_r^* \cdot \mathbf{p}_1 = \frac{1}{2}\sqrt{\frac{4\pi}{3}}\langle\phi_{M-1}^{\pi_b J_b}|\sum_i q_i Y_{-1}^1(\hat{r}_i)|\phi_M^{\pi_a J_a}\rangle$$
$$\times(\cos\theta - 1)\times(\cos\phi + i\sin\phi)$$

and

$$\frac{dN}{d\Omega\,dt} = \frac{\omega^3}{24\hbar c^3}\left|\langle\phi_{M-1}^{\pi_b J_b}|\sum_i q_i Y_{-1}^1(\hat{r}_i)|\phi_M^{\pi_a J_a}\rangle\right|^2(\cos\theta - 1)^2 \qquad (5.161)$$

Thus, a detector on the $\hat{z}$ axis ($\theta = 0$), sensitive only to right-hand circularly polarized radiation, would record no photons. This means that any photons emitted in the $+\hat{z}$ direction in the $\phi_M^{\pi_a J_a} \to \phi_{M-1}^{\pi_b J_b}$ transition are all left-hand circularly polarized. It can easily be verified that left-hand circularly polarized photons have an angular distribution given by (5.161), except that the $(\cos\theta - 1)^2$ factor is replaced by $(\cos\theta + 1)^2$.

In the $\phi_M^{\pi_a J_a} \to \phi_{M-1}^{\pi_b J_b}$ transition, the component of the angular momentum of the radiating system along the $\hat{z}$ direction decreases by $\hbar$. Angular momentum conservation requires that this angular momentum be carried away by the emitted photon. Thus, a left-hand circularly polarized photon moving in the $+\hat{z}$ direction has an angular momentum component $\hbar$ in its direction of motion. This is true in general: a left-hand circularly polarized photon *always* carries one unit ($\hbar$) of angular momentum in its direction of motion. It is said to have *positive helicity*. Conversely, a right-hand circularly polarized photon has negative helicity, and its angular momentum component in its direction of motion is $-\hbar$. A right-hand circularly polarized photon moving in the $+\hat{z}$ direction as a result of the $\phi_M^{\pi_a J_a} \to \phi_{M-1}^{\pi_b J_b}$ transition would thus imply a change of $2\hbar$ of the component of the total angular momentum of the system in the $+\hat{z}$ direction. This violation of total angular momentum conservation can be interpreted as the reason for the vanishing of (5.161) and (3.116a) when $\theta = 0$.

REVIEW PROBLEMS

5.1. Let us represent the interaction between two helium atoms by the sum of a short-range repulsive part, and a long-range attractive part:

$$V(r) = +\infty, \quad \text{for } r < a$$
$$= -|V_0| \quad \text{for } a < r < b$$
$$= 0, \quad \text{for } r > b$$

where r is the distance between the centers of the helium atoms, and a, b, and $|V_0|$ are constants.

(a) Find the condition satisfied by a, b, and $|V_0|$ for a bound state to exist in the relative motion of two helium atoms.

$$Answer: \left(|V_0|(b-a)^2 \geq \frac{\hbar^2}{m\pi^2} \right).$$

(b) Experimental evidence indicates that helium does not solidify at atmospheric pressure, even at $0°$K. What do you think this implies about the effective helium–helium interaction?

5.2. If the Schrödinger equation is written in momentum space, the momentum operator p_{op} is represented by multiplication by the variable p.

(a) Show that the canonically conjugate position operator x_{op} is represented by $-\hbar/i \cdot \partial/\partial p$.

(b) Write the Schrödinger equation for the harmonic oscillator Hamiltonian, $H = p^2/2m + m\omega^2/2 \cdot x^2$, in momentum space.

Answer:

$$\left(\left[\frac{p^2}{2m} - \frac{m\omega^2\hbar^2}{2} \frac{\partial^2}{\partial p^2} \right] \psi(p) = E\psi(p) \right).$$

(c) Find a momentum-space eigenfunction corresponding to the energy eigenvalue $\hbar\omega(n + 1/2)$.

$$Answer: \left(h_n\left(\frac{p}{\sqrt{m\omega\hbar}} \right) e^{-(p^2/2m\omega\hbar)} \right).$$

5.3. A one-dimensional harmonic oscillator is in the state

$$\phi(x, t) = \frac{\phi_0(x, t) - 2\phi_1(x, t)}{\sqrt{5}}$$

where ϕ_0 and ϕ_1 are, respectively, the normalized ground state and first excited wave functions. Find expressions for the expectation values of x, the potential energy and the total energy, all as functions of time. Useful information: the phases of ϕ_0 and ϕ_1 can be chosen so that $\langle \phi_0|x|\phi_1 \rangle = \sqrt{\hbar/2m\omega}$.

$$Answer: \left(-\frac{4}{5}\sqrt{\frac{\hbar}{2m\omega}} \cos \omega t, \frac{13}{20}\hbar\omega, \frac{13}{10}\hbar\omega \right).$$

5.4. A particle of mass m moves in the potential shown in the figure, with $V(x) = 1/2 \cdot kx^2$ for $x > 0$ and $V(x) = \infty$ for $x < 0$. What are the energies of the ground state and first excited state?

$$Answer: \mathrm{E}_{gs} = \hbar\omega\left(1 + \frac{1}{2} \right) \quad \omega = \sqrt{\frac{k}{m}}$$
$$E^* = \hbar\omega\left(3 + \frac{1}{2} \right) \quad \omega = \sqrt{\frac{k}{m}}.$$

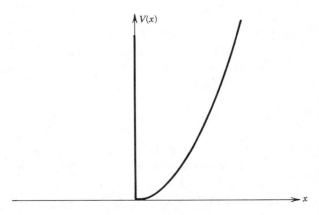

5.5. Let ψ_M^J ($M = -J, -J+1, \cdots, +J$) be the set of states related by the angular momentum operators according to (5.11) and (5.12). Prove that this set of states is closed with respect to any rotation about the origin (i.e., if any state is rotated the result can be written as a linear combination of the members of the original set.)

5.6. Prove that if ψ_M^J is an eigenstate of a spherically symmetric Hamiltonian with eigenvalue E, then ψ_{M+1}^J and ψ_{M-1}^J are also eigenstates of the Hamiltonian with the same eigenvalue (as long as $|M \pm 1| \leq J$).

5.7. The deuteron has total angular momentum $I_d = 1$, whereas the π^- has total angular momentum $I_\pi = 0$. If a π^- is captured by a deuteron from a $1s$ orbital state (so that their relative angular momentum is zero), two neutrons are produced

$$\pi^- + d \rightarrow n + n$$

(a) What is the total angular momentum of the system?   *Answer:* $(J = 1)$.

(b) What are the possible values of the total neutron spin and the neutron relative orbital angular momentum? (Remember that neutrons are $S = \frac{1}{2}$ Fermions.)
    Answer: $(S = 1, l = 1)$.

(c) Assuming that the d and n have even parity, what is the parity of the π^-?
    Answer: (odd).

5.8. Determine the rotational spectrum of the CH_4 molecule, assuming that the H nuclei lie at the vertices of a regular tetrahedron at 1.09 Å from the C nucleus.
    Answer: $(E_J = 6.5 \times 10^{-4} J(J + 1)$ eV$)$.

5.9. (a) Derive an approximate formula for the strength of the magnetic field at the hydrogen nucleus produced by an electron in the first Bohr orbit.

$$Answer: \left(\frac{e^2}{\hbar c}\right)^5 \frac{(mc^2)^2}{e^3}.$$

(b) Use your result from (a) to estimate the order of magnitude of the hyperfine splitting (in eV) of spectral lines.
    Answer: $(\sim 10^{-6}$ eV$)$.

5.10. A spectral line $(\lambda = 4800$ Å$)$ is found to have

a width of 0.015 Å. Estimate the mean life of the excited state.

$$Answer: \left(\frac{\hbar}{\Delta E} \approx 8 \times 10^{-11} \text{ s}\right).$$

5.11. What is the energy difference between electron spin-up and spin-down states in a 10 kilogauss magnetic field?
    Answer: $(1.2 \times 10^{-4}$ eV$)$.

5.12. Show that the Schrödinger equation for a particle moving in externally generated electric and magnetic fields

$$i\hbar \frac{\partial \psi}{\partial t} = \left[\frac{1}{2m}\left(\frac{\hbar}{i}\nabla - q\frac{\mathbf{A}}{c}\right)^2 + q\phi\right]\psi$$

[cf. (3.64)] is invariant under the gauge transformation

$$\mathbf{A} \rightarrow \mathbf{A}' = \mathbf{A} + \nabla\chi$$

$$\phi \rightarrow \phi' = \phi - \frac{1}{c}\frac{\partial\chi}{\partial t}$$

$$\psi \rightarrow \psi' = e^{\left(\frac{iq}{\hbar c}\right)\chi}\psi.$$

[c.f. (3.60)].

5.13. Verify that the probability density and flux
$$\rho(\mathbf{r}, t) \equiv \psi^*\psi$$
$$\mathbf{j}(\mathbf{r}, t) \equiv \frac{\hbar}{2mi}\left[\psi^*\nabla\psi - \psi\nabla\psi^* - \frac{2iq}{\hbar c}\psi^*\psi\mathbf{A}\right]$$

are invariant under the gauge transformation of Problem 5.12, and that they satisfy the probability conservation condition

$$\text{div}\,\mathbf{j} + \frac{\partial\rho}{\partial t} = 0$$

5.14. The $3p_{3/2} \rightarrow 2s_{1/2}$ transition in hydrogen has six components satisfying the dipole selection rule $\Delta M = 0, \pm 1$. What is the effect on the energies of these components of a uniform magnetic field of strength B?
    Answer:
$$\Delta E = \frac{e\hbar B}{2mc} \times \left(\frac{5}{3}, 1, \frac{1}{3}, -\frac{1}{3}, -1, -\frac{5}{3}\right).$$

5.15. Two electrons move in a one-dimensional square-well potential of width L, and with infinitely high walls.

(a) Assume that the interaction between the two electrons is negligible and calculate the energies of the ground state and of the doubly degenerate first-excited states.

$$\text{Answer: } E_{gs} = 2 \times \frac{\hbar^2}{2m}\left(\frac{\pi}{L}\right)^2$$

$$E^* = \frac{\hbar^2}{2m}\left(\left(\frac{\pi}{L}\right)^2 + \left(\frac{2\pi}{L}\right)^2\right).$$

(b) Now use first-order perturbation theory to include the effect of a weak zero-range interaction, $V_0\delta(x_1 - x_2)$. Find the perturbed energies of the states you considered in (a) above.

$$\text{Answer: } E_{gs} \approx 2 \times \frac{\hbar^2}{2m}\left(\frac{\pi}{L}\right)^2 + \frac{3V_0}{2L}$$

$$E^*(S = 1) \approx \frac{\hbar^2}{2m}\left(\left(\frac{\pi}{L}\right)^2 + \left(\frac{2\pi}{L}\right)^2\right)$$

$$E^*(S = 0) \approx \frac{\hbar^2}{2m}\left(\left(\frac{\pi}{L}\right)^2 + \left(\frac{2\pi}{L}\right)^2\right)$$
$$+ \frac{2V_0}{L}$$

5.16. The Hamiltonian for a certain particle is

$$H = \frac{p^2}{2m} - V_0\frac{e^{-r/r_0}}{r/r_0}$$

with V_0 and r_0 constant. Use the trial function $\psi_\alpha \equiv e^{-\alpha r/r_0}$ to make a variational estimate of the ground state energy.

$$\text{Answer: } E_{gs} \leq -\frac{\hbar^2\alpha^2}{2mr_0^2}\frac{2\alpha - 1}{2\alpha + 1}$$

where α satisfies $\dfrac{(2\alpha + 1)^3}{2\alpha(2\alpha + 3)} = \dfrac{2mV_0r_0^2}{\hbar^2}$.

5.17. Particles of mass m and energy E are incident from $x = -\infty$ toward a zero-range potential at the origin, $V(x) = V_0\delta(x)$. What fraction of the incident particles is reflected by this potential?

$$\text{Answer: } \left(\frac{1}{1 + \left(E\bigg/\dfrac{mV_0^2}{2\hbar^2}\right)}\right).$$

5.18. Consider low-energy scattering of an uncharged particle of mass μ by a totally absorbing potential of radius R. The energy E is low enough so that only s-waves $(l = 0)$ are affected. The absorptive character of the potential implies that at $r = R$, the $l = 0$ radial wave function is purely ingoing.

(a) Determine the scattering amplitude.

$$\text{Answer: } \left(f = -\frac{1}{2ik}, \quad k = \sqrt{\frac{2\mu E}{\hbar^2}}\right).$$

(b) Determine the total scattering cross-section

$$\text{Answer: } \left(\sigma = 4\pi|f|^2 = \frac{\pi}{k^2}\right).$$

(c) Determine the total absorption cross-section

$$\sigma_{abs} = \frac{\text{(number of particles absorbed/unit time)}}{\text{incident flux}}$$

$$\text{Answer: } \left(\sigma_{abs} = \frac{\pi}{k^2}\right).$$

5.19. A particle of mass μ and energy E is scattered elastically by a spherically symmetric potential

$$U(r) = Ae^{-br} \quad (A, B \text{ constant})$$

Use the Born approximation to calculate the scattering differential cross-section, as a function of the scattering angle, θ.

$$\text{Answer: } \frac{d\sigma}{d\Omega} = \left[\frac{Ab}{\dfrac{\hbar^2 b^2}{8\mu} + E\sin^2\dfrac{\theta}{2}}\right]^2$$

5.20. Estimate the magnetic field strength that would produce an energy density comparable to that of a chemical explosive.

$$\text{Answer: } (\sim 10^5 \text{ Gauss}).$$

CHAPTER 6
THERMAL PHYSICS

A wide variety of physical systems are considered in this chapter: gases in containers, electrons in conductors, radiation in cavities, etc. All the systems are too complex to allow exact microscopic analyses. In general, the best we can do is to use conservation principles and statistical methods to relate average values of important physical quantities. In some cases we can relate observed macroscopic quantities, like pressure or specific heat, to the detailed microscopic models of the individual components.

The notion of temperature is used throughout this chapter. It is introduced as a measure of the average translational kinetic energy of a collection of molecules. Then it plays a role in the relationship between the increase in entropy (disorder) of a system and the amount of heat added to it. Finally, the temperature serves as a parameter that governs the probability that a system will be in any one of its many allowed states. This is the most general interpretation we consider, in the sense that the other interpretations can be derived from it.

6.1 THE FIRST LAW OF THERMODYNAMICS

Consider changes taking place in a closed system. Let dQ be the heat flow *into* the system from the surroundings, dW the work done *by* the system on its surroundings, and U the internal energy of the system. U is a function of the state of the system, irrespective of how that state is reached. The first law of thermodynamics states that

$$dQ = dU + dW \tag{6.1}$$

Note that dQ is a small quantity of heat. It is *not* the differential of a heat function Q. Similarly dW is a small quantity of work. It is *not* the differential of a work function W. But dU *is* the differential of an internal energy function U.

Let $\oint$ refer to a series of changes that ends with the system returning to its original state. Since U is a function of the state of the system, a series of changes that returns the system to its original state produces no net change in U. Thus, $\oint dU = 0$. However, $\oint dQ$ and $\oint dW$ are generally not equal to zero, since there can be net transfers of work and heat between the system and the surroundings while the system goes through a cycle. The equation $\oint dU = 0 = \oint dQ - \oint dW$ is another way to state the first law of thermodynamics.

The way a system can do work on its surroundings depends on the physical situation. For example, if the system is a gas in a cylinder fitted with a frictionless piston, the gas can do work on its surroundings (the piston) by pushing it out, and the piston can do work on the gas by compressing it. In this situation, which is the most common one encountered in elementary thermodynamics, $dW = p\,dV$ with p the pressure of the gas and dV an infinitesimal volume change. If there are other forms of physical interaction between the system and its surroundings (electrical, magnetic, gravitational, etc.), then other terms have to be added to dW,

$$dW = p\,dV + d\omega \tag{6.2}$$

We will refer to $d\omega$ as "non $p\,dV$ work."

6.2 THE IDEAL GAS

When we discuss the properties of gases, the following characteristic lengths are important:

d, the range of the interaction between gas molecules (generally of the order of 10^{-8} cm)

λ, the mean distance a molecule travels before it interacts with another molecule (of the order of 10^{-5} cm for a gas under normal conditions)

L, the smallest linear dimension of the container of the gas.

In situations in which $L \gg \lambda \gg d$, the behavior of the gas is said to be *ideal*. The pressure, volume and absolute temperature of n moles of ideal gas are related by

$$pV = nRT \tag{6.3}$$

Here R is the ideal gas constant, equal to approximately 2 cal/mol K or 8.3 J/mol K. A useful way to remember R is to remember that one mole of ideal gas at 273 K and atmospheric pressure has a volume of approximately 22.4 liters.

In Section 6.8 we will see that we can derive (6.3) if we assume that the translational kinetic energy of n moles of ideal gas is given by

$$\text{KE} = \tfrac{3}{2}nRT \tag{6.4a}$$

Since each mole contains N_0 ($\approx 6.022 \times 10^{23}$) molecules, the average translational kinetic energy per molecule is

$$\overline{\text{KE}}/\text{molecule} = \frac{3}{2}\frac{R}{N_0}T = \frac{3}{2}kT \tag{6.4b}$$

The symbol k introduced in (6.4b) is called Boltzmann's constant, and has a value of 1.38×10^{-16} erg/molecule K. At 290 K, kT is approximately equal to $1/40$ eV.

A molecule consists of electrons and nuclei moving relative to one another. If the mass center of a molecule of mass m moves with speed v, the molecule has *translational* kinetic energy $\tfrac{1}{2}mv^2$. If energy is supplied to a molecule, the speed of the mass center may change, and thus there may be a change in the translational kinetic energy. But it may also happen that the added energy changes the relative motion of the parts of a single molecule, for example by causing the molecule to rotate or vibrate about its mass center. In other words, a gas can store energy as translational energy or as "excitation energy" within the molecular structure. However, if the gas consists of single atoms (such as helium or neon gas), below about 10,000 K very little excitation energy will be stored within the molecular structure, and any energy supplied to the gas will be stored as translational kinetic energy. Thus, for a *monatomic* ideal gas, the internal energy U is given by

$$U = \text{KE} = \tfrac{3}{2}nRT \qquad \text{(monatomic)} \tag{6.4c}$$

In Section 6.11 there is a discussion of the way (6.4c) must be modified when we deal with diatomic or polyatomic gases, which store excitation energy within the molecular structure.

What about a contribution to the internal energy due to the interaction between the molecules? This will certainly be present in any real gas. It leads to a dependence of the internal energy on the gas density, since changing the density changes the average intermolecular spacing. However, in the regime in which the ideal gas approximation is valid ($\lambda \gg d$), the contribution to the internal energy due to intermolecular attraction is negligible compared to the contributions due to translational kinetic energy and internal excitation energy. These depend only on the temperature. *For an ideal gas, the internal energy is a function of temperature only.* In Problem 6.6.2 below we will derive this result directly from the ideal gas law (6.3).

PROBLEM 6.2.1 Consider an atmosphere composed of an ideal gas in hydrostatic equilibrium. Suppose that the temperature $T(z)$ is given as a specified function of the height z. Find expressions for the pressure and density as functions of height, in terms of the given $T(z)$.

Hydrostatic equilibrium implies that the pressure p and the density ρ are related by

$$\frac{dp}{dz} = -g\rho \tag{6.5a}$$

This expresses the fact that the weight of a thin slab of gas is borne by the pressure difference between its upper and lower faces. If the gas is ideal and has molecular weight M, then its density is given by

$$\rho = \frac{Mn}{V} = \frac{Mp}{RT}$$

so that

$$\frac{dp}{dz} = -g\frac{M}{R}\frac{p}{T}$$

$$\frac{1}{p}\frac{dp}{dz} = \frac{d}{dz}\ln p = -g\frac{M}{R}\frac{1}{T} \tag{6.5b}$$

The solution of this differential equation is

$$\ln p(z) = \ln p(0) - \frac{gM}{R}\int_0^z \frac{dz'}{T(z')}$$

$$p(z) = p(0)e^{-(gM/R)\int_0^z(dz'/T(z'))} \tag{6.5c}$$

$$\rho(z) = \frac{Mp(z)}{RT(z)} = \frac{\rho(0)T(0)}{T(z)}e^{-(gM/R)\int_0^z(dz'/T(z'))} \tag{6.5d}$$

PROBLEM 6.2.2   Real (nonideal) gases are sometimes described by the Van der Waals equation of state. For one mole of gas, this equation is

$$\left(p + \frac{a}{V^2} \right)(V - b) = RT \qquad \text{(one mole)}$$

The constant a represents the effect of long-range attractive intermolecular forces, and the constant b represents the volume occupied by the molecules themselves.

(a) Rewrite the equation so that it applies to n moles of gas.

If one mole of gas has volume b, n moles have volume nb. The a/V^2 term in the expression for one mole shows that the correction to p is proportional to the square of the density. Thus, for n moles in volume V, the correction is $n^2 a/V^2$. To yield the correct ideal gas limit as $a \to 0$, $b \to 0$, the right-hand side should be nRT. Thus, for n moles,

$$\left(p + \frac{an^2}{V^2} \right)(V - nb) = nRT \tag{6.6}$$

(b) Which of the following is the correct expression for $\partial U/\partial V|_T$: $a/(V/n)^2$, zero, $a/(V/n - b)^2$?

For an ideal gas, $\partial U/\partial V|_T = 0$. In a Van der Waals gas, $\partial U/\partial V|_T \neq 0$ since changing the density changes the intermolecular interaction energy. Since it is the $a/(V/n)^2$ in the Van der Waals equation that expresses the effect of intermolecular forces, the correct choice will be $a/(V/n)^2$. This will be derived explicitly from the equation of state in Problem 6.6.2.

6.3 HEAT CAPACITY

If the addition of an amount of heat dQ to a system causes its temperature to change by dT, then its heat capacity C is defined by

$$C = \frac{dQ}{dT} \tag{6.7}$$

The heat capacity per mole (or sometimes per gram) is called the specific heat capacity, and is symbolized by c.

The heat capacity depends on the physical constraints effective while the heat is added. For example, suppose the system consists of one mole of ideal gas in a cylinder with a movable piston. Let c_v be the heat capacity at constant volume (the piston is kept fixed as heat is added). Then $c_v = dQ/dT = (dU + dW)/dT = dU/dT$. But we could also arrange to keep the pressure constant as the heat is added, by allowing the piston to move. The associated heat capacity is $c_p = dQ/dT = (dU + p \, dV)/dT = c_v + p \, dV/dT$. Since p is constant, $p \, dV/dT = d(pV)/dT = d(RT)/dT = R$. Thus, for an ideal gas, c_p and c_v are related by

$$c_p = c_v + R \qquad \text{(per mole)} \tag{6.8}$$

If the gas is monatomic as well as ideal, then $U = 3/2RT$, $c_v = 3/2R$, and $c_p = 5/2R$.

PROBLEM 6.3.1   A mole of paramagnetic ideal gas consists of molecules with permanent magnetic moment μ. It is contained in volume V, in a uniform constant magnetic field $\mathbf{B}$, and at temperature T. Assume that the magnetic susceptibility is given by Curie's law (6.52), and calculate the heat capacity at constant V and $\mathbf{B}$.

If the total magnetic moment of the gas is $\mathbf{M}$ ($= V \times$ magnetization density $\mathcal{M}$), its interaction with the external field $\mathbf{B}$ contributes a term $-\mathbf{B} \cdot \mathbf{M}$ to the total energy (cf. 3.57c). Thus, if we add heat dQ at constant V and $\mathbf{B}$,

$$dQ = dU + p\,dV - \mathbf{B} \cdot d\mathbf{M} = c_v dT - \mathbf{B} \cdot d\mathbf{M} \qquad (6.9)$$

Note the assumption that the gas is ideal means we neglect all interactions (even magnetic interactions) between the molecules, and so the *internal* energy U is unaffected by the magnetism. According to Curies' law (6.52),

$$\mathbf{M} = V\mathcal{M} = V\chi(T)\mathbf{B} = \frac{1}{3}N_0\frac{\mu^2}{kT}\mathbf{B}$$

$$d\mathbf{M}|_{\mathbf{B},V} = V\frac{d\chi}{dT}(T)\mathbf{B}\,dT = -\frac{1}{3}N_0\frac{\mu^2}{kT^2}\mathbf{B}\,dT$$

and so (6.9) becomes

$$dQ = \left[c_v + \frac{1}{3}N_0\frac{\mu^2 B^2}{kT^2}\right]dT$$

and the heat capacity at constant V, B is

$$c_{v,\mathbf{B}} = c_v + \frac{1}{3}N_0\frac{\mu^2 B^2}{kT^2}$$

$$= c_v + \left(\frac{\mu B}{kT}\right)^2 \bigg/ 3R \qquad (6.10)$$

Thus, the presence of $\mathbf{B}$ increases the heat capacity. When we add heat to the gas and raise its temperature, the increased molecular choas decreases the ability of the molecules to line up parallel to the external magnetic field. This causes an increase in the total energy, an increase that must be supplied by a portion of the added heat. This situation is analogous to the difference (6.8) between c_p and c_v, which is due to the part of dQ that must provide the $p\,dV$ work performed when heat is added at constant pressure.

For an ideal gas, the infinitesimal changes dp, dV, dT are related by the condition

$$d(pV) = p\,dV + V\,dp = R\,dT \qquad \text{(one mole)} \qquad (6.11a)$$

We can get another relation between dV and dT if we require that no heat enter or leave the system (an *adiabatic* process):

$$dQ = 0 = dU + p\,dV = c_v\,dT + p\,dV \qquad (6.11b)$$

If we eliminate dT between (6.11a) and (6.11b), we find that

$$p\left(1 + \frac{R}{c_v}\right)dV + V\,dp = 0$$

$$\frac{c_p}{c_v}\frac{dV}{V} + \frac{dp}{p} = 0 = d\ln\left(pV^{c_p/c_v}\right)$$

Then, when an ideal gas undergoes an adiabatic change,

$$pV^\gamma = \text{constant} \qquad (6.11c)$$

where γ represents the ratio c_p/c_v. For a monatomic ideal gas, $\gamma = (5/2R)/(3/2R) = 5/3$.

PROBLEM 6.3.2 An ideal monatomic gas at room temperature is adiabatically decompressed so that the final volume is 8 times the original. What is the final temperature?
According to (6.11c),

$$p_1 V_1^\gamma = p_2 V_2^\gamma, \qquad \frac{p_1}{p_2} = \left(\frac{V_2}{V_1} \right)^\gamma = 8^\gamma = 8^{5/3}$$

$$\frac{T_2}{T_1} = \frac{p_2 V_2}{p_1 V_1} = 8^{-5/3} \cdot 8 = 8^{-2/3}$$

$$T_2 = \frac{T_1}{8^{2/3}} = \frac{T_1}{4} = \frac{300 \text{ K}}{4} = 75 \text{ K}$$

PROBLEM 6.3.3 Suppose that the atmosphere of Problem 6.2.1 is also in equilibrium with respect to fast (adiabatic) vertical displacements of gas. What does this imply about $T(z)$?
Using the ideal gas law, we can express (6.11c) in the form $p^{1-\gamma} T^\gamma = $ constant, or equivalently

$$\ln p + \frac{\gamma}{1 - \gamma} \ln T = \text{constant}$$

Differentiating with respect to z, we get

$$\frac{d}{dz} \ln p = \frac{\gamma}{\gamma - 1} \frac{1}{T} \frac{dT}{dz}$$

If we compare this with (6.5b), we see that

$$\frac{dT}{dz} = -\frac{\gamma - 1}{\gamma} \frac{gM}{R}$$

whose solution

$$T(z) = T(0) - \frac{\gamma - 1}{\gamma} \frac{gM}{R} z$$

shows that the temperature of the atmosphere decreases linearly with height.

PROBLEM 6.3.4 Find the relation between C_p and C_v for a real (not necessarily ideal) gas.
C_v is still given by $\partial U/\partial T|_V$ since, at constant volume, $dQ = C_v \, dT = dU$. But if we add heat at constant pressure, the volume changes by an amount $dV = \partial V/\partial T|_p \, dT$. Thus, at constant pressure,

$$dQ = C_p \, dT = dU + p \, dV$$

$$= \left. \frac{\partial U}{\partial T} \right|_V dT + \left[\left. \frac{\partial U}{\partial V} \right|_T + p \right] dV$$

$$= C_v \, dT + \left[\left. \frac{\partial U}{\partial V} \right|_T + p \right] \left(\frac{\partial V}{\partial T} \right)_p dT$$

$$C_p = C_v + \left[\left. \frac{\partial U}{\partial V} \right|_T + p \right] \left(\frac{\partial V}{\partial T} \right)_p$$

This is a general relation between C_p and C_v. For an ideal gas, where $\partial U / \partial V |_T = 0$ and $\partial V / \partial T |_p = nR/P$, it reduces to $C_p = C_v + nR$. For a Van der Waals gas (6.6), we get

$$C_p = C_v + \frac{nR\left(p + \dfrac{n^2 a}{V^2}\right)}{p - \dfrac{n^2 a}{V^2} + \dfrac{2n^3 ab}{V^3}}$$

We have used here $\partial U / \partial V |_T = n^2 a / V^2$, as given in Problems 6.2.2b and 6.6.2.

6.4 ENTROPY AND THE SECOND LAW OF THERMODYNAMICS

A reversible process is one that proceeds so slowly that at any instant the system can be considered to be in equilibrium. If a series of reversible changes brings a system back to its starting point, then the *second law of thermodynamics* says that

$$\oint_{\text{rev}} \frac{dQ}{T} = 0 \tag{6.12}$$

(Remember that $\oint dQ$ is generally not equal to 0). If A and B represent two states of a system connected by two different reversible paths,

$$0 = \oint_{\text{rev}} \frac{dQ}{T} = \int_{\substack{A \\ \text{rev path 1}}}^{B} \frac{dQ}{T} + \int_{\substack{B \\ \text{rev path 2}}}^{A} \frac{dQ}{T}$$

$$= \int_{\substack{A \\ \text{rev path 1}}}^{B} \frac{dQ}{T} - \int_{\substack{A \\ \text{rev path 2}}}^{B} \frac{dQ}{T}$$

Equivalently,

$$\int_{\substack{A \\ \text{rev path 1}}}^{B} \frac{dQ}{T} = \int_{\substack{A \\ \text{rev path 2}}}^{B} \frac{dQ}{T}$$

Thus, $\int dQ/T$ between two states of a system has the same value for all reversible paths connecting the two states. Let us call state A our reference state, and keep this state fixed. For any other state B, the integral

$$S(B) \equiv \int_{\substack{A \\ \text{rev}}}^{B} \frac{dQ}{T}$$

depends only on state B, and not on the path from reference state A to B. $S(B)$ is

FIGURE 6.1 Two reversible paths connecting states A and B.

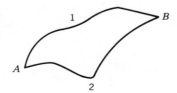

said to be the entropy of the system when it is in state B. With this choice of reference state, $S(A) = 0$. (Note: only entropy *differences* are uniquely defined). We can choose any state of the system to be our reference state. Once this state is chosen, the entropy of any other state is defined relative to it. Similar considerations were used in Section 1.4 when the potential energy function was defined for a conservative system, and again in Section 3.2 when the electrostatic potential was introduced in connection with an electric field whose curl was zero.

For a reversible process connecting infinitesimally close states,

$$\frac{dQ}{T} = dS \qquad \text{(reversible)} \qquad (6.13a)$$

For an irreversible process (and every real process is irreversible) connecting the same two states,

$$\frac{dQ}{T} < dS \qquad \text{(irreversible)} \qquad (6.13b)$$

Note that dS is the *same* in these two cases since we are going between the same initial and final states, and entropy depends *only* on the states. The difference between (6.13a) and (6.13b) is in the *left-hand sides*, dQ/T.

Perhaps the most useful set of equations in thermodynamics is obtained by combining (6.1), (6.2), and (6.13):

$$T\,dS = dU + p\,dV + d\omega \qquad \text{(reversible)} \qquad (6.14a)$$

$$T\,dS > dU + p\,dV + d\omega \qquad \text{(irreversible)} \qquad (6.14b)$$

These equations are referred to as "the combined first and second laws."

PROBLEM 6.4.1 Calculate the entropy change when a block of lead of heat capacity C is heated between absolute temperatures T_A and T_B.

Assume that the heating is done reversibly. Then

$$dS = \frac{dQ}{T} = \frac{C\,dT}{T}$$

$$\int_{T_A}^{T_B} dS = S(T_B) - S(T_A) = \int_{T_A}^{T_B} \frac{C\,dT}{T} = C \ln\left(\frac{T_B}{T_A}\right) \qquad (6.15)$$

This is the entropy change no matter how we get from the initial state T_A to the final state T_B.

PROBLEM 6.4.2 Two identical blocks of lead, each with heat capacity C, are initially at temperatures T_1 and T_2 (with $T_2 > T_1$). They are brought into thermal contact and left until they reach a common equilibrium temperature $T = (T_1 + T_2)/2$. Calculate the change in entropy of the system.

Let our reference state be the two blocks at some arbitrary temperature T_A. The initial entropy is

$$S_{\text{initial}} = S(T_1) + S(T_2) = 2S(T_A) + C\left[\ln\frac{T_1}{T_A} + \ln\frac{T_2}{T_A}\right]$$

$$= 2S(T_A) + C \ln\frac{T_1 T_2}{T_A^2}$$

The final entropy is

$$S_{\text{final}} = 2S\left(\frac{T_1 + T_2}{2}\right) = 2S(T_A) + 2C\ln\left(\frac{T_1 + T_2}{2T_A}\right)$$

and the entropy change is

$$S_{\text{final}} - S_{\text{initial}} = 2C\ln\left(\frac{T_1 + T_2}{2T_A}\right) - C\ln\frac{T_1 T_2}{T_A^2}$$

$$= C\ln\left[\frac{(T_1 + T_2)^2}{4T_1 T_2}\right] = C\ln\left[\frac{(T_1 - T_2)^2}{4T_1 T_2} + 1\right] > 0, \text{ if } T_1 \neq T_2$$

Thus, the irreversible transfer of heat between blocks at different temperatures is associated with an increase in the total entropy of the system. Alternatively, if heat dQ flows from block 2 at temperature t_2 to block 1 at lower temperatures t_1,

entropy change of block 1 is dQ/t_1 (heat flows into block 1)
entropy change of block 2 is $-dQ/t_2$ (heat flows out of block 2)

so the total entropy change is

$$\frac{dQ}{t_1} - \frac{dQ}{t_2} = dQ\left[\frac{1}{t_1} - \frac{1}{t_2}\right] > 0, \quad \text{since } t_2 > t_1$$

The expression (6.15) was derived by considering reversible changes. We can use this expression to discuss entropy changes in irreversible processes, because entropy is a function of the *state* of the system, irrespective of how this state is reached.

PROBLEM 6.4.3 Derive a formula for the entropy of a mole of ideal gas, as a function of T and V. Assume that the ranges of T and V are small enough so that c_v is constant.
We use (6.14a) with $dw = 0$ and $dU = c_v\,dT$:

$$T\,dS = c_v\,dT + p\,dV \qquad \text{(reversible)} \tag{6.16}$$

By considering separate reversible isothermal ($dT = 0$) and isovolumetric ($dV = 0$) changes, we can deduce from (6.16) that

$$\left.\frac{\partial S}{\partial V}\right|_T = \frac{p}{T} = \frac{R}{V} \qquad \text{(for an ideal gas)} \tag{6.17a}$$

$$\left.\frac{\partial S}{\partial T}\right|_V = \frac{c_v}{T} \tag{6.17b}$$

If we integrate (6.17a) from V_A to V_B, we get

$$S(V_B, T) - S(V_A, T) = R\ln\left(\frac{V_B}{V_A}\right) \tag{6.18a}$$

Similarly, integrating (6.17b) from T_A to T_B gives

$$S(V, T_B) - S(V, T_A) = c_v\ln\left(\frac{T_B}{T_A}\right) \tag{6.18b}$$

Next we set $T = T_B$ in (6.18a), and $V = V_A$ in (6.18b), and add the two equations. The result is

$$S(V_B, T_B) = S(V_A, T_A) + R \ln\left[\frac{V_B}{V_A} \cdot \left(\frac{T_B}{T_A}\right)^{c_v/R}\right] \qquad (6.19)$$

PROBLEM 6.4.4 Calculate the entropy of a mole of paramagnetic ideal gas as a function of V, T, and the magnetic moment $\mathbf{M}$. Assume $(\mu B/kT)$ is small enough so that Curie's law (6.52) is valid.

The effect of the magnetization is the addition of a term $-\mathbf{B} \cdot d\mathbf{M}$ to (6.16) (see Problem 6.3.1):

$$T\, dS = c_v\, dT + p\, dV - \mathbf{B} \cdot d\mathbf{M} \qquad \text{(reversible)}$$

If we change the components of $\mathbf{M}$ at constant V, T, we get

$$\left.\frac{\partial S}{\partial M_\alpha}\right|_{V,T} = -\frac{B_\alpha}{T} = -\frac{3k}{N_0\mu^2}M_\alpha$$

$$= -\frac{3R}{(N_0\mu)^2}M_\alpha \qquad (\alpha = x, y, z)$$

The integration with respect to each M_α proceeds as in (6.18). If we take our lower limit of $\mathbf{M}$ to be 0, then (6.19) is replaced by

$$S(V_B, T_B, \mathbf{M}) = S(V_A, T_A, 0) + R\left\{\ln\left[\frac{V_B}{V_A}\left(\frac{T_B}{T_A}\right)^{c_v/R}\right] - \frac{3}{2}\frac{\mathbf{M} \cdot \mathbf{M}}{(N_0\mu) \cdot (N_0\mu)}\right\} \qquad (6.20)$$

We see that the magnetic contribution to the entropy decreases as the magnetic order increases, and M becomes a larger fraction of its maximum value, $N_0\mu$. Note that we can only use (6.20) if $M \ll N_0\mu$, since the saturation condition $M \approx N_0\mu$ is outside the range of Curie's law [see (6.51)].

PROBLEM 6.4.5 N moles of a monatomic ideal gas are contained in volume V_0 within an insulated container. After a partition is removed, the gas expands into an adjacent evacuated volume V_0, so that the final volume of the gas is $2V_0$. What is the entropy change?

No work is done by the expanding gas, since it expands into a vacuum. Moreover, no heat flows through the thick adiabatic walls. Thus,

$$dU = dQ - dW = 0$$

Since the gas is ideal, $dU = 0$ implies that $dT = 0$. (This would not be true for a

FIGURE 6.2 A thermally insulated container of volume $2V_0$: On the left, all the gas is constrained to volume V_0 by a barrier. On the right the barrier has been removed, and the gas expands to fill the entire container.

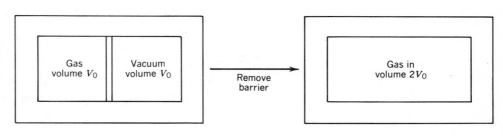

real gas.)

$$S_{\text{final}} - S_{\text{initial}} = nR \ln\left(\frac{2V_0}{V_0}\right) = nR \ln 2$$

for n moles of gas. The free expansion of the gas is an irreversible process, so the entropy is increased.

PROBLEM 6.4.6a   What are the signs of the pressure and temperature changes when an ideal gas makes an isovolumetric change in which it loses heat?

The conditions stated in the problem imply that

$$0 > dQ = dU + p\,dV = dU = d\left(\tfrac{3}{2}nRT\right)$$

so that

$$0 > dT = \frac{1}{nR}(p\,dV + V\,dp) = \frac{V}{nR}\,dp$$

Thus, we have proven that $0 > dT, dp$, so that both T and p decrease.

PROBLEM 6.4.6b   What are the signs of the pressure and entropy changes when an ideal gas makes an isothermal reversible expansion?

Here we have

$$dT = 0 = \frac{1}{nR}(p\,dV + V\,dp)$$

$$dp = -\frac{p}{V}\,dV$$

$$dQ = T\,dS = dU + p\,dV = p\,dV$$

$$dS = +\frac{p}{T}\,dV$$

Since $dV > 0$, we conclude that the pressure decreases and the entropy increases.

PROBLEM 6.4.7   A straight copper rod with a cross-section area of 5 cm^2 and a length of 10 cm is connected between boiling water and melting ice at one atmosphere pressure. The thermal conductivity of copper is 9×10^{-2} kcal/K m s. Considering this as a closed system, calculate the rate of entropy increase of the system and the rate at which the ice melts.

$$q = \text{rate of heat flow/unit area}$$

$$= \text{thermal conductivity} \times \text{temperature gradient}$$

$$= 9 \times 10^{-2} \frac{\text{kcal}}{\text{K m s}} \times \frac{100 \text{ K}}{0.1 \text{ m}} = \frac{90 \text{ kcal}}{\text{m}^2 \text{ s}}$$

For an area of 5 cm$^2 = 5 \times 10^{-4}$ m^2, the heat flow is

$$\frac{90 \text{ kcal}}{\text{m}^2 \text{ s}} \times 5 \times 10^{-4} \text{ m}^2 = .045 \frac{\text{kcal}}{\text{s}} = 45 \frac{\text{cal}}{\text{s}}$$

At the warm end, entropy is extracted from the boiling water at a rate of 45 (cal/s)/373 K. At the cold end, entropy is supplied to the ice at the rate 45 (cal/s)/273 K. The entropy of the rod is constant since we are in a steady state condition. Thus, the net rate of increase of entropy is

$$\frac{45}{273} - \frac{45}{373} = \frac{(100)(45)}{(273)(373)} \frac{\text{cal}}{\text{K s}} = .044 \frac{\text{cal}}{\text{K s}}$$

The ice melts at the rate

$$45 \frac{\text{cal}}{\text{s}} \cdot \frac{1 \text{ g}}{80 \text{ cal}} = \frac{45}{80} \frac{\text{g}}{\text{s}} = .563 \text{ g/s}$$

6.5 THE CARNOT CYCLE

A Carnot cycle is performed on an apparatus consisting of two heat reservoirs of infinite heat capacity, at respective temperatures T_1 and T_2 $(T_1 > T_2)$. A cylinder fitted with a frictionless piston contains a "working fluid." It is able to extract heat at temperature T_1 from the hotter reservoir, to dump heat at temperature T_2 into the colder reservoir, and to undergo adiabatic expansion and contraction between these two temperatures. All processes are assumed to be reversible. Figure 6.3 represents the four steps in the sequence that defines a Carnot cycle:

1. $a \to b$: Isothermal expansion at temperature T_1. Q_1 joules are removed from the heat reservoir at T_1.
2. $b \to c$: Adiabatic expansion until the working fluid reaches temperature T_2.
3. $c \to d$: Isothermal compression at temperature T_2. Q_2 joules are delivered to the heat dump at temperature T_2.
4. $d \to a$: Adiabatic compression until the working fluid reaches the initial temperature T_1.

Since all parts of the cycle are reversible, we know that $\oint dQ/T = 0$. During the adiabatic expansion and contraction, $dQ = 0$. Thus,

$$0 = \frac{Q_1}{T_1} + \frac{(-Q_2)}{T_2}, \qquad Q_2 = Q_1 \frac{T_2}{T_1}$$

The net work during the cycle is $W = Q_1 - Q_2$. The efficiency e of the cycle is the ratio of the net work done to the energy extracted from the hotter reservoir,

$$e = \frac{W}{Q_1} = \frac{Q_1 - Q_2}{Q_1} = \frac{Q_1 - Q_1 T_2/T_1}{Q_1} = 1 - \frac{T_2}{T_1} \qquad (6.21)$$

Since some heat is dumped during the $c \to d$ phase, the efficiency is always less

FIGURE 6.3 *p– V* diagram for a Carnot cycle.

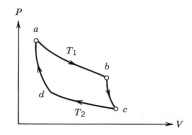

than 1. It is clear that we increase the efficiency of conversion of heat to work by *increasing* the ratio T_1/T_2.

No heat engine operating between temperatures T_1 and T_2 can convert heat to work with an efficiency exceeding that of the Carnot cycle.

The cycle can be run reversibly in the direction opposite to that indicated in Figure 6.3. It then acts as a refrigerator, extracting heat at the low temperature T_2 and dumping it at the higher temperature T_1. The "coefficient of performance" c is the ratio of the heat extracted to the mechanical work supplied,

$$c = \frac{Q_2}{W} = \frac{Q_2}{Q_1 - Q_2} = \frac{T_2}{T_1 - T_2}$$

We increase c by *decreasing* the ratio T_1/T_2.

PROBLEM 6.5.1 A helium liquefier with an inside temperature of 4 K operates in surroundings at 300 K. For each joule of energy extracted from the helium at 4 K, how many joules (at least) must be added to the surroundings as heat?

If the process is reversible,

$$\frac{W}{Q_2} = \frac{Q_1 - Q_2}{Q_2} = \frac{T_1 - T_2}{T_2} = \frac{296}{4} = 74.$$

Thus, if $Q_2 = 1$ joule, $W = 74$ joules, and $Q_1 = 74 + 1 = 75$ joules are added to the surroundings as heat.

PROBLEM 6.5.2 An ideal gas turbine may be assumed to work in a reversible Joule cycle shown in the T–S diagram. The gas is monatomic.

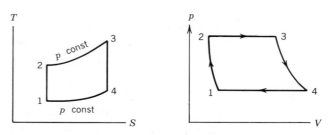

(a) Draw the corresponding p–V diagram.

Since the entropy remains constant as the temperature increases from T_1 to T_2, the step $1 \to 2$ must involve an adiabatic compression, with $pV^\gamma = pV^{5/3} =$ constant. Since p remains constant as the temperature increases from T_2 to T_3, the step $2 \to 3$ must involve an isobaric expansion, with $V = nRT/p$. Steps $3 \to 4$ and $4 \to 1$ are the reverse of steps $1 \to 2$ and $2 \to 3$, respectively.

(b) How much work is done in one cycle? Express your answer in terms of T_1, T_2, T_3, T_4.

The work done is $\oint p\,dV = \oint[T\,dS - dU] = \oint T\,dS$, which is the area within the curve representing the cycle in the T–S plane. Between points 2 and 3, $T\,dS = dQ = c_p\,dT = (5R/2)dT$,

$$\int_2^3 T\,dS = \frac{5}{2}R(T_3 - T_2)$$

Similarly, between points 4 and 1,

$$\int_4^1 T\,dS = \frac{5}{2}R(T_1 - T_4)$$

Thus, the total work done in one cycle is $(5R/2)(T_3 - T_2 + T_1 - T_4)$.

(c) Assuming fixed extreme temperatures T_1 and T_3, find the intermediate temperatures T_2 and T_4 that will result in maximum work performed.

We cannot make T_2 and T_4 arbitrarily small, because of the stipulations that $1 \rightarrow 2$ and $3 \rightarrow 4$ are adiabatic processes. But from the adiabatic conditions

$$T_1 V_1^{2/3} = T_2 V_2^{2/3}, \qquad T_3 V_3^{2/3} = T_4 V_4^{2/3}$$

and the isobaric conditions

$$\frac{T_2}{V_2} = \frac{T_3}{V_3}, \qquad \frac{T_4}{V_4} = \frac{T_1}{V_1}$$

we can deduce that $T_2 T_4 = T_1 T_3$. Thus, we need to minimize $T_2 + T_4$, subject to this restriction on $T_2 T_4$. Using the method of Lagrange multipliers, we have

$$0 = \frac{\partial}{\partial T_2}\left[T_2 + T_4 - \lambda T_2 T_4\right] = 1 - \lambda T_4$$

$$0 = \frac{\partial}{\partial T_4}\left[T_2 + T_4 - \lambda T_2 T_4\right] = 1 - \lambda T_2$$

We see that $T_2 = T_4 = \sqrt{T_1 T_3}$, and the maximum work performed is $(5R/2)[T_1 + T_3 - 2\sqrt{T_1 T_3}]$.

6.6 FREE ENERGY AND ENTHALPY

6.6.1 The Thermodynamic Potentials

Suppose that a process takes place in which a system goes from state 1 to state 2. The *thermodynamic potentials* enable us to calculate the maximum amount of work that can be extracted from the system as it undergoes the transition between these two states. There are several different potentials, corresponding to the different kinds of processes we normally encounter.

The simplest situation occurs when the process takes place in an adiabatic (thermally insulated) container. Then (6.1) and (6.2) give

$$dQ = 0 = dU + dW = dU + p\,dV + d\omega$$

$$dW = -dU \quad \text{(adiabatic)} \tag{6.22a}$$

This gives the total work dW. We may be interested in the amount of non pdV work done during this adiabatic process. Of course, if the process occurs at constant volume, $p\,dV = 0$ and all the work done is non $p\,dV$ work,

$$d\omega = -dU \quad \text{(adiabatic, constant volume)} \tag{6.22b}$$

To determine $d\omega$ in adiabatic processes in which volume is not constant, it is convenient to introduce the *enthalpy H*, defined by

$$H = U + pV \tag{6.23}$$

Then

$$dH = dU + p\,dV + V\,dp = -d\omega + V\,dp$$

$$d\omega = -dH + V\,dp \quad \text{(adiabatic)} \tag{6.24a}$$

In particular, if the process occurs at constant pressure,

$$d\omega = -dH \quad \text{(adiabatic, constant pressure)} \tag{6.24b}$$

If the container is not adiabatic, we cannot determine such strong restrictions on dW and $d\omega$, since it is possible for heat energy to flow into or out of the container as the process takes place. However, the second law of thermodynamics, (6.13), puts limits on the possible amount of heat flow. Thus, we can at least find some useful inequalities.

From (6.1) and (6.13), we have

$$dQ = dU + dW \le T\,dS \tag{6.25}$$

Now we define the *Helmholtz free energy F* by

$$F = U - TS \tag{6.26a}$$

$$dF = dU - T\,dS - S\,dT \tag{6.26b}$$

Combining (6.26b) and (6.25) yields

$$dW \le -dF - S\,dT \tag{6.27a}$$

In particular, if the process is isothermal (constant T), this becomes

$$dW \le -dF \quad \text{(constant } T) \tag{6.27b}$$

The equal signs in (6.25) and (6.27) apply if the process is reversible. Thus, if the system changes between specified states, the maximum amount of work will be extracted if that change is performed reversibly.

To get a limit on $d\omega$ in the nonadiabatic case, we define the *Gibbs free energy G* by

$$G = F + pV = U - TS + pV \tag{6.28a}$$

$$dG = dF + p\,dV + V\,dp \tag{6.28b}$$

Then (6.28b), (6.27a), and (6.2) can be combined to give

$$d\omega \le -dG + V\,dp - S\,dT \tag{6.29a}$$

$$d\omega \le -dG \quad \text{(constant } p, \text{ constant } T) \tag{6.29b}$$

Equations (6.22), (6.24), (6.27), and (6.29) imply that if a process takes place in a container, the amount of non $p\,dV$ work that can be extracted is limited by the change in U, F, H, or G of the material in the container. U and H are relevant for adiabatic processes, and F and G for isothermal ones. U and F are relevant for processes at constant volume, H and G for processes at constant pressure. U, F, H, G are called thermodynamic potentials, and are functions of the state of the system, irrespective of how that state is reached. These results are summarized in Table 6.1.

PROBLEM 6.6.1   An electric cell can be constructed that uses the chemical reaction

$$\text{Pb} + 2\text{HgCl} \rightarrow \text{PbCl}_2 + 2\text{Hg}$$

When the reaction takes place reversibly at constant volume, and at a constant temperature of 25 C, the EMF of the cell is found to be .5387 volts. What is the

TABLE 6.1 The amount of non $p\,dV$ work $d\omega$ that can be extracted from a system, in terms of changes in the thermodynamic potentials.

	Constant V	Constant p
Adiabatic	$-dU$	$-dH$
Isothermal[a]	$\leq -dF$	$\leq -dG$

[a]Note that in the isothermal cases, only inequalities can be stated. At constant V there is no distinction between non $p\,dV$ work $d\omega$ and total work dW.

change in Helmholtz free energy F when one mole of Pb is converted to $PbCl_2$ in this way?

The electrical energy produced is the EMF times the charge transported. Each lead atom transfers two electrons to mercury atoms, so that for each mole of lead consumed

$$dW = 2 \times 6.022 \times 10^{23} \text{ electrons} \times 1.602 \times 10^{-19} \frac{C}{\text{electron}} \times .5387 \text{ V}$$

$$= 1.040 \times 10^5 \text{ J} = -dF$$

It turns out that the internal energy of $PbCl_2 + 2Hg$ is lower than that of $Pb + 2HgCl$ by $.95 \times 10^5$ joules per mole of Pb. If the reaction took place in an adiabatic container at constant volume, this is the amount of electrical energy that would be produced. The extra 0.09×10^5 joules of electrical energy produced when the reaction occurs at a constant temperature of 25 C comes from heat conducted into the system as the reaction takes place. The cell acts as a refrigerator and extracts heat from its surroundings.

6.6.2 The Maxwell Relations

Several useful relations can be obtained between the thermodynamic potentials, U, F, H, G, and the thermodynamic variables p, V, S, T, if we assume that the only work done is $p\,dV$ work. These equations can be conveniently classified according to which pair of variables is chosen to be independent.

1. S and V independent.

$$dQ = T\,dS = dU + p\,dV, \qquad dU = T\,dS - p\,dV$$

$$\left.\frac{\partial U}{\partial S}\right|_V = T, \qquad \left.\frac{\partial U}{\partial V}\right|_S = -p$$

$$\left.\frac{\partial T}{\partial V}\right|_S = -\left.\frac{\partial p}{\partial S}\right|_V \tag{6.30a}$$

2. S and p independent.

$$dH = dU + p\,dV + V\,dp = T\,dS + V\,dp$$

$$\left.\frac{\partial H}{\partial S}\right|_p = T, \qquad \left.\frac{\partial H}{\partial p}\right|_S = V$$

$$\left.\frac{\partial T}{\partial p}\right|_S = \left.\frac{\partial V}{\partial S}\right|_p \tag{6.30b}$$

3. T and V independent.

$$dF = dU - T\,dS - S\,dT = -p\,dV - S\,dT$$

$$\left.\frac{\partial F}{\partial V}\right|_T = -p, \qquad \left.\frac{\partial F}{\partial T}\right|_V = -S$$

$$\left.\frac{\partial p}{\partial T}\right|_V = \left.\frac{\partial S}{\partial V}\right|_T \tag{6.30c}$$

4. T and p independent.

$$dG = dU - T\,dS - S\,dT + p\,dV + V\,dp = -S\,dT + V\,dp$$

$$\left.\frac{\partial G}{\partial T}\right|_p = -S, \qquad \left.\frac{\partial G}{\partial p}\right|_T = V$$

$$\left.\frac{\partial S}{\partial p}\right|_T = -\left.\frac{\partial V}{\partial T}\right|_p \tag{6.30d}$$

Equations (6.30a, b, c, and d) are called the Maxwell relations.

PROBLEM 6.6.2 Find an expression that will enable you to calculate $\partial U/\partial V|_T$ for a gas, given its equation of state.

Since V and T are the independent variables here, we use the relations derived in (3) above. We have

$$\left.\frac{\partial F}{\partial V}\right|_T = -p = \left.\frac{\partial(U - TS)}{\partial V}\right|_T = \left.\frac{\partial U}{\partial V}\right|_T - T\left.\frac{\partial S}{\partial V}\right|_T$$

$$= \left.\frac{\partial U}{\partial V}\right|_T - T\left.\frac{\partial p}{\partial T}\right|_V$$

Thus,

$$\left.\frac{\partial U}{\partial V}\right|_T = T\left.\frac{\partial p}{\partial T}\right|_V - p$$

We can use this expression whenever we know p as a function of T and V. For example, for an ideal gas, (6.3) implies that $T\,\partial p/\partial T|_V = p$, which yields the expected result that $\partial U/\partial V|_T = 0$. For a Van der Waals gas, we use the equation of state (6.6) to write

$$p = \frac{nRT}{V - nb} - \frac{an^2}{V^2}$$

$$T\left.\frac{\partial p}{\partial T}\right|_V = \frac{nRT}{V - nb} = p + \frac{an^2}{V^2}$$

and so

$$\left.\frac{\partial U}{\partial V}\right|_T = p + \frac{an^2}{V^2} - p = \frac{an^2}{V^2}$$

This agrees with the expressions used in Problems 6.2.2b and 6.3.4.

PROBLEM 6.6.3 Let $f(T, l)$ be the tension in a rubber band of length l and absolute temperature T. Let $S(T, l)$ be its entropy.

(a) Find a relationship between $\partial S/\partial l|_T$ and $\partial f/\partial T|_l$.

The thermodynamic potential appropriate to situations in which the independent variables are T and l is the Helmholtz free energy F. The combined first and second law applied to the rubber band is

$$T\,dS = dQ = dU - f\,dl \qquad (6.31)$$

Thus,

$$dF = dU - T\,dS - S\,dT = f\,dl - S\,dT$$

and (6.30c) is replaced by

$$\frac{\partial F}{\partial l}\bigg|_T = f, \qquad \frac{\partial F}{\partial T}\bigg|_l = -S, \qquad \frac{\partial f}{\partial T}\bigg|_l = -\frac{\partial S}{\partial l}\bigg|_T$$

(b) If the temperature of the stretched band is raised while the length is kept constant, will the tension increase of decrease? Hint: if a rubber band is stretched at constant temperature, its molecules assume a more regular arrangement.

The hint implies that $\partial S/\partial l|_T < 0$. Thus, $\partial f/\partial T|_l > 0$, and the tension increases when the band is heated at constant length.

PROBLEM 6.6.4 Consider a piece of rubber of length l with one end fixed and a force f pulling on the other end. The equation of state of the piece of rubber is

$$f = (K_0 - \alpha T)(l - l_0) \qquad (l > l_0)$$

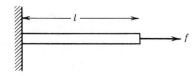

where K_0, α, l_0 are constant. The heat capacity c_l at constant length is taken to be independent of T. Suppose that this piece of rubber is adiabatically and reversibly stretched from length l_1 to length l_2. If the initial temperature was T_1, what is the final temperature?

We need $\partial T/\partial l|_S$, so let us work with S and l as independent variables. From (6.31) we have

$$\frac{\partial U}{\partial l}\bigg|_S = f, \qquad \frac{\partial U}{\partial S}\bigg|_l = T, \qquad \frac{\partial T}{\partial l}\bigg|_S = \frac{\partial f}{\partial S}\bigg|_l = \frac{\partial}{\partial S}\left[(K_0 - \alpha T)(l - l_0)\right]\bigg|_l$$

If we supply heat at constant l,

$$dQ = T\,dS = c_l\,dT, \qquad \frac{\partial T}{\partial S}\bigg|_l = \frac{T}{c_l}$$

Thus,

$$\frac{\partial T}{\partial l}\bigg|_S = -\alpha(l - l_0)\frac{T}{c_l}$$

$$\frac{1}{T}\frac{\partial T}{\partial l}\bigg|_S = \frac{\partial}{\partial l}\ln T\bigg|_S = -\alpha\frac{(l - l_0)}{c_l}$$

$$\ln\frac{T_2}{T_1} = -\frac{\alpha}{2c_l}\left[(l_2 - l_0)^2 - (l_1 - l_0)^2\right]$$

so that the final expression for T_2 is

$$T_2 = T_1 \exp\left(-\frac{\alpha}{2c_l}\left[(l_2 - l_0)^2 - (l_1 - l_0)^2\right]\right)$$

6.7 EQUILIBRIUM CONDITIONS

Consider a process occurring in a container at constant T, p, so that (6.29b) applies. Suppose that conditions are such that the Gibbs free energy G has its lowest possible value consistent with specified values of T and p. Then any change away from this state would produce $dG \geq 0$, and consequently $d\omega \leq 0$. In other words, a change can occur only if we do non $p\,dV$ work on the system (e.g., by sending in electrical energy). If we do not supply this extra energy, nothing will happen. Since the system will remain unchanged unless we stimulate it, we can say that it is at equilibrium. Thus, a sufficient condition that a system be in equilibrium at constant T and p is that G have the smallest value consistent with these values of T and p. Similarly, equilibrium at constant T and V corresponds to a minimum in F, and equilibrium under adiabatic conditions corresponds to minima in H (constant p) or U (constant V).

Suppose that two phases of the same substance are in equilibrium at some common temperature T and pressure p. Let G_1 and G_2 be their respective Gibbs free energies per mole. At equilibrium it must be that

$$G_1(T, p) = G_2(T, p) \tag{6.32a}$$

since if G_1 exceeded G_2, the total Gibbs free energy of the system could be reduced if material in phase 1 converted to phase 2, which would imply that we are not at the minimum of the total Gibbs free energy at this T and p.

Now suppose that T and p change slightly, but in a way that keeps the two phases in equilibrium. Then

$$G_1(T + dT, p + dp) = G_2(T + dT, p + dp) \tag{6.32b}$$

Subtracting (6.32a) from (6.32b) yields

$$\left.\frac{\partial G_1}{\partial T}\right|_p dT + \left.\frac{\partial G_1}{\partial p}\right|_T dp = \left.\frac{\partial G_2}{\partial T}\right|_p dT + \left.\frac{\partial G_2}{\partial p}\right|_T dp$$

$$\frac{dT}{dp} = \frac{\left.\dfrac{\partial G_2}{\partial p}\right|_T - \left.\dfrac{\partial G_1}{\partial p}\right|_T}{\left.\dfrac{\partial G_1}{\partial T}\right|_p - \left.\dfrac{\partial G_2}{\partial T}\right|_p} = \frac{V_2 - V_1}{S_2 - S_1} \tag{6.33}$$

where V_i and S_i are the volume and entropy of a mole of phase i.

Let l be the amount of heat that must be added to a mole of phase 1 to convert it to phase 2. If this heat is added reversibly at temperature T,

$$S_2 - S_1 = \frac{l}{T}$$

and (6.33) becomes

$$\frac{dT}{dp} = \frac{T(V_2 - V_1)}{l} \qquad \text{(phases 1 and 2 in equilibrium)} \tag{6.34}$$

This is the Clausius-Clapeyron equation.

If 1·represents liquid water and 2 represents water vapor, l and $V_2 - V_1$ are both positive, and so is dT/dp. For this reason the boiling temperature of water decreases as pressure decreases (it takes longer to boil an egg on a mountaintop).

If 1 represents ice and 2 represents liquid water, l is positive, but $V_2 - V_1$ is negative, since water expands when it freezes. Thus, $dT/dp < 0$, and the increased pressure under an ice-skate blade lowers the melting temperature. If this is below the ambient temperature, melting will occur, and the liquid produced lubricates the motion of the blade over the ice.

PROBLEM 6.7.1   Construct the Gibbs free energy for n moles of ideal gas, and find the equilibrium volume for specified temperature and pressure.

We use (6.28a) and the result of Problem 6.4.3 to write

$$G = U(T) - nTR \ln(VT^{3/2}) + pV$$

apart from an irrelevant added constant. To find the equilibrium volume, we must minimize G for fixed T and p,

$$0 = \left.\frac{\partial G}{\partial V}\right|_{T,p} = -\frac{nTR}{V} + p$$

$$pV = nRT$$

Thus, the equilibrium volume is the one that satisfies the ideal gas law.

PROBLEM 6.7.2   It is usually possible to assume that the molar volume of a vapor is much greater than the molar volume of the associated liquid or solid. Make the additional assumptions that the vapor obeys the ideal gas law and the heat of vaporization is approximately independent of temperature, and derive an expression for the equilibrium vapor pressure as a function of temperature.

We apply the Clausius-Clapeyron equation (6.34) to two phases, one of which is vapor and the other liquid or solid. If we neglect the liquid or solid volume, and assume that the vapor behaves like an ideal gas, we get

$$\frac{dT}{dp} = \frac{TV}{l} - \frac{T^2 R}{lp}$$

$$\frac{dp}{p} = d(\ln p) = \frac{l}{R}\frac{dT}{T^2}$$

Finally, if l is independent of T, we can integrate to get

$$\ln p = \ln p_0 - \frac{l}{RT} \qquad (p_0 = \text{constant})$$

$$p = p_0 \exp\left(-\frac{l}{RT}\right)$$

In many cases this gives a good representation of the dependence of equilibrium vapor pressure on temperature.

6.8 KINETIC THEORY OF GASES

6.8.1 Velocity Distribution

A simple but useful model of an ideal gas is a large number of molecules moving almost independently of each other in a container whose volume is much larger than the total volume of the molecules. "Almost independently" means that the energy of interaction between the molecules is small compared to their translational kinetic energy, yet this interaction is strong enough to keep the molecules in a state of dynamic equilibrium with one another. Thus, it is unlikely that one molecule would have a speed much greater than the others, since frequent collisions between the molecules lead to an almost continuous redistribution of the total kinetic energy of the gas.

Let ρ be the average number of molecules per unit volume (number density), which we assume to be constant throughout the container. We describe the distribution of molecular velocities by a function $f(\mathbf{v})$ such that $\rho f(\mathbf{v}_0)\, dv_x\, dv_y\, dv_z \equiv \rho f(\mathbf{v}_0)\, d^3v$ is the number of molecules per unit volume whose velocity $\mathbf{v}$ is within the range

$$(v_0)_x - \tfrac{1}{2}dv_x \le v_x \le (v_0)_x + \tfrac{1}{2}dv_x$$
$$(v_0)_y - \tfrac{1}{2}dv_y \le v_y \le (v_0)_y + \tfrac{1}{2}dv_y$$
$$(v_0)_z - \tfrac{1}{2}dv_z \le v_z \le (v_0)_z + \tfrac{1}{2}dv_z \tag{6.35}$$

The condition (6.35) can be described by saying that the velocity $\mathbf{v}$ is within a "velocity space" volume d^3v about $\mathbf{v}_0$. Since ρ is the total number of density, irrespective of velocity, $f(\mathbf{v})$ must be normalized so that

$$\int \rho f(\mathbf{v})\, d^3v = \rho$$
$$\int f(\mathbf{v})\, d^3v = 1 \tag{6.36}$$

Usually we can neglect the effect of external fields and of the detailed shape of the container.[1] Then all directions in space become equivalent, and we expect $f(\mathbf{v})$ to be a function of the magnitude of $\mathbf{v}$, but not of its direction. This makes it useful to introduce polar coordinates in velocity space, and to write the normalization condition (6.36) as

$$1 = \int f(v)\, d^3v = \int_{v=0}^{\infty}\int_{\theta_v=0}^{\pi}\int_{\phi_v=0}^{2\pi} f(v)v^2\, dv \sin\theta_v\, d\theta_v\, d\phi_v$$
$$= 4\pi \int_0^{\infty} f(v)v^2\, dv \tag{6.37}$$

The mean-square molecular speed can be written

$$\overline{(v^2)} = \int f(v)v^2\, d^3v = 4\pi \int f(v)v^4\, dv \tag{6.38a}$$

so that the average translational kinetic energy per molecule is

$$\overline{\mathrm{KE}}/\text{molecule} = \overline{\tfrac{1}{2}Mv^2} = 2\pi M \int_0^{\infty} f(v)v^4\, dv \tag{6.38b}$$

Since the molecules whose velocities satisfy (6.35) have number density $\rho f(\mathbf{v}_0)\, d^3v$ and average velocity $\mathbf{v}_0$, their flux density is $\rho f(\mathbf{v}_0)\, d^3v\, \mathbf{v}_0$. This means that the number of these particles crossing an infinitesimal surface area $d\mathbf{a}$ in unit time is $\rho f(\mathbf{v}_0)\, d^3v\, \mathbf{v}_0 \cdot d\mathbf{a}$. If $d\mathbf{a}$ represents a small area in the wall of the

---

[1] This will be true when the dimensions of the container are much larger than the mean free path.

FIGURE 6.4 $\mathbf{v}_0$ and $\mathbf{v}_0'$ are velocity vectors of a molecule that makes an elastic collision with surface element $d\mathbf{a}$ of the container wall.

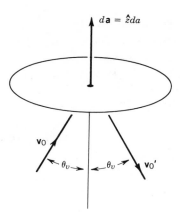

container parallel to the x–y plane ($d\mathbf{a} = \hat{z}\,da$), then the rate at which molecules satisfying (6.35) strike this area is

$$\rho f(\mathbf{v}_0)\,d^3v\,\mathbf{v}_0 \cdot \hat{z}\,da = \rho f(v_0)v_0\cos\theta_v\,da\,d^3v \qquad (\theta_v < \pi/2) \qquad (6.39a)$$

(see Figure 6.4). If we suppose that these molecules make elastic collisions with the wall, then the amount of momentum delivered to the wall by each molecule is $2Mv_0\cos\theta_v\hat{z}$. Then the rate at which the molecules satisfying (6.35) deliver momentum to $d\mathbf{a}$ is

$$\left[2M\rho f(v_0)v_0^2\cos^2\theta_v\,d^3v\right]\hat{z}\,da$$

Thus, the total force exerted on $d\mathbf{a}$ is

$$\left[2M\rho \int_0^\infty f(v_0)v_0^4\,dv_0 \int_{\theta_v=0}^\pi \cos^2\theta_v\sin\theta_v\,d\theta_v \int_{\phi_v=0}^{2\pi} d\phi_v\right]\hat{z}\,da$$

$$= \left[M\frac{4\pi}{3}\rho \int_0^\infty f(v_0)v_0^4\,dv_0\right]\hat{z}\,da$$

which implies that the upward pressure on $d\mathbf{a}$ is[2]

$$p = \frac{4\pi}{3}M\rho \int_0^\infty f(v_0)v_0^4\,dv_0 \qquad (6.39b)$$

Comparison with (6.38b) shows that

$$p = \tfrac{2}{3}\rho \cdot \overline{KE}/\text{molecule} \qquad (6.40)$$

If we now define the temperature T by

$$\overline{KE}/\text{molecule} = \tfrac{3}{2}kT \qquad (6.41)$$

then (6.40) becomes

$$p = \rho kT = \frac{nN_0}{V}kT = \frac{nRT}{V}$$

where n is the number of moles of gas. Thus, this simple model of an ideal gas,

---

[2] In Section 1.6 we also calculated the pressure that a fluid exerts on a wall by determining the rate at which the fluid delivers momentum to unit area of the wall.

together with the interpretation (6.41) of T as a measure of the average molecular kinetic energy, yields a derivation of the ideal gas law.

PROBLEM 6.8.1 Suppose that da in Figure 6.4 represents a small hole in the top of the container, and the exterior region is pumped to maintain a vacuum there. At what rate will molecules leave the container through this hole, and what will be their average kinetic energy?

We get the rate at which molecules will leave the container by integrating (6.39a) over all departing velocities

$$da \, \rho \int_{\theta_v < \pi/2} f(v) v \cos \theta_v \, d^3 v = da\rho \int_0^\infty f(v)v^3 dv \int_{\theta_v=0}^{\pi/2} \cos \theta_v \sin \theta_v \, d\theta_v \int_{\phi_v=0}^{2\pi} d\phi_v$$

$$= \left[\pi\rho \int_0^\infty f(v)v^3 \, dv \right] da \qquad (6.42a)$$

Their average kinetic energy is given by

$$(\text{KE})_{\text{avg}} = \frac{\displaystyle\int_0^\infty f(v)v^3 \frac{Mv^2}{2} \, dv}{\displaystyle\int_0^\infty f(v)v^3 \, dv} \qquad (6.42b)$$

So far no assumption has been made about the explicit form of $f(v)$. Maxwell assumed that

$$f(v) = \left(\frac{M}{2\pi kT} \right)^{3/2} e^{-Mv^2/2kT} \qquad (6.43)$$

Here M is the mass of a molecule, T is the absolute temperature, and k is Boltzmann's constant. This velocity distribution is consistent with (6.37), and with the Boltzmann probability distribution to be introduced in Section 6.9.

PROBLEM 6.8.2 Answer the questions of Problem 6.8.1 using the Maxwell distribution (6.43) for the velocity distribution of the molecules.

If we substitute (6.43) into (6.42a and b), we get[3] an outgoing particle flux of

$$da \cdot \pi\rho \left(\frac{M}{2\pi kT} \right)^{3/2} \int_0^\infty e^{-Mv^2/2kT} v^3 \, dv = da \, \pi\rho \left(\frac{M}{2\pi kT} \right)^{3/2} \frac{1}{2} \frac{\Gamma(2)}{\left(\dfrac{M}{2kT} \right)^2}$$

$$= da \cdot \rho \sqrt{\frac{kT}{2\pi M}} \qquad (6.42c)$$

---

[3] The formula

$$\int_0^\infty x^n e^{-\lambda x^2} \, dx = \frac{1}{2} \frac{\Gamma\left(\dfrac{n+1}{2} \right)}{\lambda^{\left(\frac{n+1}{2} \right)}}$$

will be found to be useful for the evaluation of averages over the Maxwell distribution. Here $\Gamma(x)$ is the gamma function, satisfying

$$\Gamma(x+1) = x\Gamma(x) \qquad \Gamma(1) = 1, \qquad \Gamma\left(\tfrac{1}{2}\right) = \sqrt{\pi}$$

and an average outgoing kinetic energy of

$$\frac{M}{2} \frac{\displaystyle\int_0^\infty e^{-Mv^2/2kT}v^5\,dv}{\displaystyle\int_0^\infty e^{-Mv^2/2kT}v^3\,dv} = \frac{M}{2}\frac{\Gamma(3)}{\Gamma(2)}\left(\frac{2kT}{M}\right)$$

$$= 2kT \qquad\qquad (6.42d)$$

The average kinetic energy within the container is $(3/2)kT$. Thus, the molecules that escape are, on the average, faster than those that stay behind.

PROBLEM 6.8.3   A space capsule whose volume is $10\ \text{m}^3$ contains air at atmospheric pressure and 20 C. It is suddenly punctured by a small meteorite, which leaves a $1\ \text{mm}^2$ hole in its side. How long will it take for half the air to leak out into the vacuum surrounding the capsule? (Assume a heat source to keep the contents at 20 C.)

If the capsule has volume V, it contains ρV molecules. According to (6.42c)

$$\frac{d}{dt}(\rho V) = V\frac{d\rho}{dt} = -da\sqrt{\frac{kT}{2\pi M}}\,\rho$$

The solution for $\rho(t)$ is

$$\rho(t) = \rho(0)e^{-da\sqrt{kT/2\pi M}\,t/V}$$

Thus, the density decreases to half its initial value in a time $t_{1/2}$ given by

$$t_{1/2} = \frac{\ln 2}{\left(\dfrac{da}{V}\sqrt{\dfrac{kT}{2\pi M}}\right)} = \frac{\ln 2}{\dfrac{10^{-2}\ \text{cm}^2}{10\times 10^6\ \text{cm}^3}\sqrt{\dfrac{1.38\times 10^{-16}\ \text{erg/K}\times 293\ \text{K}}{28\times 1.6\times 10^{-24}\ \text{g}\times 2\pi}}}$$

$$= 5.8\times 10^4\ \text{s} \approx 16\ \text{h}$$

PROBLEM 6.8.4   Figure 6.5 shows a container with a small hole of area da in one wall, with vacuum on the outside. At a distance R from the hole, at an angle θ_v to one side, is placed a particle detector with a sensitive area σ. At what rate will the detector record the arrival of particles?

The particles that leave the hole and strike the detector move in a cone of solid angle σ/R^2 whose axis is the line to the center of the detector. According to (6.39a), the rate at which such particles leave the hole with speed between v and $v + dv$ is

$$da\cdot\rho\cdot f(v)v^3\cos\theta_v\,dv\,d\Omega_v = da\cdot\rho f(v)v^3\,dv\cos\theta_v\frac{\sigma}{R^2}$$

Integrating over all speeds, we get

$$da\cdot\rho\frac{\sigma}{R^2}\cos\theta_v\int_0^\infty f(v)v^3\,dv = da\cdot\rho\cdot\frac{\sigma}{R^2}\cos\theta_v\sqrt{\frac{kT}{2\pi^3 M}}$$

If this is integrated over the entire exterior solid angle of 2π steradians, we recover (6.42c).

FIGURE 6.5 Molecules travel in straight-line paths to the detector outside the container.

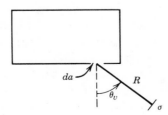

6.8.2 Mean Free Path

Suppose that a molecule within a gas has traversed a distance l since its last collision. What is the probability dp that it will make a collision in the next small interval dl? We assume that the molecular chaos is so great that there are no statistical correlations between successive collisions. Then dp will be independent of l, and we can write $dp = \gamma \, dl$, with γ constant. What does this imply about the distribution of path lengths between collisions? Let $F(l)$ be the fraction of paths that reach distance l without collision. Then $F(l) - F(l + dl)$ is the fraction that undergo collision between l and $l + dl$. Our assumption is that this is the fraction $\gamma \, dl$ of those that have reached l. Thus

$$F(l) - F(l + dl) = \gamma F(l) \, dl$$

$$\frac{dF(l)}{dl} = -\gamma F(l),$$

whose solution is

$$F(l) = F(0)e^{-\gamma l} = e^{-\gamma l} \qquad (6.44)$$

We see that short paths are more probable than long ones.

The probability of a collision between l and $l + dl$ is

$$\gamma F(l) \, dl = \gamma e^{-\gamma l} \, dl$$

The average intercollision path length, the *mean free path* λ, is given by

$$\lambda = \int_0^\infty l \gamma e^{-\gamma l} \, dl = \frac{1}{\gamma} \qquad (6.45)$$

To estimate γ, we ascribe to each molecule an effective collision cross-section area σ. As the molecule moves a distance dl through the container, this area sweeps out a volume $\sigma \, dl$. Since ρ is the molecular number density in the container, the probability that there is a molecule within volume $\sigma \, dl$ is $\rho \sigma \, dl$, and we take this to be the collision probability, $\gamma \, dl$. Thus

$$\gamma = \rho \sigma = \frac{1}{\lambda} \qquad (6.46)$$

The precise value of σ depends on the detailed interaction between the molecules, but we can get a rough estimate by taking it to be a few times larger than the molecular cross-sectional area. To make a numerical estimate, we can consider a gas at 1 atmosphere and 273 K,

$$\rho = \frac{6 \times 10^{23} \text{ molecules}}{22.4 \text{ l}} = 2.7 \times 10^{19} \text{ molecules/cm}^3$$

$$\sigma \simeq 4 \times 10^{-15} \text{ cm}^2, \qquad \gamma = \sigma \rho \approx 10^5 \text{ cm}^{-1}$$

and the mean free path is of the order of 10^{-5} cm. Thus, a molecule travels, on the average, about a thousand times the length of its diameter between collisions. For this reason the amount of internal energy associated with the molecular interactions is small compared to the translational kinetic energy of the gas.

The behavior of gas in a container depends on the ratio of the mean free path of the gas to the dimensions of the container. For example, Figure 6.6 shows a tube of small diameter D connecting two large chambers which are maintained, respectively at pressures P_1, P_2 and temperatures T_1, T_2. What will be the direction of gas flow through the tube?

Suppose first that $\lambda \ll D$. In that case, molecules in the tube are much more likely to collide with each other than with the tube. The gas behaves like a

FIGURE 6.6 Which way does the gas flow?

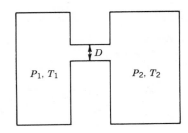

continuous fluid, and its motion is governed by the laws of classical fluid dynamics (Section 1–14e). If gravitational effects are small (as they usually are), the gas flow will be from the chamber at higher pressure toward the one at lower pressure.

Now suppose that $\lambda \gg D$. In this case a molecule is much more likely to collide with the wall of the tube than with another molecule. The molecules move through the tube essentially independently of one another. According to (6.42c), the flux from chamber 1 into the tube is proportional to $\rho_1 \sqrt{T_1} = P_1 / k\sqrt{T_1}$, and similarly, the flux from chamber 2 into the tube is proportional to $P_2 / k\sqrt{T_2}$. Thus, when $\lambda \gg D$, the direction of flow is out of the chamber with the larger value of $P/\sqrt{T}$, which is not necessarily the chamber with the higher pressure.

PROBLEM 6.8.5 The container shown in the figure has a volume $2V$ divided into two equal parts by a thin porous insulating wall with holes whose diameters are small compared to the mean free path of gas molecules. Into the container n moles of gas of molecular weight M are put, the left side being maintained at temperature T_1, and the right side at temperature T_2. Let A be the cumulative area of the holes in the wall. When equilibrium is established, what are the particle densities and pressures on each side of the wall?

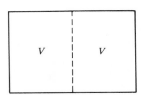

According to (6.42c), the particle flux through the holes from side i is proportional to $\rho_i \sqrt{T_i}$. At equilibrium, $\rho_1 \sqrt{T_1} = \rho_2 \sqrt{T_2}$. If there are n moles (nN_0 molecules) on both sides, then $\rho_1 V + \rho_2 V = nN_0$. These equations can be solved simultaneously to give

$$\rho_1 = \frac{\sqrt{T_2}}{\sqrt{T_1} + \sqrt{T_2}} \frac{nN_0}{V}, \qquad \rho_2 = \frac{\sqrt{T_1}}{\sqrt{T_1} + \sqrt{T_2}} \frac{nN_0}{V}$$

For an ideal gas, $P = (nN_0/V)kT = \rho kT$. Thus,

$$P_1 = \frac{kT_1 \sqrt{T_2}}{\sqrt{T_1} + \sqrt{T_2}} \frac{nN_0}{V}, \qquad P_2 = \frac{kT_2 \sqrt{T_1}}{\sqrt{T_1} + \sqrt{T_2}} \frac{nN_0}{V}$$

PROBLEM 6.8.6 The container shown in the figure on the next page has a volume $2V$ divided into two equal parts by a thin porous insulating wall with holes whose diameters are

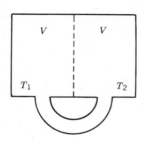

small compared to the mean free path of the gas molecules. Into the container N molecules of gas of molecular weight M are put, the left side being maintained at temperature T_1, and the right side at temperature $T_2 > T_1$. Let A be the cumulative area of the holes in the wall. A tube with a cross-section that is large compared to A connects the sides as shown. Neglecting the volume of the tube compared to V, and the viscosity of the gas, describe in detail the steady state properties of the system.

Because of the large tube connecting the two sides, the pressures on the two sides must be very nearly equal. Let N_1 and N_2 be the number of molecules on each side, and $N = N_1 + N_2$. Since $P_1 = N_1 k_1 T_1/V = P_2 = N_2 k_2 T_2/V$, it follows that

$$N_1 T_1 = N_2 T_2$$

$$N_1 = \frac{T_2}{T_1 + T_2} N, \qquad N_2 = \frac{T_1}{T_1 + T_2} N$$

According to (6.42c), the net flux of molecules across the membrane from side 1 to side 2 is

$$A\left[\frac{N_1}{V}\sqrt{\frac{kT_1}{2\pi M}} - \frac{N_2}{V}\sqrt{\frac{kT_2}{2\pi M}}\right] = \frac{AN}{V}\sqrt{\frac{k}{2\pi M}}\left[\frac{T_2\sqrt{T_1} - T_1\sqrt{T_2}}{T_1 + T_2}\right]$$

$$= \frac{AN}{V}\sqrt{\frac{kT_1 T_2}{2\pi M}}\left[\frac{\sqrt{T_2} - \sqrt{T_1}}{T_1 + T_2}\right]$$

which is greater than zero. Thus, there is a net flux across the membrane, from the cool side to the warm side. There is a return flow through the bottom tube. This flow is associated with very little pressure difference, since the connecting tube has a large cross-section.

PROBLEM 6.8.7 Most conduction electrons in a metal are kept from leaving the metal by an abrupt potential energy rise W_0 at the surface of the metal. Assume that the conduction electrons inside the metal have a Maxwellian distribution of velocity, and find the current density of thermionic emission from the metal at temperature T.

Only those electrons with $mv_x^2/2 > W_0$ and $v_x > 0$ will escape. Let ρ be the density of conduction electrons. The current density out of the metal is

$$e\rho \int_{v_x > \sqrt{2W_0/m}} \frac{\exp\left[-\dfrac{m}{2kT}\left(v_x^2 + v_y^2 + v_z^2\right)\right]}{(2\pi kT/m)^{3/2}} v_x \, dv_x \, dv_y \, dv_z$$

$$= e\rho \int_{\sqrt{2W_0/m}}^{\infty} \frac{\exp\left(-\dfrac{m}{2kT}v_x^2\right)}{(2\pi kT/m)^{1/2}} v_x \, dv_x = e\rho \sqrt{\frac{kT}{2m\pi}} e^{-W_0/kT}$$

PROBLEM 6.8.8   The gravitational acceleration at the moon's surface is 167 cm/s^2, and the radius of the moon is 1738 km. The mean temperature of the moon is 300 K.

(a) What molecular weight must a gas have so that a molecule with the rms velocity of a Maxwellian distribution cannot escape from the moon?

According to (6.41),

$$\frac{\overline{mv^2}}{2} = \frac{3}{2}kT, \qquad v_{rms} = \sqrt{\overline{v^2}} = \sqrt{\frac{3kT}{m}}$$

A molecule with the escape speed v_e has zero total energy. Thus, at the surface of the moon,

$$mv_e^2/2 - mMG/R = 0, \qquad v_e = \sqrt{\frac{2MG}{R}} = \sqrt{2gR}$$

Thus, if v_{rms} is to be less than v_e,

$$\sqrt{\frac{3kT}{m}} < \sqrt{2gR}, \qquad m > \frac{3kT}{2gR} \simeq 2.1 \times 10^{-24} \text{ g}$$

This corresponds to a molecular weight of

$$\frac{2.1 \times 10^{-24} \text{ g}}{1.6 \times 10^{-24} \text{ g/u}} \simeq 1.3 \text{ u} \qquad \text{(atomic mass unit)}$$

(b) Why has the moon no apparent atmosphere? Relate your answer to the result obtained in part (a).

A molecule like N_2 has a molecular weight of approximately 28 u, so its rms speed is less than the escape speed. However, the Maxwell distribution (6.43) has a very long tail, and enough molecules had speeds greater than v_e so that essentially all the gas escaped during the lifetime of the moon.

PROBLEM 6.8.9   Two flat plates are separated by a distance a, and are at two slightly different temperatures T_c and T_w (c for cooler, w for warmer). Helium gas at pressure p, such that the mean free path greatly exceeds a, fills the space between the plates. Assume that the velocity distribution of the molecules that have bounced off a surface is a Maxwell distribution corresponding to the temperature of that surface. Find the thermal conductivity due to the helium gas.

The thermal conductivity is the net energy flux divided by the temperature gradient. If ρ_w is the number density of molecules that have bounced off the warmer surface, the kinetic energy flux of these molecules is

$$\int dv_y\, dv_z \int_{v_x > 0} dv_x\, 2f(v_{x_1} v_{y_1} v_z)\rho_w v_x\left(\frac{m}{2}\left\{v_x^2 + v_y^2 + v_z^2\right\}\right)$$

$$= \rho_w\sqrt{\frac{(2kT_w)^3}{m\pi}}$$

We have used the Maxwell distribution (6.43) for $f(v_x, v_y, v_z)$ with an extra factor of 2 so that the normalization condition (6.36) is satisfied for $v_x > 0$ only. The net energy flux from the warm to the cold plate is

$$\sqrt{\frac{(2k)^3}{m\pi}}\left[\rho_w T_w^{3/2} - \rho_c T_c^{3/2}\right]$$

To obtain ρ_w and ρ_c, we use that fact that the net particle flux is zero:

$$\rho_w \sqrt{T_w} = \rho_c \sqrt{T_c}, \qquad \rho_w + \rho_c = \rho = \frac{p}{kT} \qquad (T \simeq T_c \simeq T_w),$$

$$\rho_w = \frac{\sqrt{T_c}}{\sqrt{T_w} + \sqrt{T_c}} \frac{p}{kT}, \qquad \rho_c = \frac{\sqrt{T_w}}{\sqrt{T_w} + \sqrt{T_c}} \frac{p}{kT},$$

and the energy flux is approximately

$$\sqrt{\frac{(2k)^3}{m\pi}} \frac{p}{kT} \frac{\sqrt{T_c}\, T_w^{3/2} - \sqrt{T_w}\, T_c^{3/2}}{2\sqrt{T}} = \sqrt{\frac{2k}{m\pi T}} [T_w - T_c]$$

Since the temperature gradient is $(T_w - T_c)/a$, the thermal conductivity due to the helium gas is $a\sqrt{2k/m\pi T}$.

6.9 STATISTICAL MECHANICS

Most of the measurements we make on a thermodynamic system tell us about averages of the physical properties of many molecules. For example, temperature is a measure of the average translational kinetic energy. Pressure is a measure of the average rate at which the molecules deliver momentum to unit area of wall. Specific heat is a measure of the average way the molecules respond to addition of energy, etc. The subject of statistical mechanics attempts to calculate such averages from the underlying microscopic structure of the system. Note that we do not attempt to find a detailed solution of the equations of motion, corresponding to some completely specified initial conditions. This would clearly be unattainable for a complicated system such as a gas of 10^{23} molecules. Fortunately, we do not need to know the detailed motion of each molecule in order to draw conclusions about the gross averages that are measured by our macroscopic experiments.

6.9.1 Classical Statistical Mechanics: Boltzmann Distribution

In classical mechanics, the instantaneous state of a system with d degrees of freedom is specified by giving the values of d generalized coordinates $q_1, \ldots, q_d$ and d generalized momenta $p_1, \ldots, p_d$. It is useful to regard the variables $q_1, \ldots, q_d$, $p_1, \ldots, p_d$ as the coordinates of a point in a $2d$-dimensional phase space (cf. the discussion on p. 52). In statistical mechanics, we visualize a large number of identical copies of our physical system. This collection of copies is called an *ensemble*. At any instant, the state of each system in the ensemble can be specified by locating its representative point in phase space. As time evolves, the collection of representative points move along trajectories determined by Hamilton's equations (1.48). If the ensemble is governed by the Boltzmann distribution, the probability that a representative phase point is within a volume $dq_1, \ldots, dp_d$ around $q_1, \ldots, p_d$ is given by

$$P(q_1, \ldots, p_d) \, dq_1, \ldots, dp_d = \frac{\exp(-H(q_1, \ldots, p_d)/kT) \, dq_1 \ldots dp_d}{\int \exp(-H(q_1', \ldots, p_d')/kT) \, dq_1' \ldots dp_d'} \qquad (6.47)$$

Here H is the Hamiltonian of the system. The denominator in (6.47)

$$Z = \int \exp\left(-\frac{H(q'_1, \ldots, p'_d)}{kT} \right) dq'_1 \ldots dp'_d \tag{6.48}$$

is called the *partition function*. It ensures that the total probability is unity:

$$\int P(q_1, \ldots, p_d) \, dq_1 \ldots dp_d = 1$$

The principle of equipartition is a useful application of the Boltzmann distribution. Suppose that

$$H(q_1, \ldots, p_d) = aq_1^2 + H'(q_2, \ldots, p_d)$$

where a is constant and H' does not depend on q_1. Then $\overline{aq_1^2}$, the average value of the aq_1^2 contribution to the energy, is given by

$$\overline{aq_1^2} = \int aq_1^2 P(q_1, \ldots, p_d) \, dq_1 \ldots dp_d$$

$$= \frac{\int aq_1^2 \exp(-H(q_1, \ldots, p_d)/kT) \, dq_1 \ldots dp_d}{\int \exp(-H(q_1, \ldots, p_d)/kT) \, dq_1 \ldots dp_d}$$

$$= \int_{-\infty}^{\infty} aq_1^2 e^{-aq_1^2/kT} \, dq_1 \bigg/ \int_{-\infty}^{\infty} e^{-aq_1^2/kT} \, dq_1$$

The numerator and denominator can be evaluated with the help of the integral given in the footnote on p. 282. Thus,

$$\overline{aq_1^2} = a\frac{\Gamma(3/2)}{\Gamma(1/2)} \frac{(a/kT)^{1/2}}{(a/kT)^{3/2}} = \frac{1}{2}\frac{\Gamma(1/2)}{\Gamma(1/2)} kT = \frac{1}{2}kT$$

We see that every generalized coordinate or momentum that contributes quadratically to the Hamiltonian has an average value of $(1/2)kT$. This is the equipartition principle.

Let us apply this result to a gas of polyatomic molecules in a container. The jth molecule contributes a translational kinetic energy

$$\frac{p_{jx}^2 + p_{jy}^2 + p_{jz}^2}{2M}$$

to the Hamiltonian. The principle of equipartition tells us that its average value is $3kT/2$. Thus, the average translational kinetic energy of a mole of gas is $3N_0 kT/2 = 3RT/2$. For an ideal monatomic gas, this represents all the energy of the gas.

Now suppose that the gas consists of finite-size molecules that can rotate as well as translate. The rotational kinetic energy of each molecule contributes three quadratic terms to the Hamiltonian (for the three orientation degrees of freedom), so the average rotational energy of a mole of gas is also $3RT/2$. If each molecule vibrates harmonically about its equilibrium configuration with n normal modes (Section 1.10), this vibrational motion contributes $2n$ quadratic terms to the Hamiltonian (a potential and a kinetic energy term for each mode). Then the average vibrational energy for each mole of gas would be $(2n)RT/2 = nRT$.

FIGURE 6.7 The path of the light beam.

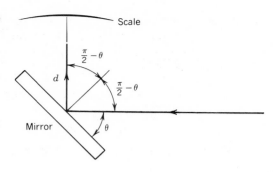

The total energy of a mole of gas would be

$$U = \tfrac{3}{2}RT + \tfrac{3}{2}RT + nRT = (3 + n)RT$$

and the specific heat[4] per mole would be $(3 + n)R$.

PROBLEM 6.9.1   A very light torsion pendulum consisting of a thin wire and a tiny mirror is suspended inside a glass-walled vessel containing gas at an absolute temperature of $T = 300$ K. A light beam is reflected by the mirror onto a scale at a distance $d = 5$ cm from the mirror. The motion of the light spot on the scale indicates vibration of the mirror. What is the root-mean square amplitude of the vibration of the light spot on the scale? The moment of inertia of the pendulum is 10^{-3} g cm^2, and its period is 4 s.

    The potential energy associated with the twisting wire is $\kappa\theta^2/2$, where κ is the torsion constant (restoring torque/angle of twist). Thus, $\overline{\kappa\theta^2/2} = kT/2$ and $\overline{\theta^2} = kT/\kappa$. It is clear from Figure 6.7 that a change $\delta\theta$ in mirror angle will produce a change $2d\,\delta\theta$ in the location of the light spot on the scale. Thus, the root-mean-square amplitude for the light spot vibration is $2d\sqrt{kT/\kappa}$. To determine κ, we use the information given about the period of oscillation of the mirror:

$$\tau = 4\text{ s} = 2\pi/\omega = 2\pi\sqrt{I/\kappa} = 2\pi\sqrt{\dfrac{10^{-3}\text{ g cm}^2}{\kappa}}$$

$$\kappa = 2.47 \times 10^{-3}\text{ g cm}^2/\text{s}^2$$

from which we calculate that

$$\text{rms amplitude} = 2\,(5\text{ cm}) \cdot \sqrt{\dfrac{1.38 \times 10^{-16}\,\dfrac{\text{erg}}{\text{K}} \times 300\text{ K}}{2.47 \times 10^{-3}\,\dfrac{\text{g cm}^2}{\text{s}^2}}} = 4.1 \times 10^{-5}\text{ cm}$$

PROBLEM 6.9.2   Consider a classical gas of N point molecules moving in a three-dimensional harmonic oscillator potential well, $V(r) = \tfrac{1}{2}Kr^2 = \tfrac{1}{2}K(x^2 + y^2 + z^2)$, at absolute temperature T. Obtain a formula for the probability that the molecule is between r and $r + dr$ from the center of attraction. Obtain a formula for the

---

[4] This is the specific heat at constant volume, since our model refers to a gas moving in a container with fixed walls.

mean square distance of the particle from the center of attraction, and check your result by comparison with the equipartition principle.

According to the Boltzmann distribution,

$$P(r, \theta, \phi)\, dv = \frac{e^{-(1/2)Kr^2/kT}\, dv}{\int e^{-(1/2)Kr^2/kT}\, dv}$$

$$= \frac{e^{-(Kr^2/2kT)}r^2 \sin\theta\, d\theta\, d\phi\, dr}{\int e^{-(Kr^2/2kT)}r^2\, dr \int \sin\theta\, d\theta\, d\phi}$$

$$= \frac{\sin\theta\, d\theta\, d\phi}{4\pi}\, \frac{e^{-Kr^2/2kT}r^2\, dr}{\Gamma(3/2) \Big/ \left(\dfrac{K}{2kT}\right)^{3/2}}$$

Thus, the probability that the molecule is between r and $r + dr$ (irrespective of θ, ϕ) is

$$\left(\frac{K}{2kT}\right)^{3/2}\frac{1}{\Gamma(3/2)}e^{-(Kr^2/2kT)}r^2\, dr = \left(\frac{K}{2kT}\right)^{3/2}\frac{1}{\sqrt{\pi}/2}e^{-(Kr^2/2kT)}r^2\, dr$$

The mean value of r^2 is

$$\overline{r^2} = \left(\frac{K}{2kT}\right)^{3/2}\frac{1}{\Gamma(3/2)}\int_0^\infty e^{-(Kr^2/2kt)}r^4\, dr$$

$$= \left(\frac{K}{2kT}\right)^{3/2}\frac{1}{\Gamma(3/2)}\frac{\Gamma(5/2)}{\left(\dfrac{K}{2kT}\right)^{5/2}} = \frac{3}{2}\cdot\frac{2kT}{K} = \frac{3kT}{K}$$

According to equipartition,

$$\overline{\frac{1}{2}K(x^2 + y^2 + z^2)} = 3\cdot\frac{1}{2}kT, \qquad \text{so} \quad \overline{r^2} = \frac{3kT}{K}$$

PROBLEM 6.9.3 Consider an ensemble of systems that have only two energy levels. Sketch the average energy as a function of temperature, and on the same graph plot the specific heat as a function of temperature.

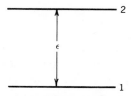

Let the energy difference between the two levels be ε. The occupation probabilities of the levels are

$$p_2 = \frac{e^{-\varepsilon/kT}}{1 + e^{-\varepsilon/kT}}$$

$$p_1 = \frac{1}{1 + e^{-\varepsilon/kT}}.$$

FIGURE 6.8 Temperature dependence of $\bar{\varepsilon}$ and c.

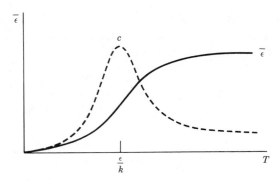

The average energy is

$$0 \cdot p_1 + \varepsilon \cdot p_2 = \frac{\varepsilon e^{-\varepsilon/kT}}{1 + e^{-\varepsilon/kT}} = \bar{\varepsilon} = \frac{\varepsilon}{e^{\varepsilon/kT} + 1} \left\{ \begin{array}{l} \xrightarrow[T \to \infty]{} \quad \dfrac{\varepsilon}{2} \\[2mm] \xrightarrow[T \to 0]{} \quad \varepsilon e^{-\varepsilon/kT} \to 0 \end{array} \right.$$

The specific heat per system will be given by

$$\frac{d\bar{\varepsilon}}{dT} = k \left[\frac{\varepsilon/kT}{e^{\varepsilon/kT} + 1} \right]^2$$

Figure 6.8 shows the T dependence of the average internal energy and the specific heat. If the two-level systems are monatomic gas molecules, then we should augment c by adding the $3k/2$ per molecule due to the translational kinetic energy. In fact this provides a very good description of c_v for atomic oxygen, which has two excited levels, at .020 eV and .028 eV, respectively, above which there is an energy gap with no levels until we reach 12.5 eV. For $kT \ll .020$ eV or $kT \gg .028$ eV, the value of c_v is approximately $3/2\ R$ per mole, but c_v shows a maximum when kT is in the vicinity of .020–.028 eV. To calculate the exact T dependence of c_v in this region, you must take account of the multiplicities of these lowest three oxygen levels (5, 3, and 1, respectively). The levels at or above 12.5 eV will not contribute to c_v until T reaches the vicinity of 100,000 K.

PROBLEM 6.9.4   Consider a mole of a paramagnetic ideal gas whose molecules each have permanent magnetic dipole moment $\boldsymbol{\mu}$. The gas is in a container of volume V, in a uniform constant magnetic field $\mathbf{B}$, and at temperature T. Find the average magnetization density $\mathcal{M}$ (magnetic moment/unit volume) and the low-field magnetic susceptibility. Neglect all interactions (even dipole-dipole interactions) between the molecules.

The degrees of freedom of the problem are the positions $\mathbf{r}_i$ of each molecule, and the polar coordinates (θ_i, ϕ_i) specifying the orientation of its dipole moment. The energy of a dipole $\boldsymbol{\mu}$ in a uniform field $\mathbf{B}$ can be taken to be $-\boldsymbol{\mu} \cdot \mathbf{B}$ (cf. 3.57c). If we choose our $\hat{z}$ direction to coincide with the direction of $\mathbf{B}$, then the magnetic contribution to the energy of molecule i is $-\mu B \cos \theta_i$. The normalized

Boltzmann distribution (6.47) reduces to

$$P\left(\mathbf{r}_1, \ldots, \mathbf{r}_{N_0}, \mathbf{p}_1, \ldots, \mathbf{p}_{N_0}, \theta_1, \phi_1, \ldots, \theta_{N_0}\phi_{N_0}\right)$$

$$= \frac{e^{-\sum_i [(\mathbf{p}_i)^2/2m - \mu B \cos\theta_i]/kt}}{\int e^{-\sum_i [(\mathbf{p}_i)^2/2m - \mu B \cos\theta_i]/kT} \prod_i d^3r_i \, d^3p_i \, d(\cos\theta_i) \, d\phi_i}$$

Because the Boltzmann factor is independent of ϕ_i, the average values of $(\mu_i)_x$ and $(\mu_i)_y$ are zero. The average value of $(\mu_i)_z = \mu \cos\theta_i$ is

$$\overline{\mu \cos\theta_i} = \frac{\displaystyle\int_{-1}^{1} e^{(\mu B/kT)(\cos\theta_i)}\mu \cos\theta_i \, d\cos\theta_i}{\displaystyle\int_{-1}^{1} e^{(\mu B/kT)(\cos\theta_i)}d\cos\theta_i}$$

$$= \mu\left[\frac{e^{\mu B/kT} + e^{-\mu B/kT}}{e^{\mu B/kT} - e^{-\mu B/kT}} - \frac{kT}{\mu B}\right] \tag{6.49a}$$

$$= \mu\left[\coth\frac{\mu B}{kT} - \frac{kT}{\mu B}\right] \tag{6.49b}$$

Since there are N_0 molecules in volume V, the magnetization density $\mathscr{M}$ is

$$\frac{N_0}{V}\mu\left[\coth\frac{\mu B}{kT} - \frac{kT}{\mu B}\right]\hat{z} \tag{6.50}$$

We can get the high- and low-field limits of (6.50) by using the properties of the exponentials in (6.49a). The results are

$$\lim_{\mu B/kT \to \infty} \mathscr{M} = \mu\frac{N_0}{V} \tag{6.51a}$$

$$\lim_{\mu B/kT \to 0} \mathscr{M} = \frac{N_0\mu}{V} \cdot \frac{1}{3}\frac{\mu B}{kT} = \frac{1}{3}\frac{N_0}{V} \cdot \frac{\mu^2 B}{kT} \tag{6.51b}$$

Equation (6.51a) represents a saturation condition, in which all the dipoles are aligned parallel to the external field. The low-field magnetic susceptibility, $\mathscr{M}/B$, is given by (6.51b):

$$\chi = \frac{1}{3}\frac{N_0}{V} \cdot \frac{\mu^2}{kT} \tag{6.52}$$

The proportionality of χ and $1/T$ is referred to as Curie's law.

PROBLEM 6.9.5 The total (electron plus proton) angular momentum quantum number for the hydrogen atom can take on the values $F = 0, 1$. The "hyperfine" interaction between the electron and proton raises the energy of the $F = 1$ level above that of the $F = 0$ level. Transitions between these levels give rise to the famous "21 cm line." At what temperature of an atomic hydrogen gas cloud will the three $F = 1$ states have a total population equal to that of the $F = 0$ ground state?

We require

$$3e^{-E_1/kT} = e^{-E_0/kT} \qquad e^{-(E_1-E_0)/kT} = 1/3, \qquad \frac{E_1 - E_0}{kT} = \ln 3$$

$$T = \frac{E_1 - E_0}{k \ln 3} = \frac{hc}{\lambda k \ln 3}$$

$$= \frac{12{,}400 \text{ eV Å} \cdot 1.6 \times 10^{-12} \text{ erg/eV}}{\ln 3 \times 21 \text{ cm} \times 1.28 \times 10^{-16} \text{ erg s} \times 10^8 \text{ Å/cm}} = .062 \text{ K}$$

PROBLEM 6.9.6 As a model of an atom that can be ionized, consider an electron that can exist either in a single bound state with energy -3 eV or in a "continuum" of states which are those of an electron confined in a box whose volume is 5 cm^3. Find an expression that determines the temperature to which the system must be heated in order for the probability of ionization to reach $\frac{1}{2}$, and give an order of magnitude estimate of this temperature.

The ionization probability (the probability that the electron is in the continuum) is

$$\frac{\displaystyle\int_0^\infty \frac{dN}{d\varepsilon} e^{-\varepsilon/kT} \, dE}{e^{3 \text{ eV}/kT} + \displaystyle\int_0^\infty \frac{dN}{d\varepsilon} e^{-\varepsilon/kT} \, d\varepsilon}$$

The density of states, $dN/d\varepsilon$, appropriate to nonrelativistic electrons in a box of volume V, is given by (5.93b). The integral we need is

$$\frac{V}{2\pi^2} \left(\frac{2m}{\hbar^2} \right)^{3/2} \int_0^\infty e^{-\varepsilon/kT} \varepsilon^{1/2} \, d\varepsilon = \frac{V}{2\pi^2} \left(\frac{2mkT}{\hbar^2} \right)^{3/2} \int_0^\infty e^{-x} x^{1/2} \, dx$$

$$= \frac{V}{2\pi^2} \left(\frac{2mkT}{\hbar^2} \right)^{3/2} \Gamma(3/2) = \frac{V}{4} \left(\frac{2mkT}{\pi\hbar^2} \right)^{3/2}$$

Thus, the ionization probability is

$$\frac{\dfrac{V}{4} \left(\dfrac{2mkT}{\pi\hbar^2} \right)^{3/2}}{e^{3 \text{ eV}/kT} + \dfrac{V}{4} \left(\dfrac{2mkT}{\pi\hbar^2} \right)^{3/2}}$$

which equals $1/2$ when

$$\frac{3 \text{ eV}}{kT} = \ln\left[\frac{V}{4} \left(\frac{2mkT}{\pi\hbar^2} \right)^{3/2} \right]$$

If we use $V = 5$ cm^3, the solution of this equation is $kT \simeq 1/14$ eV. This corresponds to a temperature of

$$T = 290 \text{ K} \times \tfrac{1}{14} \text{ eV} / \tfrac{1}{40} \text{ eV} \simeq 830 \text{ K}$$

PROBLEM 6.9.7 If a degree of freedom q_i contributes a term aq_i^2 to the Hamiltonian, the principle of equipartition says that its average contribution to the energy is $\frac{1}{2}kT$. Calculate the rms deviation of aq_i^2 about this average.

We need

$$\sigma = \sqrt{\overline{\left(aq_i^2 - \tfrac{1}{2}kT\right)^2}} = \sqrt{\overline{\left(aq_i^2\right)^2} - \left(\tfrac{1}{2}kT\right)^2}$$

$$\overline{\left(aq_i^2\right)^2} = \frac{\displaystyle\int_{-\infty}^{\infty} \left(aq_i^2\right)^2 e^{-aq_i^2/kT}\,dq_i}{\displaystyle\int_{-\infty}^{\infty} e^{-aq_i^2/kT}\,dq_i} = \frac{3}{4}\left(kT\right)^2$$

Thus,

$$\sigma_i = \sqrt{\frac{3}{4}\left(kT\right)^2 - \left(\frac{1}{2}kT\right)^2} = \frac{1}{\sqrt{2}}\,kT$$

It is easy to show that if the Hamiltonian contains N such terms $\sum_{i=1}^{N} a_i q_i^2$, with average energy $\overline{E} = N \times 1/2\,kT$, the rms deviation about this average is

$$\sigma = \sqrt{\frac{N}{2}}\,kT = \sqrt{\frac{2}{N}}\,\overline{E}$$

This implies that in a system with many degrees of freedom, such as a macroscopic collection of molecules, the average energy fluctuations about the mean energy are a very small fraction of the mean energy.

6.9.2 Connection Between Statistical Mechanics and Thermodynamics

The main link between these two subjects is provided by the equation

$$F = -kT \ln Z \tag{6.53}$$

where F is the Helmholtz free energy (6.26a), and Z is the partition function (6.48). Once we know the microscopic structure of our system, we can, in principle, calculate H and Z. Equation (6.53) then gives us F, from which we can calculate the thermodynamic properties of the system. For example, consider one mole of ideal gas in a container of volume V,

$$H = \sum_{j=1}^{N} \frac{1}{2M}\left(p_{jx}^2 + p_{jy}^2 + p_{jz}^2\right) \tag{6.54}$$

$$Z = \int \exp\left(-\sum_{j=1}^{N}\left(p_{jx}^2 + p_{jy}^2 + p_{jz}^2\right)/2MkT\right)\,dx_1\ldots dp_{Nz}$$

$$= \left[\int dx\,dy\,dz\right]^{N_0}\left[\int \exp\left(-\left\{p_x^2 + p_y^2 + p_z^2\right\}/2MkT\right)\,dp_x\,dp_y\,dp_z\right]^{N_0}$$

$$= V^{N_0}\left[\frac{\Gamma(1/2)}{1/\sqrt{2MkT}}\right]^{3N_0} = \left[V(2\pi MkT)^{3/2}\right]^{N_0}$$

$$F = kT \ln Z = -N_0 kT \ln V - \tfrac{3}{2}N_0 T \ln(2\pi MkT)$$

Using the properties of F derived on p. 275, we calculate

$$p = -\frac{\partial F}{\partial V}\bigg|_T = \frac{N_0 kT}{V} = \frac{RT}{V}, \qquad \text{the ideal gas law}$$

$$S = -\frac{\partial F}{\partial T}\bigg|_V = R\left[\ln V + \ln T^{3/2}\right] + \frac{3}{2}R\left[1 + \ln(2\pi Mk)\right]$$

$$S(B) - S(A) = R\ln\left[\frac{V_B}{V_A}\left(\frac{T_B}{T_A}\right)^{3/2}\right]$$

in agreement with the result obtained in Problem 6.4.3. Finally,

$$U = F + TS = kT\ln Z + T\frac{\partial}{\partial T}(kT\ln Z)$$

$$= kT^2\frac{\partial}{\partial T}\ln Z = kT^2 \cdot \frac{3}{2}\frac{N_0}{T} = \frac{3}{2}RT$$

Thus, we have obtained the thermodynamic properties of the gas, starting with a microscopic mechanical model (6.54).

6.9.3 Quantum Statistical Mechanics

The Heisenberg uncertainty principle implies that we cannot measure the values of a generalized coordinate and its conjugate momentum simultaneously with unlimited accuracy. Thus, it is meaningless to assert that our system is at the phase point $q_1, \ldots, p_d$, and the classical concept of phase space loses its usefulness. Instead we speak of the quantum states of the system, and if the system has d degrees of freedom, the quantum states are labeled by sets of d quantum numbers $\nu_1, \nu_2, \ldots, \nu_d$. Usually it is possible, and convenient, to modify the problem so that these quantum numbers take on only integral values. Then the integral over classical phase space is here replaced by a sum over all possible sets of quantum numbers. In order to make this correspondence dimensionally correct, it is useful to associate a phase space volume of h^d with each quantum state, where h is Planck's constant. We saw in Section 5.5 that this association gives the correct formula for the density of states of a free particle.

The Boltzmann distribution implies that the probability $P(\nu_1, \ldots, \nu_d)$ that the system is in the state labeled by quantum numbers $\nu_1, \ldots, \nu_d$ is given by

$$P(\nu_1, \ldots, \nu_d) = \frac{1}{Z}\exp\left(-E(\nu_1, \ldots, \nu_d)/kT\right) \qquad (6.55)$$

where the partition function Z is given by

$$Z = \sum_{\nu_1, \ldots, \nu_d} \exp\left(-E(\nu_1, \ldots, \nu_d)/kT\right) \qquad (6.56)$$

Here $E(\nu_1, \ldots, \nu_d)$ is the eigenvalue of H for the state with quantum numbers $\nu, \ldots, \nu_d$.

The thermodynamic properties of the system can still be obtained by using (6.53) to calculate the Helmholtz free energy from the partition function. However, the sum over quantum states in (6.56) must include only distinct, allowable states. The allowed values of $\nu_1, \ldots, \nu_d$ depend on whether we are dealing with a system of fermions, bosons, or distinguishable particles. For example, suppose that we consider a system of two particles, each of which can be in one of three single-particle states a, b, or c, with single-particle energies ε_a, ε_b, ε_c. A state of the two-particle system is specified by giving the single-particle state occupied by each particle.[5] We have the following 9 possibilities:

particle 1	a	a	b	b	a	c	c	b	c
particle 2	a	b	a	b	c	a	c	c	b
energy	$2\varepsilon_a$	$\varepsilon_a + \varepsilon_b$		$2\varepsilon_b$	$\varepsilon_a + \varepsilon_c$		$2\varepsilon_c$	$\varepsilon_b + \varepsilon_c$	

If the particles are distinguishable, each of these is a distinct two-particle state. The sum for Z has 9 terms, and equals

$$Z = e^{-2\varepsilon_a/kT} + 2e^{-(\varepsilon_a+\varepsilon_b)/kT} + e^{-2\varepsilon_b/kT} + 2e^{-(\varepsilon_a+\varepsilon_c)/kT}$$
$$+ e^{-2\varepsilon_c/kT} + 2e^{-(\varepsilon_b+\varepsilon_c)/kT}$$
$$= \left(e^{-\varepsilon_a/kT} + e^{-\varepsilon_b/kT} + e^{-\varepsilon_c/kT} \right)^2$$

If the particles are identical bosons, the second and third states in the above list are indistinguishable. All we can say is that the single-particle states a and b are occupied, but we must not distinguish between the different ways of distributing the particles among these states. Now Z has only 6 terms:

$$Z = e^{-2\varepsilon_a/kT} + e^{-(\varepsilon_a+\varepsilon_b)/kT} + e^{-2\varepsilon_b/kT} + e^{-(\varepsilon_a+\varepsilon_c)/kT}$$
$$+ e^{-2\varepsilon_c/kT} + e^{-(\varepsilon_b+\varepsilon_c)/kT}$$
$$= \tfrac{1}{2}\left[\left(e^{-\varepsilon_a/kT} + e^{-\varepsilon_b/kT} + e^{-\varepsilon_c/kT} \right)^2 + \left(e^{-2\varepsilon_a/kT} + e^{-2\varepsilon_b/kT} + e^{-2\varepsilon_c/kT} \right) \right]$$

If the particles are identical fermions, we again do not distinguish between different distributions of the particles over the same states. But now we must also include the effect of the Pauli exclusion principle, which disallows any two-particle state in which more than one particle occupies the same single-particle state. The sum for Z now has only 3 terms:

$$Z = e^{-(\varepsilon_a+\varepsilon_b)/kT} + e^{-(\varepsilon_a+\varepsilon_c)/kT} + e^{-(\varepsilon_b+\varepsilon_c)/kT}$$
$$= \tfrac{1}{2}\left[\left(e^{-\varepsilon_a/kT} + e^{-\varepsilon_b/kT} + e^{-\varepsilon_c/kT} \right)^2 - \left(e^{-2\varepsilon_a/kT} + e^{-2\varepsilon_b/kT} + e^{-2\varepsilon_c/kT} \right) \right]$$

6.10 BOSONS AND FERMIONS

In general, exact evaluation of Z for fermions or bosons is difficult. However, many of the important results can be obtained by using the following rule:

In a gas of weakly interacting indistinguishable particles, the average number of particles in the single-particle state α is

$$n_\alpha = \frac{1}{e^{(\varepsilon_\alpha - \mu)/kT} \pm 1} \qquad \left\{ \begin{array}{l} \text{for bosons} - \\ \text{for fermions} + \end{array} \right\} \qquad (6.57)$$

Here ε_α is the energy of the single-particle state α, and μ is the "chemical potential." It is obtained from the requirement that the total number of particles have some specified value, N,

$$N = \sum_\alpha n_\alpha = \sum_\alpha \frac{1}{e^{(\varepsilon_\alpha - \mu)/kT} \pm 1} \qquad (6.58)$$

---

[5] This assumes that the two particles interact weakly (see the footnote on page 223).

For particles of zero rest mass (photons, neutrinos), $\mu = 0$. Equation (6.57) reduces to the Boltzmann distribution in situations in which most of the particles are in states for which $(\varepsilon_\alpha - \mu)/kT \gg 1$.

6.10.1 Electromagnetic Radiation in a Cavity

The state of the radiation in the cavity is determined by specifying the number of quanta of excitation (photons) for each cavity mode. If there are n quanta in a mode of frequency ν, the contribution to the energy of the system is $nh\nu$. The situation is formally identical to that for a gas of independent bosons. The single-particle states for each boson are the cavity modes, and the single-particle energy for a mode of frequency ν is $h\nu$. According to (6.57), the average number of photons in a mode of frequency ν is

$$\frac{1}{e^{h\nu/kT} - 1}$$

and since each photon has energy $h\nu$, the average energy in this mode is

$$\frac{h\nu}{e^{h\nu/kT} - 1}$$

The number of such modes with frequency between ν and $\nu + d\nu$ is $8\pi V\nu^2\, d\nu/c^2$ if V is the volume of the cavity (cf. 5.93c). Thus, the average energy per unit volume in modes with frequency between ν and $\nu + d\nu$ is

$$u_\nu = \frac{8\pi h\nu^3}{c^3\left[e^{h\nu/kT} - 1\right]}\, d\nu \qquad \text{(Planck's law)} \qquad (6.59)$$

6.10.2 Debye Theory of the Specific Heat of Crystals

We first analyze the vibrational motion of a crystal into its normal modes. The number of vibrational modes per unit frequency interval is given by the same formula as for the electromagnetic radiation in a cavity, except that there are three vibrational modes associated with each allowed k-value, rather than the two states of electromagnetic polarization. Thus, there are $12\pi V\nu^2 d\nu/c^3$ vibrational modes with frequency between ν and $\nu + d\nu$. Here c is the speed of propagation of vibrational waves in the crystal, which we assume to be independent of ν and the state of polarization.

A crystal with N atoms has $3N$ vibrational normal modes. Thus, in the Debye model there is a maximum vibrational frequency $\nu_{\max}$, given by

$$3N = \int_0^{\nu_{\max}} \frac{12\pi V}{c^3}\nu^2\, d\nu, \qquad \nu_{\max} = \left[\frac{3}{4\pi}\frac{N}{V}\right]^{1/3}$$

The total vibrational energy of the crystal is then

$$U = \int_0^{\nu_{\max}} \frac{12\pi V}{c^3}\frac{h\nu^3}{e^{h\nu/kT} - 1}\, d\nu = \frac{9NkT^4}{\Theta^3}\int_0^{\Theta/T} \frac{z^3\, dz}{e^z - 1}$$

where Θ is the Debye temperature, defined by

$$\Theta = \frac{h\nu_{\max}}{k}$$

We can now calculate c_v by differentiating U with respect to T. We give the result here in the low- and high-temperature limits:

$$c_v = \frac{\partial U}{\partial T} \;\rightarrow\; \frac{12Nk\pi^4}{5\Theta^3}T^3, \qquad \text{if } T \ll \Theta$$

$$\rightarrow 3Nk, \qquad\qquad \text{if } T \gg \Theta$$

Note that the high-temperature limit is the result predicted by the equipartition principle, applied to the potential and kinetic energy associated with each mode.

PROBLEM 6.10.1 The lattice heat capacity of a certain form of carbon has a temperature dependence T^2, instead of the more common T^3 dependence for solids. What can you infer about the structure of this particular phase of carbon?

The T^3 dependence in the c_v of the Debye law is due to the T^4 dependence of U at low temperatures. This arises when we extract the T-dependence of the integral

$$\int \frac{\nu^3\, d\nu}{e^{h\nu/kT} - 1} = \left(\frac{kT}{h}\right)^4 \int \frac{x^3\, dx}{e^x - 1}$$

The $\nu^3\, d\nu$ factor came from the $h\nu$ energy of each mode times the $\nu^2\, d\nu$ factor from the density of states (5.93c). If the solid were two-dimensional, the density of states would have a factor $\nu\, d\nu$, the integrand a factor $\nu^2\, d\nu$, so U at low T would be proportional to T^3 and c_v to T^2.

PROBLEM 6.10.2 Show that, in the Debye approximation, the heat capacity of a monatomic lattice in one dimension is proportional to T/Θ for low temperature ($T \ll \Theta$). The quantity Θ is the effective Debye temperature, given by $\Theta = \hbar\pi v/Ka$, with a the interatomic spacing and v the sound velocity.

For a lattice of length L, the density of states is $dN/dk = L/2\pi$. With lattice spacing a, the number of modes is L/a. Thus, k_{max} is given by

$$\frac{L}{a} = \int_0^{k_{max}} \frac{dN}{dk}\, dk = \frac{L}{2\pi}k_{max}, \qquad k_{max} = \frac{2\pi}{a}$$

The mode k has frequency $\omega = kv$ and energy $\hbar\omega = \hbar kv$. Thus,

$$U = \frac{L}{\pi}\int_0^{k_{max}} \frac{\hbar kv}{\exp(\hbar kv/kT) - 1}\, dk = \frac{Lk}{2a}\frac{T^2}{\Theta}\int_0^{\Theta/T} \frac{x\, dx}{e^x - 1}$$

$$c_L = \frac{dU}{dT} \approx \frac{Lk}{a}\frac{T}{\Theta}\int_0^{\Theta/T} \frac{x\, dx}{e^x - 1}.$$

For $T \ll \Theta$ the integral is nearly independent of T, and c_L is proportional to T/Θ.

6.10.3 The Fermi Gas

A Fermi gas is a collection of weakly interacting fermions. It has been used to model many different phenomena, such as the conduction electrons in a solid, the protons and neutrons in a nucleus, the neutrinos that fill the universe. The concepts that are used in these different applications are rather similar. We choose as our illustration the conduction electrons in a metal, such as copper. We

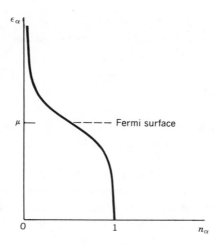

FIGURE 6.9 The occupation probability n_α of single-particle state α, as a function of the single-particle energy.

assume that these form a gas of fermions that move freely throughout the metal. Each copper atom contributes one conduction electron; the remaining 28 atomic electrons remain tightly bound to the $Z = 29$ copper nucleus. These Cu^+ ions form a lattice which provides the mechanical rigidity of the metal and a background of positive charge which neutralizes, on the average, the negative charge of the conduction electrons.

We get the lowest state of the conduction electron Fermi gas by filling the lowest available single-particle states, consistent with the Pauli principle. We can put two electrons (with oppositely directed intrinsic spins) into every single-particle orbital state and the number of single-particle orbital states in the energy interval ε to $\varepsilon + d\varepsilon$ is $V(2\pi/h^3)(2m)^{3/2}\varepsilon^{1/2}\,d\varepsilon$. Thus, if there are N conduction electrons in volume V, the condition analogous to (6.58) which determines the "Fermi energy" (the name given to the chemical potential in this application) is

$$N = 2\int_0^\infty V\frac{\dfrac{2\pi}{h^3}(2m)^{3/2}\varepsilon^{1/2}\,d\varepsilon}{e^{(\varepsilon-\mu)/kT}+1} \tag{6.60}$$

We will soon see that for temperatures less than about 10,000 K, the occupation probability (6.57) changes from nearly 1 to nearly 0 in an energy interval that is small compared to the Fermi energy (see Figure 6.9). This interval is called the "Fermi surface." In order to simplify the evaluation of the integral in (6.60) let us assume that the Fermi surface is infinitely sharp, so that

$$\frac{1}{e^{(\varepsilon-\mu)/kT}+1} \simeq \begin{cases} 1 & \text{for } \varepsilon < \mu \\ 0 & \text{for } \varepsilon > \mu \end{cases}$$

in which case (6.60) becomes

$$\frac{N}{2} = \int_0^\mu V\frac{2\pi}{h^3}(2m)^{3/2}\varepsilon^{1/2}\,d\varepsilon = V\frac{2\pi}{h^3}(2m)^{3/2}\frac{2}{3}\mu^{3/2}$$

$$\mu = \frac{h^2}{2m}\left[\frac{3}{8\pi}\frac{N}{V}\right]^{2/3} \tag{6.61}$$

Copper has a molecular weight of about 63 g, and a mass density of about 9 g/cm^3. Since N is approximately equal to the number of copper atoms, the Fermi energy is equal to

$$\mu = \frac{\left(6.63 \times 10^{-27}\,\text{erg s}\right)^2}{2 \times 9.1 \times 10^{-28}\,\text{g}} \times \left[\frac{3}{8\pi}\frac{9\,\text{g}}{\text{cm}^3} \cdot \frac{6 \times 10^{23}\,\text{atom}}{63\,\text{g}} \cdot \frac{1\,\text{electron}}{\text{atom}}\right]^{2/3}$$

$$= 1.14 \times 10^{-11}\,\text{erg} = 7.1\,\text{eV}$$

Next we estimate the fraction of the electrons whose states lie "in" the Fermi surface. We take these states to be the ones with energies between $\mu \pm 2kT$. As ε goes through this interval, n_α in (6.57) drops from 0.88 to 0.12. There are approximately

$$V \cdot \frac{2\pi}{h^3}(2m)^{3/2}\mu^{1/2} \times 4kT$$

orbital states in this region, compared with $N/2$ occupied orbital states all together. Thus, the fraction of the conduction electrons whose states are in the Fermi surface is

$$\frac{V\dfrac{2\pi}{h^3}(2m)^{3/2}\mu^{1/2} \times 4kT}{V\dfrac{2\pi}{h^3}(2m)^{3/2}(2/3)\mu^{3/2}} = \frac{6kT}{\mu} \tag{6.62}$$

For copper at 300 K, this is about $(6 \times 0.025)/7.1 = 0.02$. We see that a very small fraction of the conduction electrons are in the region of the Fermi surface. These are the only electrons that can absorb small amounts of energy. Hence, at temperatures less than about 10,000 K, the electrons make a small contribution to the specific heat. From (6.62) we can conclude that this small contribution varies linearly with T.

PROBLEM 6.10.3 Estimate a bound, in K, on the temperatures at which electrons in a metal with 10^{22} electrons/cm^3 could be treated using Maxwell-Boltzmann statistics.

We need

$$kT \gg \varepsilon_F = \frac{h^2}{2m}\left[\frac{3}{8\pi}\rho\right]^{2/3}$$

$$T \gg \frac{h^2}{2mk}\left[\frac{3}{8\pi}\rho\right]^{2/3} = \frac{h^2c^2}{2mc^2k}\left[\frac{3}{8\pi}\rho\right]^{2/3}$$

$$= \frac{(12,400)^2 \times 300}{2 \times 0.51 \times 10^6 \times 0.025}\left[\frac{3}{8\pi} \times 10^{22}\right]^{2/3} \times 10^{-16}\,\text{K}$$

$$= 20,000\,\text{K}$$

PROBLEM 6.10.4 Metallic sodium crystallizes in a *bcc* lattice, with 2 conduction electrons per unit cell, and with lattice constant (cube edge) 4.28 Å. Determine the Fermi temperature in K and the Fermi energy in eV for the conduction electrons, using the free-electron model.

$$E_F = \left(\frac{3}{8\pi} \frac{N}{V} \right)^{2/3} \frac{h^2}{2m} = \left(\frac{3}{8\pi} \frac{2}{(4.28 \text{ Å})^3} \right)^{2/3} \frac{(hc)^2}{2mc^2}$$

$$= \left(\frac{3}{8\pi} \frac{2}{(4.28 \text{ Å})^3} \right)^{2/3} \frac{(12{,}400 \text{ eV Å})^2}{2 \times .51 \times 10^6 \text{ eV}} = 3.2 \text{ eV}$$

$$T_F = 300 \text{ K} \times \frac{3.2}{.025} \simeq 38{,}000 \text{ K}$$

PROBLEM 6.10.5 N noninteracting spin-$\frac{1}{2}$ fermions move in one dimension in a common simple harmonic oscillator potential. Each particle has a magnetic moment μ. Assume that N is even.

(a) Find the specific heat of the system near $T = 0$.

Near $T = 0$ the Fermi function $1/(e^{(\varepsilon - \varepsilon_F)/kT} + 1)$ falls very sharply near $\varepsilon = \varepsilon_F$, where ε_F is the Fermi energy (chemical potential). The states of the system have energies separated by $\hbar\omega$ (harmonic oscillator). Near $T = 0$ the occupation of the states just above ε_F and the number of holes in the states just below ε_F are equal, and these numbers drop by a factor $e^{-\hbar\omega/kT}$ for each step away from ε_F. Thus, near $T = 0$ we can consider only the two states closest to ε_F on either side, and ε_F will lie *midway* between them. For the state just below ε_F, the average occupation is

$$(2s + 1)\frac{1}{e^{-\hbar\omega/2kT} + 1} \simeq 2\left(1 - e^{-\hbar\omega/2kT}\right)$$

since $s = \frac{1}{2}$. For the state just above ε_F, the average occupation is

$$\frac{2}{1 + e^{\hbar\omega/2kT}} \simeq 2e^{-\hbar\omega/2kT}$$

The contribution to the energy of the system due to these states is

$$2\left(\varepsilon_F + \frac{\hbar\omega}{2}\right)e^{-\hbar\omega/2kT} + 2\left(\varepsilon_F - \frac{\hbar\omega}{2}\right)\left(1 - e^{-\hbar\omega/2kT}\right) = \text{const.} + 2\hbar\omega e^{-\hbar\omega/2kT}$$

while, in this approximation, all other states remain either filled ($\varepsilon < \varepsilon_F$) or empty ($\varepsilon > \varepsilon_F$). Thus the specific heat near $T = 0$ is

$$\frac{d}{dT}\left(2\hbar\omega e^{-\hbar\omega/2kT}\right) = \frac{1}{k}\left(\frac{\hbar\omega}{T}\right)^2 e^{-\hbar\omega/2kT}$$

(b) Find the specific heat of the system at high temperatures.

At high temperatures, the electron occupation probabilities will follow the Boltzmann distribution (approximately), and we can use the principle of equipartition. For each electron there are two quadratic terms in the Hamiltonian (kinetic and potential energy), so

$$c_v = 2N\frac{R}{2} = NR$$

(c) Find the magnetic susceptibility of the system at $T = 0$.

In the absence of a magnetic field, at $T = 0$, the magnetic moment of the system is zero, since each single-particle oscillator state has two electrons in it with oppositely directed spins. If we turn on a small magnetic field, the energies of the spin-up states will be shifted slightly relative to the energies of the spin-down states, but at $T = 0$ the relative numbers of spin-up and spin-down states will not be changed. Thus, the total magnetic moment will still be zero, even in the presence of a small magnetic field, and the magnetic susceptibility of the system will be zero at $T = 0$.

PROBLEM 6.10.6   It has been speculated that the universe is filled with neutrinos that are continually being emitted and absorbed to maintain thermal equilibrium. Neutrinos are spin-$\frac{1}{2}$ fermions, whose energy is given by $\varepsilon = cp$, where p is the momentum and c is the speed of light. Determine the average number of neutrinos in the universe if the temperature is 3 K and the volume of the universe is 10^{81} cm^3. Express your answer in terms of the integral $I_n = \int_0^\infty x^n \, dx/(e^x + 1)$.

The number of neutrino states per unit energy interval is given by (5.84c). To get the total number of neutrinos in the universe we must integrate the occupation probability (6.57) over all possible states:

$$N = \frac{V}{\pi^2(\hbar c)^3} \int_0^\infty \frac{\varepsilon^2 \, d\varepsilon}{e^{\varepsilon/kT} + 1} = \frac{V}{\pi^2}\left(\frac{kT}{\hbar c}\right)^3 \int_0^\infty \frac{x^2 \, dx}{e^x + 1}$$

The integral can be done numerically; it equals approximately 1.803. N equals

$$\frac{10^{81} \text{ cm}^3}{\pi^2}\left[\frac{1.38 \times 10^{-16} \, \dfrac{\text{erg}}{\text{K}} \times 3 \text{ K}}{1.05 \times 10^{-27} \text{ erg s} \times 3 \times 10^{10} \text{ cm/s}}\right]^3 \times 1.803 = 4 \times 10^{83} \text{ neutrinos}$$

PROBLEM 6.10.7   The conduction electrons in Problem 6.8.7. are fermions. Use Fermi-Dirac statistics to find the thermionic emission current density as a function of the temperature T and the work function $\phi = W_0 - W_F$, W_F being the Fermi energy. Note that, for ordinary conditions, $\phi \gg kT$.

The average number of electrons in the momentum interval between $\mathbf{p}$ and $\mathbf{p} + d\mathbf{p}$ is

$$\frac{2V \, dp_x \, dp_y \, dp_z}{h^3} \frac{1}{1 + \exp\left[\dfrac{(p_x^2 + p_y^2 + p_z^2)}{2m} - W_F\right]\Big/ kT}$$

Equivalently, in velocity space we have

$$\frac{2Vm^3}{h^3} dv_x \, dv_y \, dv_z \frac{1}{1 + \exp\left[\dfrac{m}{2}\left(v_x^2 + v_y^2 + v_z^2\right) - W_F\right]\Big/ kT}$$

The thermionic current density is

$$\frac{2m^3}{h^3} e \int_{\substack{\sqrt{2W_0/m} < v_x < \infty \\ -\infty < v_y, \, v_z < \infty}} dv_x \, dv_y \, dv_z \frac{v_x}{1 + \exp\left[\dfrac{m}{2}\left(v_x^2 + v_y^2 + v_z^2\right) - W_F\right]\Big/ kT}$$

Everywhere in the region of integration, the exponential is much larger than unity. Thus, we can approximate the current by

$$\frac{2m^3}{h^3}e\int_{\sqrt{2W_0/m}}^{\infty}dv_x\cdot v_x\exp\left(-\left[\frac{mv_x^2}{2}-W_F/kT\right]\right)\left[\int_{-\infty}^{\infty}dv_y\exp-\frac{mv_y^2}{2kT}\right]^2$$

$$=\frac{2m^3}{h^3}e\frac{kT}{m}\exp\left[-\frac{(W_0-W_F)}{kT}\right]\times\frac{2\pi kT}{m}$$

$$=\frac{4\pi me}{h^3}(kT)^2\exp\left[-\frac{(W_0-W_F)}{kT}\right]$$

6.11 QUANTUM CORRECTIONS TO THE PRINCIPLE OF EQUIPARTITION

A degree of freedom makes its full contribution of $kT/2$ to the average energy only when kT is large compared to the excitation energy of the quantum states associated with that degree of freedom (cf. Problem 6.9.3.). Thus, the details of the energy spectra of the individual molecules in a system can have an effect on the variation with temperature of the specific heat of the system.

A striking example of the way molecular spectra affect specific heat is afforded by a diatomic gas. For a typical diatomic molecule, the rotational level spacing in the vicinity of the ground state is approximately $1/400$ eV, the vibrational level spacing is about $1/4$ eV and the electronic level spacing is about 1 eV. Remember that $T = 300$ K corresponds to $kT \approx 1/40$ eV. Thus, if $T < 30$ K, so that $kT < 1/400$ eV, rotational, vibrational, and electronic degrees of freedom make a very small contribution to the average energy. The average energy is almost all due to translational kinetic energy, and $U = 3nRT/2$, $c_v = 3R/2$ per mole.

If 30 K $< T < 3000$ K, so that $1/400$ eV $< kT < 1/4$ eV, translational and rotational degrees of freedom are excited and contribute significantly to the average energy. Thus, $U = 3nRT/2 + nRT = 5nRT/2$, $c_v = 5R/2$ per mole. Note that the rotational energy contributes only two degrees of freedom. These correspond to rotation of the molecule about any two axes perpendicular to its symmetry axis. The moment of inertia of a diatomic molecule about its symmetry

FIGURE 6.10 The molar specific heat of an ideal diatomic gas as a function of temperature.

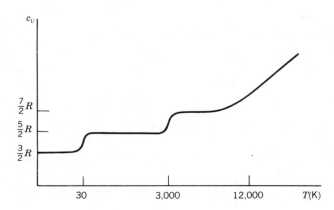

axis is so small that the excitation energy of quantum states of rotation about this axis would be much higher than $1/4$ eV.

If 3000 K $< T < 12{,}000$ K, so that $1/4$ eV $< kT < 1$ eV, translational, rotational, and vibrational degrees of freedom are excited and contribute significantly to the average energy. The vibrational motion contributes terms in r^2 and p_r^2 to the Hamiltonian (r is the interatomic separation). Thus, $U = 3nRT/2 + nRT + nRT = 7nRT/2$, $c_v = 7R/2$ per mole. For $T > 12{,}000$ K, c_v generally continues to increase as the electronic motion of the atoms is disrupted.

6.12 THE STEFAN-BOLTZMANN LAW

Planck's law (6.59) gives the electromagnetic energy density (per unit volume, per unit frequency interval) in a cavity at temperature T. Suppose the wall of the cavity contains a small hole of area da. At what rate does electromagnetic energy stream out through this hole? We can answer this question by using the formula (6.42a) derived for particle flux leaving through a hole if we identify the particle density ρ with Planck's energy density u_ν, and use

$$f(v) = \frac{\delta(v - c)}{4\pi v^2}$$

for the velocity distribution. This choice for $f(v)$ is correctly normalized according to (6.37), and guarantees that all the photons have speed c. If these ρ and $f(v)$ are used in (6.42a) we get

$$\frac{\pi u_\nu c}{4\pi} = \frac{2\pi h \nu^3}{c^2 \left[e^{h\nu/kT} - 1 \right]} \tag{6.63a}$$

for the rate, per unit frequency interval, at which electromagnetic energy leaves through a hole of unit area in the cavity wall. We get the total radiant emission by integrating (6.63a) over the entire frequency spectrum

$$\int_0^\infty \frac{2\pi h \nu^3}{c^2 \left[e^{h\nu/kT} - 1 \right]} \, d\nu = \frac{2\pi h}{c^2} \left(\frac{kT}{h} \right)^4 \int_0^\infty \frac{x^3 \, dx}{e^x - 1} = \sigma T^4 \tag{6.63b}$$

with

$$\sigma = \frac{2\pi^5}{15 c^2} \frac{k^4}{h^3} \approx 5.67 \times \frac{10^{-5} \text{ erg}}{\text{cm}^2 \text{ s K}^4}$$

Equation (6.63) expresses the Stefan-Boltzmann law. We have derived it by calculating the rate at which energy leaves through a hole in the wall of a cavity at temperature T. The same expression, σT^4, governs the rate at which energy is radiated from unit area of an ideal "black body" at temperature T. An ideal black body is one which absorbs all radiant energy that falls on it.

PROBLEM 6.12.1 A small metal sphere painted flat black is situated in interplanetary space, at a point where the sun appears to have an angular diameter $\theta \ll 1$ radian. Assume that the sun radiates like a black-body at temperature $T_s = 6{,}000$ K, and estimate the equilibrium temperature of the sphere. Evaluate your result for $\theta = 0.57°$. Assume that the small sphere is a perfect absorber and a perfect heat conductor.

Let the radius of the sun be r_s, and its surface temperature be T_s. It radiates

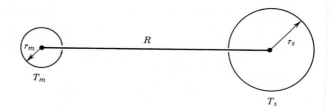

energy at the rate $4\pi r_s^2 \sigma T_s^4$. If the metallic sphere has radius r_m, and is at distance R from the sun, it catches a fraction $\pi r_m^2 / 4\pi R^2$ of the radiation emitted by the sun, so it absorbs solar energy at the rate

$$4\pi r_s^2 \sigma T_s^4 \times \pi r_m^2 / 4\pi R^2$$

If the metallic sphere has temperature T_m, it radiates at a total rate

$$4\pi r_m^2 \sigma T_m^4$$

Thus, an energy balance for the sphere yields

$$\frac{\pi r_s^2 r_m^2}{R^2} \sigma T_s^4 = 4\pi r_m^2 \sigma T_m^4$$

$$T_m = \left[\frac{r_s}{2R} \right]^{1/2} T_s$$

If θ is the angular diameter of the sun, as seen at the sphere, then

$$\theta = \frac{2r_s}{R} \text{ rad}, \qquad T_m = [\theta/4]^{1/2} T_s$$

Using $T_s = 6{,}000$ K, $\theta = 0.57° = 0.57 \times \pi/180$ rad, we find $T_m = 300$ K, which gives a good account of the average temperature of the earth.

PROBLEM 6.12.2 Large flat sheets of material are arranged parallel to one another, separated by vacua. For example, this might be an unusually elaborate insulating window. The spacing between the sheets is much smaller than the lateral dimensions of each sheet. Assume that the sheets are perfect absorber-radiators, and perfect thermal conductors, while the side support structures are perfect thermal insulators. The first and last sheets have fixed temperatures T_1, T_N. Find the temperatures $T_2, T_3, \ldots, T_{N-1}$ and the heat flux Q.

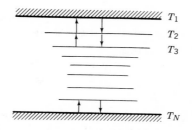

When the system is in a steady state, the net heat flux between each pair of plates is the same:

$$\frac{Q}{\sigma} = T_1^4 - T_2^4 = T_2^4 - T_3^4 = \cdots = T_{N-1}^4 - T_N^4$$

Thus, the heat flux can be calculated from,

$$Q = \frac{\sigma}{N-1}\left(T_1^4 - T_N^4\right)$$

and the intermediate temperatures are given by

$$T_2^4 = T_1^4 - \frac{Q}{\sigma} = \frac{(N-2)T_1^4 + T_N^4}{N-1}$$

$$T_3^4 = T_2^4 - \frac{Q}{\sigma} = \frac{(N-3)T_1^4 + 2T_N^4}{N-1}$$

$$\vdots \quad \vdots \quad \vdots \qquad \vdots$$

$$T_r^4 = T_{r-1}^4 - \frac{Q}{\sigma} = \frac{(N-r)T_1^4 + (r-1)T_N^4}{N-1}$$

If $T_N = T_1 - \Delta T$ with $\Delta T \ll T$, Q is approximately linear in ΔT:

$$Q = \frac{\sigma}{N-1}\left[T_1^4 - (T_1 - \Delta T)^4\right] \approx \frac{4\sigma}{N-1}T_1^3\Delta T$$

The R-value $(Q/\Delta T)$ for this system is $4\sigma T_1^3/(N-1)$.

PROBLEM 6.12.3 The diagram shows a source of light, a lens, and a thin black heat-conducting disk. The light source is 1 mm in diameter and emits 100 watts of radiation isotropically . The lens is 2 cm in diameter, has a focal length of 10 cm, and passes all radiation that impinges on it. The diameter of the disk is 1/2 mm, and the image of the source is just large enough to cover the disk completely. What is the equilibrium temperature of the disk?

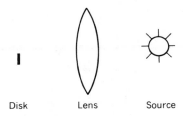

Disk   Lens   Source

Let d_s and d_d be the distances between the lens and source and lens and disk, respectively. Their ratio is equal to the ratio of source and image diameters [see (4.40)]

$$\frac{d_s}{d_d} = \frac{1 \text{ mm}}{\frac{1}{2} \text{ mm}} = 2$$

If we combine this with the lens equation (4.39)

$$\frac{1}{d_d} + \frac{1}{d_s} = \frac{1}{f} = \frac{1}{10 \text{ cm}}$$

we find that $d_s = 30$ cm and $d_d = 15$ cm. Thus, the rate at which energy falls on the lens is approximately

$$100 \text{ W} \times \frac{\frac{\pi}{4}(2 \text{ cm})^2}{4\pi(30 \text{ cm})^2} = \frac{100}{3600}\text{W}$$

This is also the rate at which energy falls on the disk. The rate at which the (two-sided) disk radiates is

$$2 \times \frac{\pi \, (.05 \text{ cm})^2}{4} \times \sigma T^4 = \frac{100}{3600} \text{ W}$$

so that T is given by

$$T = \left[\frac{\dfrac{100}{3600} \dfrac{\text{J}}{\text{s}} \times \dfrac{10^7 \text{erg}}{\text{J}}}{2 \cdot \dfrac{\pi}{4} \, (.05 \text{ cm})^2 \times 5.67 \times 10^{-5} \, \dfrac{\text{erg}}{\text{cm}^2 \text{ s K}^4}} \right]^{1/4}$$

$$= 1057 \text{ K}$$

PROBLEM 6.12.4   If the sun behaves like a black body at 6000 K with a diameter of 10^6 km, what is its total microwave emitted power per megacycle width at 3 cm?

The amount of radiant energy emitted per unit area per unit frequency interval is given by (6.63a). Thus, a black body of radius R emits into bandwidth W Hz an amount of energy equal to

$$4\pi R^2 \frac{2\pi h}{c^2} \frac{\nu^3}{e^{h\nu/kT} - 1} W = 4\pi R^2 \frac{2\pi h}{c^2} \frac{\left(\dfrac{c}{\lambda}\right)^3}{e^{hc/\lambda kT} - 1} W \text{ erg/s}$$

For $R = 0.5 \times 10^6$ km, $\lambda = 3$ cm, $W = 10^6$ Hz, and $T = 6000$ K, this is equal to 1.75×10^{16} erg/s.

REVIEW PROBLEMS

6.1. In cold weather we heat rooms. How does this affect the total energy of the air in the room?
Answer: It doesn't.

6.2. A Carnot cycle operates as a heat engine between two bodies of equal mass and equal specific heat, until their temperatures are equal. If their initial temperatures were T_1 and T_2 (with $T_1 > T_2$), find their common final temperature and the total work done by the engine.   *Answer*: $T_f = \sqrt{T_1 T_2}$
$$W = c[T_1 + T_2 - 2\sqrt{T_1 T_2}\,].$$

6.3. Compare the entropy of 10^{20} He molecules in a one-liter vessel plus 10^{20} O_2 molecules in a similar one-liter vessel, with the entropy of all these molecules in a single one-liter vessel, at the same temperature. *Answer*: No difference.

6.4. Derive the following relationships obeyed by c_p and c_v, the specific heats of a gas at con-stant volume and pressure,

$$\left.\frac{\partial c_v}{\partial v}\right)_T = T\left.\frac{\partial^2 P}{\partial T^2}\right)_v$$

$$\left.\frac{\partial c_p}{\partial P}\right)_T = -T\left.\frac{\partial^2 V}{\partial T^2}\right)_p$$

Answer: Start with $c_v = \left.T\frac{\partial S}{\partial T}\right)_v$, $c_p = \left.T\frac{\partial S}{\partial T}\right)_p$

6.5. One mole of an ideal monatomic gas is expanded adiabatically and reversibly from temperature T_0 and volume V_0 to a volume $2V_0$. It is then compressed reversibly and iso-thermally back to its original volume.

(a) Calculate the change in internal energy of the gas in terms of T_0, V_0, and the gas constant R.   *Answer*: $\dfrac{3}{2}RT_0\left[\dfrac{1}{2^{2/3}} - 1\right]$

(b) Calculate the change in entropy of the gas.   *Answer*: $-R \ln 2$.

6.6. A spoonful of water is placed in a bottle, the cork is inserted, and the bottle is heated to 100 C. What is the pressure in the bottle?

$$Answer: \frac{646}{273} \text{ atm.}$$

6.7. Find a relation between the equilibrium vapor pressure of water and its absolute temperature, by making the following assumptions:

(a) The latent heat of vaporization depends upon absolute temperature according to the relation

$$L = a - bT$$

where a and b are constants.

(b) The volume occupied by the liquid is negligible compared to the volume occupied by the vapor.

(c) The vapor behaves like an ideal gas.

$$Answer: P = \frac{\text{constant}}{e^{a/RT}T^{b/R}}.$$

6.8. Acoustic frequencies are low enough so that every point in a gas through which a sound wave is traveling can be regarded as being in thermodynamic equilibrium. On the other hand, acoustic frequencies are high enough so that there isn't sufficient time for heat flow to have much effect on the instantaneous temperature distribution. Thus, it is reasonable to treat the processes going on at every point in the gas as adiabatic. Derive an expression for the speed of propagation, c, of an acoustic wave through an ideal gas at pressure p and temperature T. You can start with the expression for c given in (1.115c).

$$Answer: c = \sqrt{\gamma \frac{RT}{M}}.$$

6.9. A dilute gas consisting of molecules with a permanent electric dipole moment **p** is brought into a uniform electric field. Find the Helmholtz free energy of the system, and the average electric dipole moment per unit volume of gas.

$$Answer: F = -NkT \ln (2m\pi kT)^{3/2}V$$
$$\cdot \frac{2kT}{Ep} \sinh\left(\frac{Ep}{kT}\right)\Big\}$$
$$P = Np\left\{ \coth\left(\frac{Ep}{kT}\right) - \frac{BT}{Ep} \right\}.$$

6.10. A gas has an isotropic velocity distribution. Show that the particle flux (number per unit time crossing unit area in one direction) is given by $1/4 \cdot \rho\bar{v}$, where ρ is the particle number density and v is the mean particle speed.

6.11. The neutron flux at the center of a reactor is 4×10^{10} neutrons/m² sec. Assume that the neutrons have a Maxwell velocity distribution corresponding to a temperature of $300°K$.

(a) Find the number of neutrons/m³.
    Answer: 6.4×10^{13}.

(b) Find the partial pressure at the neutron gas.   *Answer*: $2.4 \times 10^{-7} N/m^2$.

6.12. If a closed system is in contact with a reservoir, and is in thermal equilibrium, the temperature is, by definition, the temperature of the reservoir. However, the energy of the system will fluctuate. Using the canonical ensemble, show that the magnitude of these fluctuations is given by

$$\overline{(\delta E)^2} = \overline{E^2} - (\overline{E})^2 = kT^2c_v$$

where c_v is the heat capacity of the system at constant volume.

6.13. Find the total energy of an electron gas in which N electrons are in volume V at $T = 0$ K.

$$Answer: \frac{3^{5/3}}{80\pi^{2/3}} \cdot \frac{h^2N^{5/3}}{mV^{2/3}}.$$

6.14. At 100 K the molar specific heat at constant volume equals 3 cal/mole C for H_2, whereas all other diatomic molecules have molar specific heats of 5 cal/mole C. Explain.
Answer: H_2 has a low moment of inertia, and rotational kinetic energy $> kT$.

6.15. (a) The momentum p and kinetic energy E of a particle with zero rest mass are related by

$$p = \frac{E}{c}$$

[cf. (2.15a)]. Consider a gas of photons in a perfectly reflecting container. Show that the pressure P exerted on the walls of the container is related to the energy density u by

$$P = \tfrac{1}{3}u$$

(b) Show that the ideal gas law can be written in an analogous form, if u is interpreted as the kinetic energy density of the molecules, and the constant $1/3$ is replaced by a different constant. What is this different constant?   *Answer*: $\tfrac{2}{3}$.

6.16. Consider an enclosure filled with black-body radiation at temperature T. The enclosure is fitted with a movable piston with a reflecting surface. Use the relationship between pressure and energy density derived in Problem 6.15, together with the combined first and second laws of thermodynamics (6.14a), to derive formulae for the radiation energy density and entropy as functions of T and V.

Answer: $u = \text{constant} \times T^4$
$S = \text{constant} \times \tfrac{4}{3}VT^3$.

6.17. Suppose that the bulb of an ordinary 100-watt incandescent lamp is blackened to absorb all the radiation emitted by the filament. The lamp is surrounded by vacuum. How hot will the bulb get?   *Answer*: 562 K.

6.18. A small flat heat-conducting disk is placed a distance D away, and parallel to, an infinitely large plane surface maintained at temperature T. All surfaces radiate as if they were perfect black bodies. What is the equilibrium temperature of the disk?

Answer: $2^{-1/4}T$.

6.19. A cubic ice crystal is at rest relative to the sun at the same distance from the sun as the earth is. It is balanced there under the simultaneous influences of gravitational attraction toward the sun and outward pressure due to electromagnetic radiation from the sun. How big is the crystal? Assume that the solar constant at the top of the atmosphere is 1 kw/m^2, and that the effective gravitational field strength of the sun at the earth's orbit is $6 \times 10^{-4}g$.

Answer: 10^{-4} cm.

6.20. Make order-of-magnitude estimates of the following quantities:

(a) Number of watts emitted by 1 cm^2 of tungsten at 3000 K.   *Answer*: 460 Watts.

(b) Number of molecules/cm^3 in a vacuum of 10^{-9} torr (mm Hg).

Answer: 3.5×10^7/cm^3.

(c) Velocity of an electron at the Fermi surface in a metal.

Answer: 1.6×10^8 cm/sec.

(d) Pressure (in atmospheres) of a gas at room temperature required to make its dielectric constant appreciably different from 1 (e.g., 1.2 or 1.3).   *Answer*: 1000 atm.

BIBLIOGRAPHY

The following list contains books that we have used in our own studies and teaching. More complete bibliographies can be found in the books by Goldstein, French, Jackson, Fowles, Gasiorowicz and Kittel and Kroemer.

General

R. Feynman, *Lectures on Physics*, Reading, Massachusetts: Addison-Wesley, 1963. Probably the best general physics text written this century. Every physics student or physicist, no matter how experienced, will profit by reading these volumes.

Chapter 1

Fetter and Walecka, *Theoretical Mechanics of Particles and Continua*, New York: McGraw-Hill, 1980.

H. Goldstein, *Classical Mechanics*, 2nd ed., Reading, Massachusetts, Addison-Wesley, 1980.

Landau and Lifshitz, *Mechanics*, 3rd ed., New York: Pergamon, 1976

K. Symon, *Mechanics*, 3rd ed., Reading, Massachusetts: Addison-Wesley, 1971.

Chapter 2

A.P. French, *Special Relativity*, New York: Norton, 1968

R. Sard, *Mechanics, Special Relativity, and Classical Particle Dynamics*, New York: Benjamin, 1970

Taylor and Wheeler, *Spacetime Physics*, San Francisco: Freeman, 1963

Chapter 3

J.D. Jackson, *Classical Electrodynamics*, New York: Wiley, 1975

Panofsky and Phillips, *Classical Electricity and Magnetism*, Reading, Massachusetts Addison-Wesley, 1962

E. Purcell, *Berkeley Physics Course*, vol. 2, New York: McGraw-Hill, 1963

Reitz, Milford and Christie, *Foundations of Electromagnetic Theory*, 3rd ed., New York: Addison-Wesley, 1979

V. Rojansky, *Electromagnetic Fields and Waves*, New York: Prentice-Hall, 1971

Chapter 4

R. Ditchburn, *Light*, 3rd ed., 2 vols. New York: Academic Press, 1976

G. Fowles, *Introduction to Modern Optics*, New York: Holt, Reinhart and Winston, 1968

Jenkins and White, *Fundamentals of Optics*, 3rd ed., McGraw-Hill, 1957

F. Sears, *Optics*, Reading, Massachusetts: Addison-Wesley, 1938

Chapter 5

G. Baym, *Lectures on Quantum Mechanics*, New York: Benjamin, 1969

Beard and Beard, *Quantum Mechanics and Applications*, Boston: Allyn Bacon, Inc., 1970

S. Gasiorowicz, *Quantum Mechanics*, New York: Wiley, 1974

E. Merzbacher, *Quantum Mechanics*, New York: Wiley, 1970

L.I. Schiff, *Quantum Mechanics*, New York: McGraw-Hill, 1968

Chapter 6

C. Kittel and H. Kroemer, *Thermal Physics*, San Francisco: W. H. Freeman and Company, 1983

A. Pippard, *Elements of Classical Thermodynamics*, New York: Cambridge University Press, 1957

F. Reif, *Fundamentals of Statistical and Thermal Physics*, New York: McGraw-Hill, 1965

ter Haar and Wergeland, *Elements of Thermodynamics*, Reading, Massachusetts: Addison-Wesley, 1966

INDEX